Benjamin Fine, Anthony Gaglione, Alexei Myasnikov,
Gerhard Rosenberger, Dennis Spellman

The Elementary Theory of Groups

De Gruyter Expositions in Mathematics

Volume 60

Benjamin Fine, Anthony Gaglione, Alexei Myasnikov,
Gerhard Rosenberger, Dennis Spellman

The Elementary Theory of Groups

A Guide through the Proofs of the Tarski Conjectures

DE GRUYTER

Mathematics Subject Classification 2010
20F05, 20F06, 20F10, 20F65, 20F67, 20F69, 20E05, 20E06, 20E08, 20E10, 20A15, 03C95, 03C98, 03C68

Authors
Prof. Dr. Benjamin Fine
Fairfield University
Department of Mathematics
1073 North Benson Road
Fairfield, CT 06430
USA
ben1902@aol.com

Prof. Dr. Anthony Gaglione
United States Naval Academy
Department of Mathematics
9E Mail Stop
Annapolis, MD 21401
USA
amg@usna.edu

Alexei Myasnikov
McGill University
Dept. of Mathematics and Statistics
805 Sherbrooke St. West
Montréal, QC H3A 2K6
Canada
alexeim@math.mcgill.ca

Prof. Dr. Gerhard Rosenberger
Heinrich-Barth-Str. 1
20146 Hamburg
Germany
rosenber@fim.uni-passau.de

Prof. Dr. Dennis Spellman
Temple University
Dept. of Mathematics
Broad and Montgomery
Philadelphia, PA 19122
USA

ISBN 978-3-11-034199-7
e-ISBN (PDF) 978-3-11-034203-1
Set-ISBN 978-3-11-034204-8
ISSN 0938-6572

Library of Congress Cataloging-in-Publication Data
A CIP catalog record for this book has been applied for at the Library of Congress.

Bibliographic information published by the Deutsche Nationalbibliothek
The Deutsche Nationalbibliothek lists this publication in the Deutsche Nationalbibliografie; detailed bibliographic data are available on the Internet at http://dnb.dnb.de.

Typesetting: le-tex publishing services GmbH, Leipzig
Printing and binding: CPI books GmbH, Leck
♾ Printed on acid-free paper
Printed in Germany

www.degruyter.com

Preface

In 1940, Alfred Tarski, the noted logician, asked three major questions about the elementary or first-order theory of the class of non-Abelian free groups. These were subsequently formalized into conjectures. The first of these *Tarski conjectures* about non Abelian free groups is that all non-Abelian free groups have exactly the same first-order theory. The second is that the natural embedding of one free group into another is an elementary embedding. This second conjecture implies the first. Finally Tarski asked if the elementary theory of the non-Abelian free groups is decidable, that is does there exist an algorithm to determine if a first order sentence is true or not within the class of non-Abelian free groups. The conjectures remained open for over fifty years.

In a series of papers from 1998–2006 the first two Tarski conjectures were answered in the affirmative by Olga Kharlampovich and Alexei Myasnikov [152, 153, 154, 155, 156] and independently by Zil Sela [233, 234, 235, 236, 237]. Kharlampovich and Myasnikov also proved Tarski conjecture 3. The proofs of both Kharlampovich and Myasnikov and of Sela involve long and complicated applications of algebraic geometry over free groups (Sela calls this Diophantine geometry) as well as an extension of methods to solve equations over free groups originally developed by Makanin and Razborov. The material necessary to understand these proofs is quite daunting even for accomplished group theorists and logicians. This book is an examination of the material on group theory and logic and on the general elementary theory of groups that is necessary to begin to understand the proofs. This material includes a complete exposition of the theory of fully residually free groups or limit groups as well a complete description of the algebraic geometry of free groups. Also included are introductory material on combinatorial and geometric group theory and first-order logic. There is then a short outline of the proof of the Tarski conjectures. We found that in many cases, group theorists don't know enough logic to understand the proof while the same is true for logicians, that is the logicians for the most part don't understand enough of the group theory. Part of our goal in this book is to correct this.

We first introduce some basic ideas and give some history.

The elementary theory of groups is tied to first-order logic and to model theory. We will look at elementary logic and model theory in more detail in Chapter 4. We start with a first-order language appropriate for group theory. This language, which we denote L_0, is the first-order language with equality containing a binary operation symbol $\cdot$ a unary operation symbol $^{-1}$ and a constant symbol 1. A *sentence* in L_0 is a logical expression using variables, the operational symbols above, equality, logical connectives and quantifiers. A *universal sentence* of L_0 is one of the form $\forall\overline{x}\{\phi(\overline{x})\}$ where $\overline{x}$ is a tuple of distinct variables, $\phi(\overline{x})$ is a formula of L_0 containing no quantifiers and containing at most the variables of $\overline{x}$. For example

$$\forall(x, y)\{xy = yx\}$$

is a universal sentence describing an Abelian group. Similarly an *existential sentence* is one of the form $\exists\overline{x}\{\phi(\overline{x})\}$ where $\overline{x}$ and $\phi(\overline{x})$ are as above. For example

$$\exists(x,y)\{xy \neq yx\}$$

is an existential sentence describing a non-Abelian group. A *universal-existential sentence* is one of the form $\forall\overline{x}\exists\overline{y}\{\phi(\overline{x},\overline{y})\}$. Similarly defined is an *existential-universal sentence*. It is known that every sentence of L_0 is logically equivalent to one of the form $Q_1x_1\ldots Q_nx_n\phi(\overline{x})$ where $\overline{x} = (x_1,\ldots,x_n)$ is a tuple of distinct variables, each Q_i for $i = 1,\ldots,n$ is a quantifier, either $\forall$ or $\exists$, and $\phi(\overline{x})$ is a formula of L_0 containing no quantifiers and containing freely at most the variables $x_1,\ldots,x_n$. Further vacuous quantifications are permitted. Finally a *positive sentence* is one logically equivalent to a sentence constructed using (at most) the connectives $\vee, \wedge, \forall, \exists$.

If G is a group then the *universal theory* of G consists of the set of all universal sentences of L_0 true in G. Since any universal sentence is equivalent to the negation of an existential sentence it follows that two groups have the same universal theory if and only if they have the same *existential theory*. The set of all sentences of L_0 true in G is called the *first-order theory* or the *elementary theory* of G. We denote this by $Th(G)$. We note that being *first-order* or *elementary* means that in the intended interpretation of any formula or sentence all of the variables (free or bound) are assumed to take on as values only individual group elements – never, for example, subsets of, nor functions on, the group in which they are interpreted.

We say that two groups G and H are *elementarily equivalent* (symbolically $G \equiv H$) if they have the same first-order theory, that is $Th(G) = Th(H)$.

Group monomorphisms which preserve first-order formulas are called elementary embeddings. Specifically, if H and G are groups and

$$f\colon H \to G$$

is a monomorphism then f is an *elementary embedding* provided whenever $\phi(x_0, \ldots, x_n)$ is a formula of L_0 containing free at most the distinct variables $x_0,\ldots,x_n$ and $(h_0,\ldots,h_n) \in H^{n+1}$ then $\phi(h_0,\ldots,h_n)$ is true in H if and only if $\phi(f(h_0),\ldots,f(h_n))$ is true in G. If H is a subgroup of G and the inclusion map $i\colon H \to G$ is an elementary embedding then we say that G is an *elementary extension* of H.

The genesis of the Tarski problems is the observation that most theorems concerning free groups are independent of the rank of the free group. As an example we note the Nielsen–Schreier Theorem (see Chapter 2) which says that any subgroup of a free group is itself a free group (independent of the rank of the overgroup). Another example is the result that an Abelian subgroup of a free group, again of any rank, must be cyclic. Proceeding further, suppose that $n < m$ are positive integers. From the Nielsen–Schreier Theorem it is clear that a free group of rank n can be embedded isomorphically into a free group of rank m. Hence F_n can be embedded into F_m. Further it can be shown that a free group of any countable rank can be embedded isomorphically into a free group of rank 2. It follows that F_m can be embedded into F_n. Therefore F_n and F_m must be *similar*.

There was some initial early success on the Tarski conjectures. Vaught showed that the Tarski conjectures 1,2 are true if G and H are both free groups of infinite rank. Combining his result with the elementary chain theorem (see Chapter 4) reduced the conjectures to free groups of finite rank. He also provided a criterion, now called the Tarski–Vaught criterion, to determine if an embedding of one group into another is an elementary embedding.

The next significant progress was due to Merzljakov. The *positive theory* of a group G consists of all the positive sentences true in G. Merzljakov [185] showed that the non-Abelian free groups have the same positive theory. A proof of Merzljakov's [185] result can be given involving generalized equations and a quantifier elimination process. This was a precursor to the methods used in the eventual solution of the overall Tarski problems.

Work following Merzljakov centered on restricted theories of free groups. It is fairly straightforward to show that any two non-Abelian free groups satisfy the same universal theory. Sacerdote [226] proved that this could be extended to *universal-existential sentences* or π_2-sentences (see Chapter 4).

Despite this early sucessful work, the conjectures remained open for over fifty years after Tarski initially proposed them. In a 1988 paper surveying combinatorial group theory [175] Roger Lyndon called the Tarski problems, which he described as folklore, among the outstanding open problems (at that time) in the field. After this point the pieces in the big puzzle began to placed together. First a result of Gaglione and Spellman and independently V. Remeslennikov and building on results of Gilbert Baumslag and Benjamin Baumslag showed that finitely generated groups that share the same universal theory as the class of non-Abelian free groups are precisely the class of finitely generated fully residually free groups. This shifted the focus to the class of finitely generated fully residually free groups in the search for the class of groups that share the same complete first-order theory as the non-Abelian free groups. Further from the Tarski–Vaught criterion the concentration was on the solution of equations in groups.

Dealing with systems of equation over free groups, it was clear from the beginning that to deal with the Tarski's conjectures a precise description of solution sets of equations (and inequations) over free groups was needed. Therefore in analogy with the classical solutions of polynomial equations over fields what was needed was a translation of classicial algebraic geometry to an algebraic geometry over groups. In the late 1990s Gilbert Baumslag, Olga Kharlampovich, Alexei Myasnikov, and Vladimir Remeslennikov developed the basics of algebraic geometry over groups introducing analogs of the standard algebraic geometric notions such as algebraic sets, the Zariski topology, Noetherian domains, irreducible varieties, radicals and coordinate groups.

The first general results on equations in groups appeared in the 1960s. Roger Lyndon developed several extremely important ideas. He considered completions of a given group G by allowing exponents in various rings and then used these completions to parameterize solutions of equations in G. Along these lines he introduced the free exponential group $F^{\mathbb{Z}[t]}$ with exponents in the integral polynomial ring $\mathbb{Z}[t]$. Subsequently it was shown that the finitely generated subgroups of this free exponential group coincides with the class of finitley generated fully residually free groups and hence with the class of universally free groups.

Another idea originating with Lyndon, in addition to generalizing the ring of exponents to $\mathbb{Z}[t]$, is to consider groups with free length functions with values in some ordered Abelian group. This allows one to axiomatize the classical Nielsen technique based on the standard length function in free groups and apply it to "non-standard" extensions of free groups, for instance, to ultrapowers of free groups. A link with the Tarski problems comes here by the Keisler–Shelah theorem, that states that two groups are elementarily equivalent if and only if their ultrapowers (with respect to a non-principal ultrafilter) are isomorphic.

In the eighties new crucial concepts were introduced. Makanin proved the algorithmic decidability of the Diophantine problem over free groups, and showed that both, the universal theory and the positive theory of a free group are algorithmically decidable. He created an extremely powerful technique (the Makanin elimination process) to deal with equations over free groups. Shortly afterwards, Razborov then described the solution set of an arbitrary system of equations over a free group in terms of what is known now as Makanin–Razborov diagrams.

A few years later Edmunds and Commerford and Grigorchuck and Kurchanov described solution sets of arbitrary quadratic equations over free groups. These equations came to group theory from topology and their role in group theory was not initially clear. However it was subsequently proved, and it became fundamental to the proof of the Tarski conjectures, that an arbitrary system of equations is equivalent to a collection of a special type of quadratic systems.

This book is laid out in the following manner. In Chapter 2 we present the necessary material from Combinatorial Group Theory. This will include the material on free groups and group amalgams.

Over the past twenty years, building on work of Gromov, Rips, Bass and Serre and others, geometric ideas have gained prominence. This has been given the overall name Geometric Group Theory and includes hyperbolic group theory and the theory of groups acting on various types of trees. In Chapter 3 we describe the essential ideas in Geometric Group Theory.

In Chapter 4 we will formally introduce the ideas from first-order languages and model theory, most of which are not as well-known to group theorists as they should be. We will also review the concepts of filters, ultra-filters and ultra-products which are essential tools in the study of elementary properties.

In Chapter 5 we will give a more formal description of the Tarski problems as well as a survey of Tarski-like results for other classes of groups.

In Chapters 6 and 7 we describe the vast body of results on fully residually free groups. In Chapter 6 we introduce fully residually free groups, and related concepts, and present the basic properties of such groups including the equivalence with universally free groups. We also describe the general structure theory of these groups.

In Sela's approach the class of fully residually free groups arises as the class of limiting groups from homomorphisms into free groups. Sela terms these limit groups. In Chapter 7 we describe the equivalences for various interpretations of the class of limit groups including some topological interpretations.

In Chapter 8 we present the basic framework of algebraic geometry over groups. This includes algebraic sets, the Zariski topology, Noetherian domains, irreducible varieties, radicals and coordinate groups. We also prove that the coordinate groups of irreducible algebraic varieties over free groups are the limit groups.

In Chapter 9 we outline the Kharlampovich–Myasnikov proof of the Tarski problems. This involves an induction on the number of quantifiers in a logical sentence and a quantifier elimination process that they call the elimination process.

As part of the solution of the Tarski conjectures both Kharlampovich–Myasnikov and Sela provide a complete description of the class of groups that share the elementary theory of the non-Abelian free groups. These extend beyond the class of non-Abelian free groups and are called the elementary free groups. In Chapter 10 we consider the general theory of the elementary free groups and present some properties that go beyond what is true in free groups. It is known that the surface groups of high enough genus are elementary free and this provides a method to prove results in surface groups that are otherwise quite inaccesible and difficult. We discuss this also in Chapter 10.

Finally in Chapter 11 we discuss a large body of results concerning discriminating groups, a class of groups introduced by G.Baumslag, Myasnikov and Remeslennikov as a by-product of the development of algebraic geometry.

We hope that the book will find good use among the group theoretic and logic community.

We would like to thank people who have looked over portions of the book and helped us with the preparation including Gilbert Baumslag and Olga Kharlampovich. Especially we thank Anja Moldenhauer who carefully looked over the chapters and prepared the diagrams and figures.

Ben Fine
Anthony Gaglione
Alexei Myasnikov
Gerhard Rosenberger
Dennis Spellman

Contents

1 Group theory and logic: introduction

1.1 Group theory and logic

The algebraic concept of a group and eventually the discipline of group theory arose in the early nineteenth century initially from the solution by Galois to the problem of solvability by radicals. This early work led primarily to finite groups and specifically to permutation groups, although as Cayley's Theorem points out this is really no restriction. Later, infinite groups became prominent through their use in Geometry and Klein's Erlanger Program (1876). Continuous groups were introduced by Lie and others to extend the methods of Galois for algebraic equations to the solutions of differential equations. Almost concurrently with the introduction of continuous groups, infinite discrete groups arose as tools in the study of combinatorial topology as introduced by Poincare. In addition, infinite discrete groups also became prominent in complex analysis via the work of Fricke and Klein on discrete groups of motions of the hyperbolic plane (Fuchsian groups). As these group objects were introduced through algebra, topology, geometry and analysis it became clear that there were strong interactions with formal logic. Each of the concrete examples of groups, mentioned above, permutation groups, matrix groups, groups of geometric transformations etc. are *models* in the sense of formal logic of abstract logical structures and languages. These ties became even clearer in the twentieth century as the study of mathematical logic became formalized.

This book will concentrate on the interactions between group theory and logic and will focus primarily on infinite discrete groups. We will deal with ideas and extensions of concepts arising around the *Tarski Problems* and their solution. The statements of the Tarski problems will be explained in the next section. First however we introduce some material that is needed to describe these problems.

The study of infinite discrete group theory essentially uses *combinatorial group theory*. This subdiscipline, within group theory, can roughly be described as the study of groups via *group presentations*. A presentation for a group G consists of a set of generators $\{g_\nu\}$ for G from which any element of G can be generated as a word or expression in the $\{g_\nu\}$ together with a set of relations on these generators from which any part of the group table can be constructed. In Chapter 2 we will examine combinatorial group theory in detail. Although a group presentation is a succinct way to express a group, it was clear from the beginnings of the discipline, that working with group presentations required some detailed algorithmic knowledge and certain decision questions.

In 1910 Max Dehn, as part of his work with group presentations for the fundamental groups of orientable surfaces, presented the three most *fundamental decision problems*. The first of these is the *word problem* or *identity problem*. This is given as follows:

(1) **Word Problem:** Suppose G is a group given by a finite presentation. Does there exist an algorithm to determine if an arbitrary word W in the generators of G defines the identity element of G?

Specifically if

$$G = \langle g_i; i = 1, \ldots, n; R_j = 1, j = 1, \ldots, m\rangle$$

is a finite presentation for G and $W(g_\nu)$ is an arbitrary word in the generators of G, can one decide algorithmically, in a finite number of steps, whether $W(g_\nu)$ represents the identity in G or not. If such an algorithm exists we say that G has a *solvable word problem*. If not, G has an *unsolvable word problem*. Dehn presented a geometric method to show that the fundamental group of an orientable surface of genus $g \geq 2$, which we denote by S_g, has a solvable word problem. In particular he gave an algorithm which systematically reduced the length of any word equal to the identity in $\pi(S_g)$. If a particular word's length is greater than 1 and cannot be reduced then that word does not represent the identity. Such an algorithm is now called a *Dehn algorithm*. Subsequently small cancellation theory, (see Chapter 3), was developed to determine additional groups that have Dehn algorithms. More recently it was shown that finitely presented groups with Dehn algorithms are precisely the word-hyperbolic groups of Gromov (see Chapter 2). In 1955 Novikov [205] and independently Boone [34] proved that, in general. the word problem is unsolvable, that is, there exist finitely presented groups with unsolvable word problems. Hence questions about word problems now focus on which particular classes of groups have solvable word problems. As decribed by Magnus (see [178]), given the Novikov–Boone result, any solution of the word problem is actually a triumph over nature.

The second fundamental decision problem is the *conjugacy problem* given by:

(2) **Conjugacy Problem:** Suppose G is a group given by a finite presentation. Does there exist an algorithm to determine if an arbitary pair of words u, v in the generators of G define conjugate elements of G?

A solution of the conjugacy problem implies a solution of the word problem. Hence it follows from the Novikov–Boone result that the conjugacy problem is unsolvable in general. It has been shown further (see [178] and [186]) that there do exist finitely presented groups with solvable word problems and unsolvable conjugacy problems. As for the word problem, results on the conjugacy problem, concentrate on which groups, or classes of groups, do have solvable conjugacy problems. In particular many small cancellation groups also have solvable conjugacy problems. Geometric techniques were introduced into small cancellation theory to mimic Dehn's original approach and prove these results for small cancellation groups.

The final fundamental decision problem is the most difficult. It is the *isomorphism problem*.

(3) **Isomorphism Problem:** Does there exist an algorithm to determine given two arbitary finite presentations whether the groups they present are isomorphic or not?

A solution to the isomorphism problem would imply a solution to the word problem so again as a consequence of the Novikov and Boone theorem the isomorphism problem is unsolvable in general. We digress a bit here to show a tie between the isomorphism problem and model theory. There is a fundamental first-order language (see Chapter 4) that we will denote by L_0 describing basic group theory. We will make these ideas precise in Chapter 4. A group presentation can be considered as an augmented language L of L_0. A group G, given by a presentation, is then a model of the language L. The isomorphism problem can be phrased as whether it can be algorithmically decided whether models of two different languages describe isomorphic groups or not.

Since the isomorphism problem is so complicated we restrict ourselves mostly to certain classes of group presentations and ask for the solution of the isomorphism problem within these classes. This means, given two groups from a class $\mathcal{C}$ of group presentations, can we decide algorithmically whether they are isomorphic or not. We then say the isomorphism problem is solvable or unsolvable for the class $\mathcal{C}$. Further, given a fixed group presentation for a group G within the class $\mathcal{C}$ can we decide algorithmically for G whether or not G is isomorphic to a group H with a different group presentation from $\mathcal{C}$. We then say the isomorphism problem is solvable or unsolvable for G within the class $\mathcal{C}$. Examples of the types of group presentations we have in mind are the class of free presentations, the class of one-relator presentations and the class of finite presentations defining torsion-free hyperbolic groups.

From the standpoint of combinatorial group theory the simplest type of group is a *free group*, and for free groups all the decision problems are solvable. Recall that a free group is constructed combinatorially in the following way. Let X be a set of letters. Then the group F is free on X if X is a set of generators for F and there are no nontrivial relations. In particular a free group F has a presentation of the form $F = \langle X; \ \rangle$. We say that X is a *free basis* for F and in general a *free group* is a group with a free basis. The cardinality of a free basis, $|X|$, for a free group F is unique, and is called the *rank of* F. If $|X| < \infty$, F is of *finite rank*. If F has rank n and $X = \{x_1, \dots, x_n\}$ we say that F is free on $\{x_1, \dots, x_n\}$. We denote this by $F(x_1, x_2, \dots, x_n)$ or by just F_n.

There is also a categorical algebraic description of a free group. From this point of view a group F is *free on a subset* X if every map $f: X \to G$ can be extended to a unique homomorphism $f: F \to G$. We denote by F_X a group with free basis X. From this algebraic characterization it is easy to show that the cardinality of a free basis $|X|$ for a free group F is unique. That is $F_X \cong F_Y$ if and only if $|X| = |Y|$. This solves the isomorphism problem for *free presentations*, that is, presentations with no relations. The cardinality of a free basis is called the *rank of* F. If $|X| < \infty$, F is of *finite rank*. If F has rank n and $X = \{x_1, \dots, x_n\}$ we say that F is free on $\{x_1, \dots, x_n\}$. We denote this by $F(x_1, x_2, \dots, x_n)$. This categorical formulation allows one to show that every group is a homomorphic image of a free group. It is relatively easy to see that a free group is Abelian if and only if it is cyclic, that is has rank zero or one.

Besides the solution of the isomorphism problem for free presentations there are straightforward algorithms, within free groups, to solve both the word problem and the conjugacy problem. We will discuss these in Chapter 2. Many classes of groups were introduced because they behaved algorithmically (relative to the decision problems) like non-Abelian free groups. Small cancellation groups and torsion-free hyperbolic groups fit into this context (see Chapter 3).

The genesis of the Tarski problems is the observation that most theorems concerning free groups are independent of the rank of the free group. As an example we note the Nielsen–Schreier Theorem (see Chapter 2) which says that any subgroup of a free group is itself a free group (independent of the rank of the overgroup). Another example is the result that an Abelian subgroup of a free group, again of any rank, must be cyclic. Proceeding further, suppose that $n < m$ are positive integers. From the Nielsen–Schreier Theorem it is clear that a free group of rank n can be embedded isomorphically into a free group of rank m. Hence F_n can be embedded into F_m. Further it can be shown that a free group of any countable rank can be embedded isomorphically into a free group of rank 2. It follows that F_m can be embedded into F_n. Therefore F_n and F_m must be *similar*. The main *Tarski problem* or *Tarski conjecture* is that any two countable free groups have the same *first-order* or *elementary theory*. We introduce some terminology to put these questions more formally and then look at them more carefully in Chapter 5.

1.2 The elementary theory of groups

The elementary theory of groups is tied to first-order logic and to model theory. In this section we briefly introduce the necessary material from these areas in order to state the Tarski problems. We will look at elementary logic and model theory in more detail in Chapter 4. The starting point is a first-order language appropriate for group theory. This language, which we denote L_0, is the first-order language with equality containing a binary operation symbol $\cdot$ a unary operation symbol $^{-1}$ and a constant symbol 1. A *sentence* in L_0 is a logical expression using variables, the operational symbols above, equality, logical connectives and quantifiers. A *universal sentence* of L_0 is one of the form $\forall \overline{x}\{\phi(\overline{x})\}$ where $\overline{x}$ is a tuple of distinct variables, $\phi(\overline{x})$ is a formula of L_0 containing no quantifiers and containing at most the variables of $\overline{x}$. For example

$$\forall(x, y)\{xy = yx\}$$

is a universal sentence describing an Abelian group. Similarly an *existential sentence* is one of the form $\exists \overline{x}\{\phi(\overline{x})\}$ where $\overline{x}$ and $\phi(\overline{x})$ are as above. For example

$$\exists(x, y)\{xy \neq yx\}$$

is an existential sentence describing a non-Abelian group. A *universal-existential sentence* is one of the form $\forall \overline{x} \exists \overline{y}\{\phi(\overline{x}, \overline{y})\}$. Similarly defined is an *existential-universal sentence*. It is known that every sentence of L_0 is logically equivalent to one of the form $Q_1x_1 \ldots Q_nx_n\phi(\overline{x})$ where $\overline{x} = (x_1, \ldots, x_n)$ is a tuple of distinct variables, each Q_i for $i = 1, \ldots, n$ is a quantifier, either $\forall$ or $\exists$, and $\phi(\overline{x})$ is a formula of L_0 containing no quantifiers and containing freely at most the variables $x_1, \ldots, x_n$. Further vacuous quantifications are permitted. Finally a *positive sentence* is one logically equivalent to a sentence constructed using (at most) the connectives $\vee, \wedge, \forall, \exists$, that is no negations are allowed.

If G is a group then the *universal theory* of G consists of the set of all universal sentences of L_0 true in G. Since any universal sentence is equivalent to the negation of an existential sentence it follows that two groups have the same universal theory if and only if they have the same *existential theory*. The set of all sentences of L_0 true in G is called the *first-order theory* or the *elementary theory* of G. We denote this by $Th(G)$. We note that being *first-order* or *elementary* means that in the intended interpretation of any formula or sentence all of the variables (free or bound) are assumed to take on as values only individual group elements – never, for example, subsets of, nor functions on. the group in which they are interpreted.

We say that two groups G and H are *elementarily equivalent* (symbolically $G \equiv H$) if they have the same first-order theory, that is $Th(G) = Th(H)$.

Group monomorphisms which preserve first-order formulas are called elementary embeddings. Specifically, if H and G are groups and

$$f : H \to G$$

is a monomorphism then f is an *elementary embedding* provided whenever $\phi(x_0, \ldots, x_n)$ is a formula of L_0 containing free at most the distinct variables $x_0, \ldots, x_n$ and $(h_0, \ldots, h_n) \in H^{n+1}$ then $\phi(h_0, \ldots, h_n)$ is true in H if and only if $\phi(f(h_0), \ldots, f(h_n))$ is true in G. If H is a subgroup of G and the inclusion map $i : H \to G$ is an elementary embedding then we say that G is an *elementary extension* of H.

With these basic ideas we can formally state the Tarski conjectures. The concept of a free factor will be introduced in Chapter 2.

Conjecture (Tarski Conjecture 1). *Any two non-Abelian free groups are elementarily equivalent. That is any two non-Abelian free groups satisfy exactly the same first-order theory.*

Conjecture (Tarski Conjecture 2). *If the non-Abelian free group H is a free factor in the free group G then the inclusion map $H \to G$ is an elementary embedding.*

Clearly the second conjecture is stronger than the first and implies the first. If the second conjecture were true then the theory of the non-Abelian free groups would be *complete*, that is given a sentence ϕ of L_0 then either ϕ is true in every non-Abelian free group or ϕ is false in every non-Abelian free group.

In addition to the completeness of the theory of the non-Abelian free groups the question of its *decidability* also arises. The *decidability* of the theory of non-Abelian free groups means the question of whether there exists a recursive algorithm which, given a sentence ϕ of L_0 decides whether or not ϕ is true in every non-Abelian free group. Tarski further conjectured that the theory of the non-Abelian free groups is decidable.

Conjecture (Tarski Conjecture 3). *The elementary theory of the non-Abelian free groups is decidable.*

As an outgrowth of the development of algebraic geometry over groups initiated by G. Baumslag, A. Myasnikov and V. Remeslennikov [16, 17] significant progress has been made on the Tarski problems. Kharlampovich and Myasnikov [152, 153, 154, 155, 156, 157], using this algebraic geometry, and independently Sela [232, 233, 234, 235, 236, 237], using what he termed diophantine geometry over groups, proved that all of the above Tarski conjectures are indeed true. Both proofs grew out of work on equations over free groups, fully residually free groups (in Sela's terminology limit groups) and on the above mentioned algebraic geometry of groups. Their methods are the genesis of these notes. We will outline the method of proof in Chapter 9. The following outlines the fundamental results of Kharlampovich and Myasnikov and Sela.

Theorem 1.2.1. *If the non-Abelian free group H is a free factor in the free group G then the inclusion map $H \to G$ is an elementary embedding. Hence any two non-Abelian free groups are elementarily equivalent. Moreover the theory of the non-Abelian free groups is decidable.*

1.3 Overview of this monograph

This monograph is laid out in the following manner. In Chapter 2 we will go over the necessary material from Combinatorial Group Theory. This will include the material on free groups and group amalgams. In Chapter 3 we describe the essential ideas involved in Geometric Group Theory. In Chapter 4 we will introduce formally the ideas from first-order languages and model theory, most of which are not as well-known to group theorists as they should be. We will also review the concepts of filters, ultra-filters and ultra-products which are essential tools in the study of elementary properties. In Chapter 5 we will give a more formal description of the Tarski problems. In Chapters 6, 7 and 8 we lay the groundwork for the solution of the Tarski problems. In Chapters 6 and 7 we introduce residually free groups and related concepts and talk extensively about their structure. In Chapter 7, the basic framework of algebraic geometry over groups will be devloped. In Chapter 9 we outline the Kharlampovich–Myasnikov proof of the Tarski problems. This involves what is termed the implicit function theorem for free groups. In Chapter 10 we look at discriminating, co-discriminating and squarelike groups, concepts which arose from the techniques of algebraic geometry but which have developed a life of their own. Finally in Chapter 11 we list open questions and make some further comments.

2 Combinatorial group theory

2.1 Combinatorial group theory

Most of the techniques, as well as the motivation for the study of the elementary theory of groups arise, in combinatorial group theory, especially the theory of free groups. In this chapter we review the basic concepts from this discipline.

Combinatorial group theory is roughly that branch of group theory which studies groups via their presentations, that is by generators and relations. Recall that a free group can be considered as a group with a free generating system for which there are no nontrivial relations. Free groups arise in many areas of mathematics as well as in pure mathematical logic and serve as motivating examples for many of the important questions concerning the elementary theory of groups. As a starting point then for our study of elementary theory we must look at the theory of free groups and the related idea of group amalgams.

Group products or *group amalgams* are the key constructions in infinite group theory. These products will play a pivotal role in the elementary theory of groups. The general idea of group amalgams is to decompose (if possible) an infinite group G into an amalgam (in a way which we will describe shortly) of some of its subgroups. These subgroups are then called the *factors* of G. Information about G can then be deduced from the corresponding information on the factors. Thus amalgam decompositions play a role in infinite group theory similar to a prime factorization theorem – although the amalgam decomposition of a group G need not be unique.

There are essentially two different types of group amalgams – *free products with amalgamation* and *HNN groups*. An infinite group, however, may decompose as both a free product with amalgamation or in a different manner as an HNN group. Free products are a special case of free products with amalgamation, so we discuss these first.

Before beginning we note that there are two main approaches to the theory of group amalgams. The first is a classical combinatorial approach which deals primarily with presentations for the group and its factors. The second approach is a geometric-topological technique which depends on how the group acts (as a group of isometries) on a graph. The second method is due to Bass and Serre.

2.2 Free groups and free products

There are several standard references on group amalgams. For most of the proofs in this chapter we refer to these. A complete treatment of the combinatorial approach is given in the books by Magnus–Karrass–Solitar [178] and Lyndon–Schupp [176]. The Bass–Serre Theory can be found in Serre's book Trees [238]. Other approaches to Bass–

Serre theory can be found in the articles by G. P. Scott and C. T. C. Wall [228], M. Tretkoff [248], in the Springer lecture notes by Dicks [66] and the book by Cohen [56].

Much of the theory of group amalgams arises from the theory of free groups so we discuss these first.

Definition 2.2.1. A group F is *free on a subset* X if every map $f: X \to G$ can be extended to a unique homomorphism $f: F \to G$. X is a *free basis* for F and in general a *free group* is a group with a free basis.

The cardinality of a free basis $|X|$ for a free group F is unique and is called the *rank of* F. If $|X| < \infty$, F is of *finite rank*. If F has rank n and $X = \{x_1, \ldots, x_n\}$ we say that F is free on $\{x_1, \ldots, x_n\}$. We denote this by $F(x_1, x_2, \ldots, x_n)$ or by just F_n.

Free groups F exist and can be realized combinatorially by taking a set X as a set of generators for F and having no non-trivial relations. In particular:

Theorem 2.2.1. *F is a free group if and only if F has a presentation of the form $F = \langle X; \ \rangle$.*

This theorem depends on the concept of a *freely reduced word*. If F is free on X, a *freely reduced word* in F on X is a word of the form $w = x_{v_1}^{e_1} x_{v_2}^{e_2} \ldots x_{v_n}^{e_n}$ where $x_{v_i} \in X$, $x_{v_i} \neq x_{v_{i+1}}$ if $e_i = -e_{i+1}$ and $e_i = \pm 1$. The *free length* of this word w is then $L(w) = |w| = n$. The word 1, that is the word with $n = 0$, is also considered a reduced word called the empty word. Theorem 2.2.1 is equivalent to:

Theorem 2.2.2. *F is a free group if and only if there is a generating set X such that every element of F has a unique representation as a freely reduced word on X.*

Reduced words provide *normal forms* for elements in F. An important concept is the following: a freely reduced word $W = x_{v_1}^{e_1} x_{v_2}^{e_2} \ldots x_{v_n}^{e_n}$ is *cyclically reduced* if $v_1 \neq v_n$ or if $v_1 = v_n$ then $e_1 \neq -e_n$. Clearly then every element of a free group is conjugate to an element given by a cyclically reduced word.

The significance of free groups stems from the following result, which is easily deduced from the definition. Let G be any group and F the free group on the elements of G considered as a set. The identity map $f: G \to G$ can be extended to a homomorphism of F onto G, therefore:

Theorem 2.2.3. *Every group G is a homomorphic image of a free group.*

We note that we may state this alternatively as follows. Let G be any group and X a generating systems for G. Let F be the free group on X. Then the identity map $f: G \to G$ can be extended to a homomorphism of F onto G.

This provides another view of group presentations. Suppose

$$G = \langle X; R \rangle$$

and let F be a free group on X. Then

$$G = F/N$$

for some normal subgroup N of F. Each element of N must map onto the identity in G and hence the corresponding word in N defines a relator in G. Hence each word in N is a consequence of the complete set of relators R. Setting this in the free group F this says that N is the normal closure within F, that is the smallest normal subgroup within F containing R.

The theory of free groups has a large and extensive literature. We state several simple properties.

Theorem 2.2.4. *A free group is torsion-free.*

Next, it can be shown that in a free group, two elements g_1, g_2 are conjugate if and only if a cyclically reduced word for g_1 is a cyclic permutation of a cyclically reduced word for g_2. From this we deduce:

Theorem 2.2.5. *An Abelian subgroup of a free group must be cyclic.*

Finally a celebrated theorem of Nielsen and Schreier states that a subgroup of a free group must be free. As we will see there is a corresponding type of subgroup theorem for each of the amalgam structures.

Theorem 2.2.6 (Nielsen–Schreier). *A subgroup of a free group is itself a free group.*

There are several different proofs of this result (see [178]) with the most straightforward being topological in nature. We give a simple topological proof in the next section. A proof using Bass–Serre Theory is found in [238].

Nielsen, using a technique now called Nielsen transformations in his honor, first proved this theorem about 1920 for finitely generated subgroups. Shreier shortly after found a combinatorial method to extend this to arbitrary subgroups. A complete version of the original combinatorial proof appears in [178] and [42] and in the notes by Johnson [141]. We will sketch a proof using Nielsen transformations in Section 2.6.

Scheier's combinatorial proof also allows for a description of the free basis for the subgroup. In particular, let F be free on X, and $H \subset F$ a subgroup. Let $T = \{t_\alpha\}$ be a complete set of right coset representatives for $F \bmod H$ with the property that if $t_\alpha = x_{v_1}^{e_1} x_{v_2}^{e_2} \dots x_{v_n}^{e_n} \in T$, with $\epsilon_i = \pm 1$ then all the initial segments $1, x_{v_1}^{e_1}, x_{v_1}^{e_1} x_{v_2}^{e_2}$, etc. are also in T. Such a system of coset representatives can always be found and is called a *Schreier system* or *Schreier transversal* for H. If $g \in F$ let $\overline{g}$ represent its coset representative in T and further define for $g \in F$ and $t \in T$, $S_{tg} = tg(\overline{tg})^{-1}$. Notice that $S_{tg} \in H$ for all t, g. We then have:

Theorem 2.2.7 (Explicit Form of Nielsen–Schreier). *Let F be free on X and H a subgroup of F. If T is a Schreier transversal for $F \bmod H$ then H is free on the set $\{S_{tx}; t \in T, x \in X, S_{tx} \neq 1\}$.*

Example 2.2.1. Let F be free on $\{a, b\}$ and $H = F(X^2)$ the normal subgroup of F generated by all squares in F.

Then $F/F(X^2) = \langle a, b; a^2 = b^2 = (ab)^2 = 1\rangle = \mathbb{Z}_2 \times \mathbb{Z}_2$. It follows that a Schreier system for F mod H is $\{1, a, b, ab\}$ with $\overline{a} = a, \overline{b} = b$ and $\overline{ba} = ab$. From this it can be shown that H is free on the generating set

$$x_1 = a^2, \quad x_2 = bab^{-1}a^{-1}, \quad x_3 = b^2, \quad x_4 = abab^{-1}, \quad x_5 = ab^2a^{-1}.$$

The theorem also allows for a computation of the rank of H given the rank of F and the index. Specifically:

Lemma 2.2.1. *Suppose F is free of rank n and $|F : H| = k$. Then H is free of rank $nk - k + 1$.*

From the example we see that F is free of rank 2, H has index 4 so H is free of rank $(2)(4) - 4 + 1 = 5$.

As described in Chapter 1, the genesis of the Tarski problems is the observation that most theorems concerning free groups are independent of the rank of the free group. This observation follows for the most part from the Nielsen–Schreier Theorem. Suppose that $n < m$ are positive integers. It is clear that a free group of rank n can be embedded isomorphically into a free group of rank m. Hence F_n can be embedded into F_m. From the Nielsen–Schreier Theorem and the Reidemeister–Schreier Method, which we will describe below, it can be shown that a free group of any countable rank can be embedded isomorphically into a free group of rank 2. It follows that F_m can be embedded into F_n. Therefore F_n and F_m must be *similar*. Hence the main *Tarski problem* or *Tarski conjecture* is that any two countable free groups have the same *first-order* or *elementary theory*. We will make this precise as we proceed.

The second version of the Nielsen–Schreier theorem is the basis for the *Reidemeister–Schreier process*. This is a method to find presentations for subgroups of a group G when given a presentation of G. We give a brief description. A complete description and a verification of its correctness is found in [178] and in [42].

Let G be a group with the presentation $\langle a_1, \ldots a_n; R_1 = \cdots = R_k = 1\rangle$. Let H be a subgroup of G and T a Schreier system for G mod H defined analogously as above.

Process (Reidemeister–Schreier Process). Let G, H and T be as above. Then H is generated by the set

$$\{S_{ta_v}; t \in T, a_v \in \{a_1, \ldots, a_n\}; S_{ta_v} \neq 1\}$$

with a complete set of defining relations given by conjugates of the original relators rewritten in terms of the subgroup generating set.

In order to actually rewrite the relators in terms of the new generators we use a mapping τ on words on the generators of G called the *Reidemeister rewriting process*. This map is defined as follows:

If $\quad W = a_{v_1}^{e_1} a_{v_2}^{e_2} \ldots a_{v_j}^{e_j}$ with $e_i = \pm 1$ defines an element of H

then $\quad \tau(W) = S_{t_1, a_{v_1}}^{e_1} S_{t_2, a_{v_2}}^{e_2} \ldots S_{t_j, a_{v_j}}^{e_j}$

where t_i is the coset representative of the initial segment of W preceding a_{v_i} if $e_i = 1$ and t_i is the representative of the initial segment of W up to and including $a_{v_i}^{-1}$ if $e_i = -1$. The complete set of relators rewritten in terms of the subgroup generators is then given by

$$\{\tau(tR_it^{-1})\} \quad \text{with } t \in T \text{ and } R_i \text{ runs over all relators in } G\,.$$

We present two examples; one with a finite group and then an important example with a free group which shows that a countable free group contains free subgroups of arbitrary ranks.

Example 2.2.2. Let $G = A_4$ be the alternating group on 4 symbols. Then a presentation for G is

$$G = A_4 = \langle a, b; a^2 = b^3 = (ab)^3 = 1\rangle\,.$$

Let $H = A_4'$ the commutator subgroup. We use the above method to find a presentation for H. Now

$$G/H = A_4/A_4' = \langle a, b; a^2 = b^3 = (ab)^3 = [a, b] = 1\rangle = \langle b; b^3 = 1\rangle\,.$$

Therefore $|A_4 : A_4'| = 3$. A Schreier system is then $\{1, b, b^2\}$. The generators for A_4' are then

$$X_1 = S_{1a} = a, \quad X_2 = S_{ba} = bab^{-1}, \quad X_3 = S_{b^2a} = b^2ab$$

while the relations are

(1) $\tau(aa) = S_{1a}S_{1a} = X_1^2$
(2) $\tau(baab^{-1}) = X_2^2$
(3) $\tau(b^2aab^{-2}) = X_3^2$
(4) $\tau(bbb) = 1$
(5) $\tau(bbbbb^{-1}) = 1$
(6) $\tau(b^2bbbb^{-2}) = 1$
(7) $\tau(ababab) = S_{1a}S_{ba}S_{b^2a} = X_1X_2X_3$
(8) $\tau(bababab b^{-1}) = S_{ba}S_{b^2a}S_{1a} = X_2X_3X_1$
(9) $\tau(b^2abababb^{-2}) = S_{b^2a}S_{1a}S_{ba} = X_3X_1X_2$

Therefore after eliminating redundant relations and using that $X_3 = X_1X_2$ we get as a presentation for A_4',

$$\langle X_1, X_2; X_1^2 = X_2^2 = (X_1X_2)^2 = 1\rangle\,.$$

Example 2.2.3. Let $F = \langle x, y; \rangle$ be the free group of rank 2. Let H be the commutator subgroup. Then

$$F/H = \langle x, y; [x, y] = 1\rangle = \mathbb{Z} \times \mathbb{Z}$$

a free Abelian group of rank 2. It follows that H has infinite index in F. As Schreier coset representatives we can take

$$t_{m,n} = x^m y^n, \quad m = 0, \pm1, \pm2\dots, \quad n = 0, \pm1, \pm2\dots$$

The corresponding Schreier generators for H are

$$x_{m,n} = x^m y^n x^{-m} y^{-n}, \quad m = 0, \pm 1, \pm 2, \ldots, \quad n = 0, \pm 1, \pm 2 \ldots$$

The relations are only trivial and therefore H is free on the countable infinitely many generators above. It follows that a free group of rank 2 contains as a subgroup a free group of countably infinite rank. Since a free group of countable infinite rank contains as subgroups free groups of all finite ranks it follows that a free group of rank 2 contains as a subgroup a free subgroup of any arbitrary finite rank.

Theorem 2.2.8. *Let F be free of rank 2. Then the commutator subgroup F' is free of countable infinite rank. In particular a free group of rank 2 contains as a subgroup a free group of any finite rank n.*

Lemma 2.2.2. *Let n, m be any pair of positive integers $n, m \geq 2$ and F_n, F_m free groups of ranks n, m respectively. Then F_n can be embedded into F_n and F_m can be embedded into F_m.*

Closely related to free groups in both form and properties are free products of groups. Let $A = \langle a_1, \ldots; R_1 = 1, \ldots \rangle$ and $B = \langle b_1, \ldots; S_1 = 1, \ldots \rangle$ be two groups. We consider A and B to be disjoint. Then:

Definition 2.2.2. The *free product* of A and B denoted $A * B$ is the group G with the presentation $\langle a_1, \ldots, b_1, \ldots; R_1 = 1, \ldots, S_1 = 1, \ldots \rangle$, that is the generators of G consist of the disjoint union of the generators of A and B with relators taken as the disjoint union of the relators R_i of A and S_j of B. A and B are called the *factors* of G.

In an analogous manner the concept of a free product can be extended to an arbitrary collection of groups.

Definition 2.2.3. If $A_\alpha = \langle \text{gens } A_\alpha; \text{rels } A_\alpha \rangle$, $\alpha \in \mathcal{I}$, is a collection of groups, then their free product $G = *A_\alpha$ is the group whose generators consist of the disjoint union of the generators of the A_α and whose relators are the disjoint union of the relators of the A_α.

Free products exist and are non-trivial if the factors are non-trivial. We have:

Theorem 2.2.9. *Let $G = A * B$. Then the maps $A \to G$ and $B \to G$ are injections. The subgroup of G generated by the generators of A has the presentation $\langle$generators of A; relators of $A\rangle$, that is, is isomorphic to A. Similarly for B. Thus A and B can be considered as subgroups of G. In particular $A * B$ is non-trivial if A and B are.*

Free products share many properties with free groups. First of all there is a categorical formulation of free products. Specifically we have:

Theorem 2.2.10. *A group G is the free product of its subgroups A and B if A and B generate G and given homomorphisms $f_1: A \to H$, $f_2: B \to H$ into a group H there exists a unique homomorphism $f: G \to H$ extending f_1 and f_2.*

Secondly each element of a free product has a *normal form* related to the reduced words of free groups. If $G = A * B$ then a *reduced sequence* or *reduced word* in G is a sequence $g_1 g_2 \dots g_n$ with $g_i \neq 1$, each g_i in either A or B and g_i, g_{i+1} not both in the same factor. Then:

Theorem 2.2.11. *Each element $g \in G = A * B$ has a unique representation as a reduced sequence. The length n is unique and is called the syllable length. The case $n = 0$ is reserved for the identity.*

A reduced word $g_1 \dots g_n \in G = A * B$ is called *cyclically reduced* if either $n \leq 1$ or $n \geq 2$ and g_1 and g_n are from different factors. Certainly every element of G is conjugate to a cyclically reduced word.

From this we obtain several important properties of free products which carry over to more general amalgams.

Theorem 2.2.12. *An element of finite order in a free product is conjugate to an element of finite order in a factor. In particular a finite subgroup of a free product is entirely contained in a conjugate of a factor.*

Theorem 2.2.13. *If two elements of a free product commute then they are both powers of a single element or are contained in a conjugate of an Abelian subgroup of a factor.*

Finally a theorem of Kurosh extends the Nielsen–Schreier theorem to free products.

Theorem 2.2.14 (Kurosh). *A subgroup of a free product is also a free product. Explicitly if $G = A * B$ and $H \subset G$ then*

$$H = F * (*A_\alpha) * (*B_\beta)$$

where F is a free group and $(*A_\alpha)$ is a free product of conjugates of subgroups of A and $(*B_\beta)$ is a free product of conjugates of subgroups of B.

We note that the rank of F as well as the number of the other factors can be computed. A complete discussion of these is in [178], [42] and [176].

2.3 Group complexes and the fundamental group

Combinatorial group theory has its origins in topology and complex analysis. Especially important in the development is the theory of the fundamental group. This connection is so deep that many people consider combinatorial group theory as the study of the fundamental group – especially the fundamental group of a low-dimensional complex. This connection proceeds in both directions. The fundamental group provides methods and insights to study the topology. In the other direction the topology can be used to study the groups.

Recall that if X is a topological space then its *fundamental group* based at a point x_0, denoted $\pi(X, x_0)$, is the group of all homotopy classes of closed paths at x_0. If X is path connected then the fundamental groups at different points are all isomorphic and we can speak of the fundamental group of X which we will denote $\pi(X)$. Historically group presentations were developed to handle the fundamental groups of spaces which allowed simplicial or cellular decompositions. In these cases the presentation of the fundamental group can be read off from the combinatorial decomposition of the space.

An (abstract) *simplicial complex* or *cell complex* K is a topological space consisting of a set of points called the *vertices*, which we will denote $V(K)$ and collections of subsets of vertices called *simplexes* or *cells* which have the property that the intersection of any two simplices is again a simplex. If n is the number of vertices in a cell then $n - 1$ is called its *dimension*. Hence the set of vertices are the 0-dimensional cells and a simplex $\{v_1, \ldots, v_n\}$ is an $(n - 1)$-dimensional cell. The 1-dimensional cells are called *edges*. These have the form $\{u, v\}$ where u and v are vertices. One should think of the cells in a geometric manner so that the edges are really edges, the 2-cells are filled triangles, that are equivalent to disks and so on. The maximum dimension of any cell in a complex K is called the dimension of K. From now on we will assume that our simplicial complexes are connected.

A *graph* Γ is just a 1-dimensional simplicial complex. Hence Γ consists of just vertices and edges. If K is any complex then the set of vertices and edges is called the *1-skeleton* of K. Similarly all the cells of dimension less than or equal to 2 comprise the 2-skeleton. A connected graph with no closed paths in it is called a *tree*. If K is any complex then a *maximal tree* in K is a tree that can be contained in no other tree within K.

From the viewpoint of combinatorial group theory what is relevant is that if K is a complex then a presentation of its fundamental group can be determined from its 2-skeleton and read off directly. In particular:

Theorem 2.3.1. *Suppose that K is a connected cell complex. Suppose that T is a maximal tree within the 1-skeleton of K. Then a presentation for $\pi(K)$ can be determined in the following manner:*

Generators: all edges outside of the maximal tree T

Relations: (a) $\{u, v\} = 1$ *if* $\{u, v\}$ *is an edge in* T

(b) $\{u, v\}\{v, w\} = \{u, w\}$ *if* u, v, w *lie in a simplex of* K.

From this the following is obvious:

Lemma 2.3.1. *The fundamental group of a connected graph is free. Further its rank is the number of edges outside a maximal tree.*

A connected graph is homotopic to a wedge or bouquet of circles. If there are n circles in a bouquet of circles then the fundamental group is free of rank n. The converse is also true. A free group can be realized as the fundamental group of a wedge of circles.

An important concept in applying combinatorial group theory is that of a *covering complex*.

Definition 2.3.1. Suppose that K is a cell complex. Then a cell complex K_1 is a covering complex for K if there exists a surjection $p\colon K_1 \to K$ called a *covering map* with the property that for any cell $S \in K$ the inverse image $p^{-1}(S)$ is a union of pairwise disjoint cells in K_1 and p restricted to any of the preimage cells is a homeomorphism.

That is for each cell S in K we have

$$p^{-1}(S) = \cup S_i'$$

with s_i' a cell in K_1 and $p\colon S_i' \to S$ is a bijection for each i.

The following then becomes clear.

Lemma 2.3.2. *If K_1 is a connected covering complex for K then K_1 and K have the same dimension.*

What is crucial in using covering complexes to study the fundamental group is that there is a *Galois theory* of covering complexes and maps. The covering map p induces a homomorphism of the fundamental group which we will also call p. Then:

Theorem 2.3.2. *Let K_1 be a covering complex of K with covering map p. Then $p(\pi(K_1))$ is a subgroup of $\pi(K)$. Conversely to each subgroup H of $\pi_(K)$ there is a covering complex K_1 with $\pi(K_1) = H$. Hence there is a one-to-one correspondence between subgroups of the fundamental group of a complex K and covers of K.*

A topological space X is *simply connected* if $\pi(X) = \{1\}$. Hence the covering complex of K corresponding to the identity in $\pi(K)$ is simply connected. This is called the *universal cover* of K since it covers any other cover of K.

Based on Theorem 2.3.2 we get a very simple proof of the Nielsen–Schreier theorem.

Theorem 2.3.3 (Nielsen–Schreier). *Any subgroup of a free group is free.*

Proof. Let F be a free group. Then $F = \pi(K)$ where K is a connected graph. Let H be a subgroup of F. Then H corresponds to a cover K_1 of K. But a cover is also 1-dimensional and hence $H = \pi(K_1)$ where K_1 is a connected graph and hence H is also free. □

Using a technique, now called Nielsen transformations, Nielsen first proved this theorem about 1920 for finitely generated subgroups. Schreier found a method to extend this to arbitary subgroups. The proof above was done maybe twenty years later. We will look at the Nielsen and Schreier proofs in the next chapter.

The fact that a presentation of a fundamental group of a simplicial complex is determined by its 2-skeleton goes in the other direction also. That is given an arbitrary presentation there exists a 2-dimensional complex whose fundamental group has that presentation. Essentially given a presentation $\langle X; R\rangle$ we consider a wedge of circles with cardinality $|X|$. We then paste on a 2-cell for each relator W in R bounded by the path corresponding to the word W.

Theorem 2.3.4. *Given an arbitrary presentation $\langle X; R\rangle$ there exists a connected 2-complex K with $\pi(K) = \langle X; R\rangle$.*

We note that the books by Rotman [225] and Camps, Kühling and Rosenberger [43] have very nice detailed and accessible descriptions of groups and complexes.

Cayley, and then Dehn, introduced for each group G, a graph, now called its *Cayley graph*, as a tool to apply complexes to the study of G. The Cayley graph is actually tied to a presentation and not to the group itself. Gromov reversed the procedure and showed that by considering the geometry of the Cayley graph one could get information about the group. This led to the development of the theory of hyperbolic groups which we will introduce in Chapter 4.

Definition 2.3.2. Let $G = \langle X; R\rangle$ be a presentation. We form a graph $\Gamma(G, X)$ in the following way. Let $A = X \cup X^{-1}$. For the vertex set of $\Gamma(G, X)$ we take the elements of G, that is $V(\Gamma) = \{g\colon g \in G\}$. The edges of Γ are given by the set $\{(g, x)\colon g \in G, x \in A\}$. We call g the initial point and gx is the terminal point. That is two points g, g_1 in the vertex set are connected by an edge if $g_1 = gx$ for some $x \in A$. We have $(g, x)^{-1} = (gx, x^{-1})$. This gives an oriented graph called the *Cayley graph* of G on the generating set X.

Call x the *label* on the edge (g, x). Given a $g \in G$ then G is represented by at least one word W in A. This represents a path in the Cayley graph. The length of the word W is the length of the path. This is equivalent to making each edge have length one. If we take the distance between 2 points as the minimum path length we make the Cayley graph a metric space. This metric is called the *word metric*. If we extend this metric to all pairs of points in the Cayley graph in the obvious way (making each edge a unit real interval) then the Cayley graph becomes a geodesic metric space. Each closed path in the Cayley graph represents a relator.

By left multiplication the group G acts on the Cayley graph as a group of isometries. Further the action of G on the Cayley graph is without inversion, that is, $ge \neq e^{-1}$ if e is an edge.

If we sew in a 2-cell for each closed path in the Cayley graph we get a simply connected 2-complex called the *Cayley complex*.

2.4 Group amalgams

By a *group amalgam* we mean either a *free product with amalgamation* or an *HNN group* which we define below. As mentioned earlier, these constructions are the essential building tools for infinite groups. They generalize the concept of a free product while at the same time preserving many of the basic properties.

In this section we introduce these amalgam constructions. We first discuss free products with amalgamation. Let $A = \langle a_1, \ldots; R_1 = 1, \ldots \rangle$ and $B = \langle b_1, \ldots; S_1 = 1, \ldots \rangle$ be two groups with $H \subset A, K \subset B$ proper subgroups and $f: H \to K$ an isomorphism. Again we assume that A and B are disjoint. Then:

Definition 2.4.1. The *free product of A and B amalgamating H to K* is the group G with the presentation

$$G = \langle a_1, \ldots, b_1, \ldots; R_1 = 1, \ldots, S_1 = 1, \ldots, H = f(H) \rangle,$$

that is, the group G has as generators the disjoint union of the generators of A and B and has as relations the disjoint union of the relations of A and B together with an additional set of relations giving the subgroup isomorphism. Identifying H with its isomorphic image we say that G is the *free product of A and B with H amalgamated* denoted

$$G = A *_H B.$$

The groups A and B are called the *factors* of G.

A group G is a (non-trivial) *free product with amalgamation* or *amalgamated free product* if $G = G_1 *_H G_2$ for some groups G_1 and G_2 both non-trivial and some proper non-trivial subgroup H in G_1 and also in G_2.

Taking $H = \{1\}$ we obtain a free product. Therefore free products are just special cases of free products with amalgamation. As with free products, the factors inject into G and their intersection, as a subgroup of G, is H, the amalgamated subgroup.

Theorem 2.4.1. *Let $G = A *_H B$. Then $A \to G$ and $B \to G$ are injections. The subgroup of G generated by the generators of A has the presentation* $\langle \text{gens } A; \text{rels } A \rangle$. *Similarly for B. Thus A and B can be considered as subgroups of G and $A \cap B = H$.*

The proof of this theorem depends upon a *normal form* for elements of free products with amalgamation. Let $G = A *_H B$ and let L_1 be a set of left coset representatives for A mod H and let L_2 be a set of left coset representatives for B mod H, normalized in both cases by taking 1 to represent H. Then a *reduced sequence* or *reduced word* or *normal form* in $G = A *_H B$ is a sequence of the form

$$g_1 g_2 \cdots g_n h$$

where $h \in H, 1 \neq g_i \in L_1 \cup L_2$ and $g_1 \ldots g_n$ is a reduced word in the free product $A * B$, that is $g_{j+1} \notin L_i$ if $g_j \in L_i$.

Theorem 2.4.2 (Normal Form Theorem for Free Products with Amalgamation). *If $G = A *_H B$ then every $g \in G$ has a unique representation as a reduced sequence.*

Extending the concept from free products, a reduced word $g_1 \dots g_n h$ in $G = A *_H B$ is called *cyclically reduced* if either $n \leq 1$ or $n \geq 2$ and g_1 and g_n are from different factors. Certainly every element of G is conjugate to a cyclically reduced word. From this we obtain properties analagous to those in free groups and free products. Specifically:

Theorem 2.4.3.
(1) *An element of $G = A *_H B$ of finite order must be conjugate to an element of finite order in one of the factors. Thus a finite subgroup or more generally a bounded subgroup must be entirely contained in a conjugate of a factor.*
(2) *An Abelian subgroup of $G = A *_H B$ is*
  (a) *a conjugate of an Abelian subgroup of A or B or*
  (b) *a countable ascending union of conjugates of subgroups of H or*
  (c) *a direct product of an infinite cyclic group and a conjugate of a subgroup of H.*

The concept of a free product with amalgamation can be extended in a straightforward manner to more than two factors.

Definition 2.4.2. Let $\{G_i\}, i \in I$, be a family of groups. Let A be a group and for each i, $f_i \colon A \to G_i$ a monomorphism. Then the *free product of the G_i amalgamating A* is the quotient group of the free product $G = *_i G_i$ modulo the normal subgroup of $*_i G_i$ generated by all relations $f_i(a) = f_j(a)$ with $a \in A$ and $i, j \in I$.

As in the case of two factors, each G_i injects into G, and each element can be expressed as a normal form.

Before moving on to HNN groups we mention that there is also a categorical formulation of free products with amalgamation. Specifically

Theorem 2.4.4. *Suppose G is a group, G_1, G_2 subgroups and A a group together with injections $\theta_1 : A \to G_1, \theta_2 : A \to G_2$. Then $G = G_1 *_A G_2$ if for every group H and every pair of homomorphisms $f_1 \colon G_1 \to H$, $f_2 \colon G_2 \to H$ making the following diagram commute*

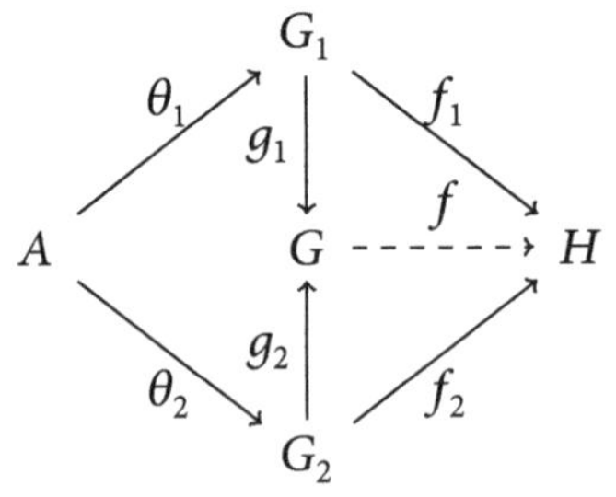

there exists a unique homomorphism $f : G \to H$ extending f_1, f_2.

Our second basic amalgam construction is that of an HNN group. This construction has properties very nearly parallel to those of free products with amalgamation. As pointed out in [238] they are really two different aspects of the same idea.

Definition 2.4.3. Let G be a group, $\{A_i\}, i \in I$, a family of subgroups of G, and for each $i \in I$, $f_i\colon A_i \to G$ a monomorphism. Then an *HNN extension of* G is a group G^* of the form

$$G^* = \langle t_i, i \in I, \text{gens}\, G; \text{rels}\, G, t_i^{-1}A_it_i = f_i(A_i), i \in I\rangle.$$

G is called the *base*, $\{t_i\}_{i\in I}$ the *free part* or *stable letters* and $\{A_i, f_i(A_i)\}$ the *associated subgroups*.

G^* is an *HNN group* if it can be expressed as an HNN extension of some base.

The base group G embeds in the HNN extension in the obvious manner.

Theorem 2.4.5. *Let G^* be an HNN extension of base G. Then G is embedded in G^* by $g \to g$, that is the subgroup of G^* generated by the generators of G, has the presentation* $\langle \text{gens}\, G; \text{rels}\, G\rangle$. *Further the free part $\{t_i\}$ is a basis for a free subgroup of G^*.*

As in all previous cases the proof of this depends upon a normal form for elements of HNN groups. However this is somewhat more intricate than that for free products with amalgamation.

Suppose that G^* is an HNN extension of G with the associated subgroups $\{A_i, f_i(A_i)\}$ and suppose we choose a fixed set of left coset representatives for A_i and $f_i(A_i)$ in G where all A_i and $f_i(A_i)$ are represented by 1. Then a *normal form* in G^* is a sequence

$$g_1t_{i_1}^{e_1}g_2t_{i_2}^{e_2}\dots t_{i_k}^{e_k}g_{k+1}, e_j = \pm 1\,,$$

where $g_1, \dots, g_{k+1}$ are elements of G such that for $j \le k$ if $e_j = 1$ then g_j is a left coset representative for A_i in G while if $e_j = -1$ then g_j is a left coset representative for $f_{i_j}(A_{i_j})$ in G and $e_j = e_{j+1}$ whenever $g_{j+1} = 1$ and $i_j = i_{j+1}$.

Theorem 2.4.6. *Every element w in G^* has a unique representation as a normal form.*

From this, as before, we obtain a classification of torsion elements as well as a classification of Abelian subgroups.

Theorem 2.4.7. *Let G^* be an HNN extension of G.*
(1) *Elements of finite order in G^* are conjugate to elements of finite order in the base G. Further finite subgroups must be contained in conjugates of the base.*
(2) *An Abelian subgroup H of G^* is one of the following:*
   (a) *A subgroup of a conjugate of the base.*
   (b) *A countable ascending union of subgroups of conjugates of the associated subgroups.*
   (c) *An HNN group with presentation*

$$\langle t', H'; \text{rels}\, H', t'^{-1}H't' = H''\rangle,$$

with $H'' \subset H'$ and H' is the intersection of the Abelian subgroup H with finitely many conjugates of the associated subgroups.

We note that it is possible for a group to be both an HNN group and a free product with amalgamation. Consider the group

$$G = \langle a, t, u; a^2 = (at)^3 = [t, u] = 1 \rangle .$$

Let $G_1 = \langle a, t; a^2 = (at)^3 = 1 \rangle$. This is a free product of a cyclic group of order 2 generated by a and a cyclic group of order 3 generated by at. Therefore t has infinite order in G. Further let $G_2 = \langle t, u; [t, u] = 1 \rangle$ a free Abelian group of rank 2. The identification $t \to t$ is then an isomorphism and G is a free product of G_1 and G_2 with the infinite cyclic subgroup generated by t amalgamated.

Now write G as $\langle u, a, t; a^2 = (at)^3 = 1, u^{-1}tu = t \rangle$. Again let $G_1 = \langle a, t; a^2 = (at)^3 = 1 \rangle = \mathbb{Z}_2 * \mathbb{Z}_3$. Then G is an HNN extension of G_1 with the single pair of associated subgroups $\langle t \rangle$ and $\langle f(t) \rangle$ where $f(t) = t$.

We note that the above presentation is a presentation for the groups $PE_2(O_d)$, the two dimensional projective elementary matrix group with entries in O_d, where O_d is the ring of integers in the quadratic imaginary number field $Q(\sqrt{-d})$ and $d \neq 1, 2, 3, 7, 11$ (see [72]).

HNN groups were originally developed by G. Higman, H. Neumann and B. Neumann (whence the name) in order to prove several important embedding theorems. In particular:

Theorem 2.4.8 (HNN). *Every countable group can be embedded in a two generator group.*

The following concept called SQ-universality was developed from the above theorem.

Definition 2.4.4. A group G is *SQ-universal* if every countable group can be embedded isomorphically as a subgroup of a quotient of G.

Thus the Higman, Neumann, Neumann theorem above says that a free group of rank 2 is SQ-universal. Many linear groups as well as groups arising from low-dimensional topology are SQ-universal. SQ-universality might be thought of as a measure of "largeness" of an infinite group.

Amalgam structures can also be mirrored in terms of cell complexes and covering space theory. This approach is based on a well-known topological result due to Seifert and Van Kampen. In 1931 Seifert and Van Kampen proved the following theorem.

Theorem 2.4.9 (Seifert-van Kampen). *Let K be a complex having connected subcomplexes L_1 and L_2 such that $L_1 \cup L_2 = K$ and $L_1 \cap L_2$ is nonempty and connected. Let*

v_0 be a vertex in $L_1 \cap L_2$. Then the fundamental group $\pi(K\colon v_0)$ is a pushout of the diagram

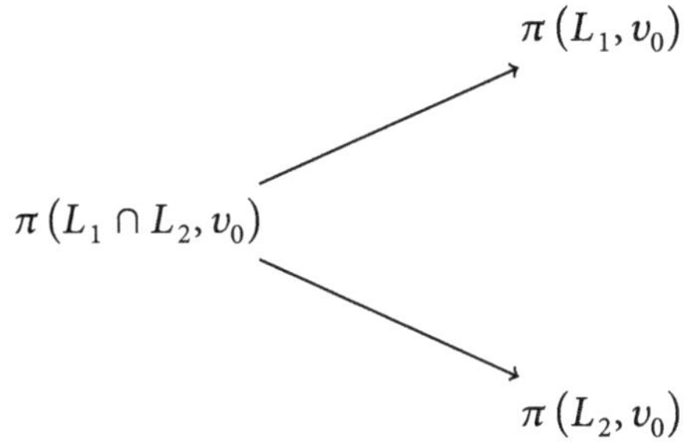

where the arrows are induced by inclusion maps of $L_1 \cap L_2$ into L_1 and L_2 respectively.

Recall that a *pushout* of groups is defined in the following way. Suppose that B, A_1, A_2 are groups and $f_1\colon B \to A_1$ and $f_2\colon B \to A_2$ are homomorphisms. We have the diagram

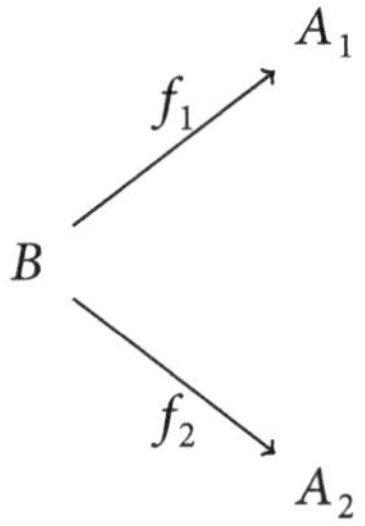

A *solution* of this diagram is a group C and homomorphisms g_1, g_2 making the following diagram commute.

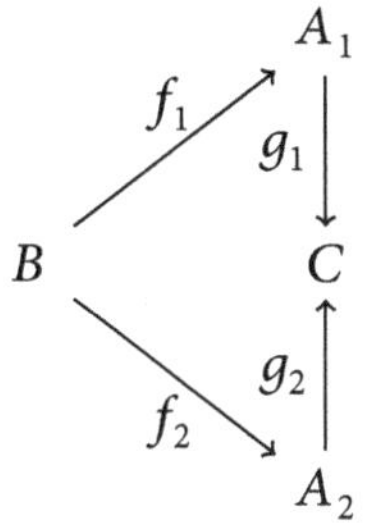

A *pushout* of the original diagram is a solution (C, g_1, g_2) such that for any other solution (D, h_1, h_2) there is a unique homomorphism $f: C \to D$ making the following diagram commute.

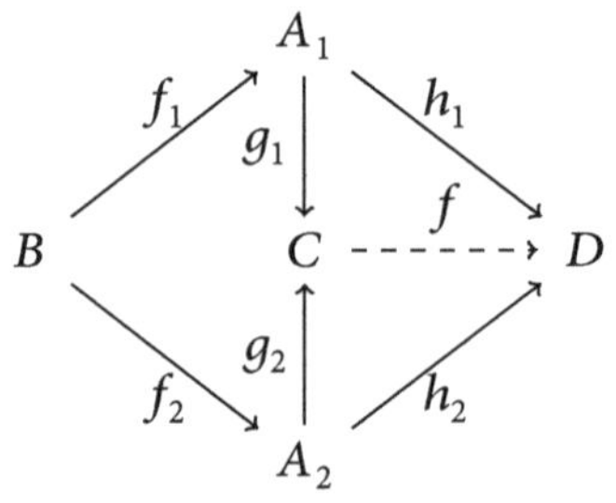

Notice that the free product of A_1 and A_2 amalgamating B is a pushout. However the fundamental group defined by the Seifert–Van Kampen Theorem is not yet the free product with amalgamation since the homomorphisms of $\pi(L_1 \cap L_2)$ in $\pi(L_1)$ and of $\pi(L_1 \cap L_2)$ in $\pi(L_2)$ may not result in isomorphic subgroups. However if these homomorphisms are both injective then the resulting group is the free product with amalgamation. Further, if $L_1 \cap L_2$ is simply connected, then the group obtained is the free product of the two fundamental groups.

Theorem 2.4.10. *Let K be a complex having connected subcomplexes L_1 and L_2 such that $L_1 \cup L_2 = K$ and $L_1 \cap L_2$ is simply connected. Then for a vertex v_0 in $L_1 \cap L_2$,*

$$\pi(K, v_0) = \pi(L_1, v_0) * \pi(L_2, v_0).$$

In particular if $L_1 \cap L_2$ is a tree this result follows.

This formulation provides a relatively easy proof of the Kurosh subgroup theorem.

HNN extensions can also be described in terms of cell complexes. Let K be a connected cell complex with disjoint homeomorphic cell subcomplexes A and B with $\phi: A \to B$ being the homeomorphism. We construct a cell complex $\overline{K}$ where A and B are identified via this isomorphism. Specifically we define $\overline{K}$ as the quotient space of $K \cup (A \times I)$ where I is the unit interval and we identify $a \in A$ with $(a, 0)$ and $\phi(a)$ with $(a, 1)$. We say that we have *added a handle* to K.

Theorem 2.4.11. *Let K be a connected cell complex with disjoint homeomorphic subcomplexes A and B. Let $\overline{K}$ be the complex formed by adding a handle to K. Then $\pi(\overline{K})$ is an HNN extension with base group $\pi(K)$.*

Although in this geometric context we assume that the subcomplexes are disjoint, in the algebraic setting, the isomorphic subgroups may intersect or even coincide. Later we will see a construction called extensions of centralizers where it is crucial that the associated subgroups coincide. As we will see from Bass–Serre theory, free products with amalgamation and HNN extensions are variations on the same idea.

A more thorough treatment of this cell complex approach as well as the proofs can be found in [225].

2.5 Subgroup theorems for amalgams

The Nielsen–Schreier theorem says that subgroups of free groups are themselves free groups while the Kurosh Theorem says that subgroups of free products are free products. There are similar results for the other amalgam constructions. In particular we have that subgroups of free products with amalgamation and subgroups of HNN groups both have a special type of HNN construction which we will describe shortly.

There are several versions of these results. In terms of combinatorial group theory the most straightforward versions are due to Karrass and Solitar [146, 147], [149]. The subgroup theorems, as well as the Kurosh Theorem, are actually special cases of a structure theorem for groups acting on trees. We refer the reader to Serre's book [238] for a complete discussion of this. We point out that it requires a detailed analysis in each case to actually recover the Karrass and Solitar results from the general structure theorem.

In order to present the Karrass–Solitar subgroup theorems we first need to extend the concept of a free product with amalgamation.

Definition 2.5.1. Let T be a tree, that is, a connected oriented graph without closed paths. Suppose that for each vertex P of T there exists a group G_P and for each edge y of T a group G_y together with an injection $G_y \to G_{t(y)}$ where $t(y)$ is the terminal vertex of y. Suppose further that $G_y = G_{\overline{y}}$ where $\overline{y}$ is the oppositely oriented edge to y. Then the group G formed by taking the free product of the groups G_P for all vertices P of T modulo the identifications induced by the edge groups is called the *tree product of the* $\{G_P\}$ *amalgamated along the* G_y. A group G is a *tree product* if it can be decomposed as above. The groups $\{G_P\}$ are called the *vertex groups*.

Notice that an ordinary free product with amalgamation $G = G_1 *_A G_2$ can be considered as a tree product with two vertex groups G_1 and G_2. If the number of vertices and edges of the tree is finite we call G a finite tree product. Tree products possess many of the same properties as ordinary free products with amalgamation. For our purposes what is important is that each vertex group G_P injects into the tree product.

We now state the Karrass–Solitar result for subgroups of free products with amalgamation.

Theorem 2.5.1. *Let $G = G_1 *_A G_2$ be a non-trivial free product with amalgamation. If H is a subgroup of G then H is an HNN group*

$$H = \langle t_1, \ldots, t_n, \ldots, S; \text{rels } S, t_1^{-1} S_1 t_1 = f_1(S_1), \ldots \rangle$$

whose base S is a tree product. Each vertex group in the base is a conjugate of $G_1 \cap H$ or $G_2 \cap H$ and each amalgamated subgroup is a conjugate of A intersected with H. Further the associated subgroups $\{S_i, f_i(S_i)\}$are also conjugates of A intersected with H and each associated subgroup is contained in a vertex group.

An HNN group whose base is a tree product and where each associated subgroup is a subgroup of a vertex group is called a *treed HNN group*. The above theorem says that subgroups of free products with amalgamation are treed HNN groups. We note also that the free part might be trivial so the subgroup itself is just a tree product.

Karrass and Solitar also explicitly describe the vertex groups, amalgamated subgroups and associated subgroups in terms of double coset representatives for H (see [146]).

Notice that if the amalgamated subgroup A is trivial then H is the free product of the free part and conjugates of subgroups of the factors. This recovers the Kurosh Theorem.

Karrass and Solitar have also proven a corresponding subgroup theorem for HNN groups.

Theorem 2.5.2. *Suppose G is an HNN group with a presentation*

$$G = \langle t_1, \ldots, t_n, \ldots, K; \text{rels } K, t_i^{-1} L_i t_i = f_i(L_i), i = 1, \ldots \rangle.$$

Then any subgroup H of G is a treed HNN group. Further the vertex groups of the base of H are conjugates of the base K intersected with H while the amalgamated subgroups in the base of H are conjugates of the L_i intersected with H and the non-trivial associated subgroups in H are conjugates of K or of the L_i intersected with H.

2.6 Nielsen transformations

One of the main tools in the study of free groups and related constructions involving infinite groups is the linear cancellation method using *Nielsen transformations*. Introduced originally by Nielsen [203] to prove the subgroup theorem for free groups, they can be considered as the basic transformations in moving from one generating system of any finitely generated group to another generating system. Along these lines, Nielsen also proved that the Nielsen transformations generate the automorphism group for any free group of finite rank. Nielsen transformations can be considered as the non-commutative analogs of row reduction of matrices and have proved to be indispensible in the theory of free groups.

The theory of Nielsen transformations has been extended to free products with amalgamation by H.Zieschang [257], and this theory has been further refined by Rosenberger [219, 220, 221], Kalia and Rosenberger [143] and Collins and Zieschang [59]. It has also been extended to HNN groups by Peczynski and Reiwer [210]. Using the theory many subgroup results on these constructions can be deduced. In addition several important results, such as the Kurosh theorem, can be reproved using the Nielsen technique.

In this section we first describe the basic notation of Nielsen transformations for free groups and some applications. We then show how the theory can be extended to group amalgams. Details can be found in either the original papers cited above or in the survey paper by Fine, Rosenberger and Stille [105]. Other results on the Nielsen method which we will use in this book will be introduced when needed.

For our purposes, we consider all groups G to be countable and only consider finite subsets $\{x_1, \ldots, x_n\}, n \geq 1$, in G. For a finite subset $\{x_1, \ldots, x_n\}, n \geq 1$, in G we define an *elementary Nielsen transformation* as a transformation of one of the following five types:

(N1) replace $(x_1, x_2, x_3, \ldots, x_n)$ by $(x_2, x_1, x_3, \ldots, x_n)$;
(N2) replace $(x_1, \ldots, x_{n-1}, x_n)$ by $(x_n, x_1, \ldots, x_{n-1})$;
(N3) replace $(x_1, x_2, \ldots, x_n)$ by $(x_1^{-1}, x_2, \ldots, x_n)$;
(N4) replace $(x_1, x_2, \ldots, x_n)$ by $(x_1 x_2, x_2, \ldots, x_n)$;
(N5) delete some x_i where $x_i = 1, 1 \leq i \leq n$.

A *Nielsen transformation* is a finite product of elementary Nielsen transformations. It is a *regular* Nielsen transformation if there is no factor of type (N5) otherwise it is *singular*. Each elementary Nielsen transformation of type (N1), (N2), (N3) or (N4) has an inverse which is a regular Nielsen transformation. It follows then that the regular Nielsen transformations form a group which contains every permutation of the set $\{x_1, \ldots, x_n\}$. For a subset $\{x_1, \ldots, x_n\}$ we let $\langle x_1, \ldots, x_n \rangle$ denote the subgroup of G that they generate. If $\{x_1, \ldots, x_n\}$ is carried by a Nielsen transformation into $\{y_1, \ldots, y_m\}, 1 \leq m \leq n$, then clearly $\langle x_1, \ldots, x_n \rangle = \langle y_1, \ldots, y_m \rangle$, that is they generate the same subgroup of G. If this Nielsen transformation is regular then we must have that $m = n$ in which case we say that $\{x_1, \ldots, x_n\}$ is *Nielsen equivalent* to $\{y_1, \ldots, y_m\}$.

In connection with Nielsen transformations in finite generating sets of a subgroup of a group, we always assume that this subset never contains both an element and its inverse, unless they are equal.

Now let F be a free group with fixed basis A. The length $L(w)$ of an element from F is the length of the reduced word for w. This clearly depends on the basis A. We write $u_1 \ldots u_q \equiv v_1 \ldots v_m$ for the equality together with the fact that $L(v_1 \ldots v_m) = L(v_1) + L(v_2) + \cdots + L(v_m)$, all $u_i, v_j \in F$.

Let $X = \{x_1, \ldots, x_n\}, n \geq 1$, be a finite subset of the free group F. X is called *Nielsen reduced* if for all triples of elements u, v, w from $X^{\pm 1}$ of the form $x_i^{\epsilon_i}, \epsilon_i = \pm 1$, the following conditions hold:

(R1) $u \neq 1$;
(R2) $uv \neq 1$ implies $L(uv) \geq L(u), L(v)$;
(R3) $uv \neq 1$ and $vw \neq 1$ implies that

$$L(uvw) > L(u) - L(v) + L(w) .$$

Being Nielsen reduced implies that there is not too much cancellation in multiplying elements from X.

Using in F, any fixed order relative to the basis A, which does not distinguish between inverse elements then if $X = \{x_1, \ldots, x_n\}$ is finite then X can be carried by a Nielsen transformation into some $Y = \{y_1, \ldots, y_m\}, 1 \leq m \leq n$, with Y Nielsen reduced.

If $X = \{x_1, \ldots, x_n\}, n \geq 1$, is Nielsen reduced then for each $u \in X^{\pm 1}$ there are words p_u, q_u, k_u with $k_u \neq 1$ such that $u \equiv p_u k_u q_u$ is reduced and such that if $w = u_1 \ldots u_q, q \geq 0, u_i \in X^{\pm 1}$, all $u_i u_{i+1} \neq 1$, then $k_{u_1}, \ldots, k_{u_q}$ remain uncancelled in the reduced form of w and $L(w) \geq q$. From this it is straightforward that if X is Nielsen reduced then $\langle X \rangle$ is free with X as a basis. Hence every finitely generated subgroup of a free group is free. This was Nielsen's original proof. It can be extended to remove the finitely generated restriction {see [176]}. Further if F has finite rank m and $X = \{x_1, \ldots, x_n\}, 1 \leq n \leq m$, is Nielsen reduced and generates F then $m = n$ and X is a basis for F. If $X = \{x_1, \ldots, x_n\}$ is Nielsen reduced then $x_1, \ldots, x_n$ are the shortest generators of $\langle X \rangle$ which exist, that is if $y_1, \ldots, y_n$ are other free generators of $\langle X \rangle$ and both sets are ordered according to the length L then $L(y_i) \geq L(x_i), i = 1, \ldots, n$.

It is clear that a Nielsen transformation applied to a basis of a finitely generated free group defines an automorphism. Nielsen also proved that the Nielsen transformations generate the automorphism group of a free group of finite rank.

We now describe an application which we used frequently. This application is related to the study of surface groups and co-compact Fuchsian groups. Recall again that a surface group has a presentation

$$\langle a_1, b_1, \ldots, a_g, b_g; [a_1, b_1] \ldots [a_g, b_g] = 1 \rangle \text{ in the orientable case}$$

or

$$\langle a_1, \ldots, a_g; a_1^2 \ldots a_g^2 = 1 \rangle \text{ in the non-orientable case} .$$

Any automorphism of a surface group moves the relator to a Nielsen equivalent word. What is of interest then is how free group words of the form of the surface group relator behave under Nielsen transformations. This was studied by Rosenberger [222, 223] and what was obtained was the following:

Theorem 2.6.1 ([222], [223]). *Let F be the free group on $a_1, \ldots, a_n$ and*

$$P(a_1, \ldots, a_n) = a_1^{\alpha_1} \ldots a_p^{\alpha_p} [a_{p+1}, a_{p+2}] \ldots [a_{n-1}, a_n] \in F$$

with $0 \le p \le n$, $n-p$ *even and* $\alpha_i \ge 1$ *for* $i = 1, \dots, n$. *Let* $X = \{x_1, \dots, x_m\}$ *be any finite system in* F *and let* $H = \langle X \rangle$. *Suppose that* H *contains some conjugate of* $P(a_1, \dots, a_n)$. *Then:*

(a) $\{x_1, \dots, x_m\}$ *can be carried by a Nielsen transformation into a free basis for* H *which contains a conjugate of* $P(a_1, \dots, a_n)$; *or*

(b) $\{x_1, \dots, x_m\}$ *can be carried by a Nielsen transformation into a free basis* $\{y_1, \dots, y_k\}$ *for* H *with* $m \ge k \ge n$, $y_i = za_i^{\gamma_i}z^{-1}$, $1 \le \gamma_i < \alpha_i$, $\gamma_i | \alpha_i$ *for* $i = 1, \dots, p$, $y_j = za_jz^{-1}$ *for* $j = p+1, \dots, n$ *and* $z \in F$.

A proof of this theorem using the Nielsen method can be found in either [222] or [105]. In these references are further extensions of the above result.

Lemma 2.6.1. *Let* F, $P(a_1, \dots, a_n)$, X *and* H *be as in Theorem 2.6.1. Suppose that all* α_i *are primes (not necessarily different) and that* H *contains* $P(a_1, \dots, a_n)$. *Then either* $P(a_1, \dots, a_n)$ *is a member of a basis of* H *or* $H = F$.

H. Zieschang [257] developed the Nielsen cancellation method in free products with amalgamation. This was refined by G.Rosenberger [219, 220, 221] and R. N. Kalia and Rosenberger [143]. A further refinement of this technique was given by D.Collins and H. Zieschang in [59] which we do not consider here in detail. We restrict ourselves in this book to the free product of two groups with an amalgamated subgroup, although the method works more generally.

Let $G = H_1 *_A H_2, H_1 \ne A \ne H_2$, denote the non-trivial free product of the groups H_1 and H_2 with the amalgamated subgroup $A = H_1 \cap H_2$. If $A = \{1\}$ then G is just the free product $G = H_1 * H_2$ of H_1 and H_2. We choose in each H_i, $i = 1, 2$, a system L_i of left coset representatives of A in H_i normalized by taking 1 to represent A. Each $x \in G$ has a unique representation $x = h_1 \dots h_n a$ with $a \in A$, $1 \ne h_j \in L_1 \cup L_2$ and $h_{j+1} \notin L_i$ if $h_j \in L_i$. The *length* of x denoted $L(x)$ is then defined to be n and G is then (partially) ordered by length.

In order to obtain results analagous to those in free groups it is found that the ordering defined by the length L is too coarse. Therefore, as in the free group case, we need a finer pre-ordering of G. For this purpose we define a symmetric normal form for elements $x \in G$. Take the inverses L_i^{-1} of the left coset representatives as a system of right coset representatives. Then each $x \in G$ has a unique representation

$$x = l_1 \dots l_m k_x r_m \dots r_1$$

with $m \ge 0$, $k_x \in H_1 \cup H_2$, $1 \ne l_j \in L_1 \cup L_2$, $1 \ne r_j \in L_1^{-1} \cup L_2^{-1}$ and $l_{j+1} \notin L_i$ if $l_j \in L_i$, $r_{j+1} \notin L_i^{-1}$ if $r_j \in L_i^{-1}$. Further if $k_x \in A$ then l_m and r_m belong to different H_i (if $m \ge 1$), and if $k_x \in H_i \setminus A$ then $l_m \notin H_i$, $r_m \notin H_i$ (if $m \ge 1$).

We then have $L(x) = 2m$ if $k_x \in A$ and $L(x) = 2m + 1$ if $k_x \notin A$. We call $l_1 \dots l_m$ the *leading half*, $r_m \dots r_1$ the *rear half* and k_x the *kernel* of x. One advantage of this symmetric normal form is that in forming products, cancellations can usually be reduced to free cancellations.

We now introduce an ordering $\leq$ on G. We assume that for each H_i, the system L_i of left coset representatives has a strict total order. For our applications we may assume that the groups are countable. This is no restriction if one considers a given finitely generated subgroup of G or a given finite system in G. If G is countable then just enumerate the system L_i and order it correspondingly. Let the elements of L_1 precede those of L_2. Then we order for each m the product $l_1 \dots l_m$ of left coset representatives (where $1 \neq l_j \in L_1 \cup L_2$ and $l_{j+1} \notin L_i$ if $l_j \in L_i$), first by length and second lexigraphically. Hence if $l_1 \dots l_m < l'_1 \dots l'_m$ then for any permitted l_{m+1}, l'_{m+1} we have $l_1 \dots l_{m+1} < l'_1 \dots l'_{m+1}$. If G is countable then further each product $l_1 \dots l_m$ has only finitely many predecessors of the form $l_1 \dots l_{m-1} l'_m$ (where $l'_m \in L_i$ if $l_m \in L_i$). Thus without loss of generality we assume that G is countable.

We define an ordering on the products of right coset representatives in the L_i^{-1} by taking inverses. We now extend this ordering to the set of pairs $\{g, g^{-1}\}$, $g \in G$, where the notation is chosen so that the leading half of g precedes that of g^{-1} with respect to ordering $\leq$. Then we set $\{g, g^{-1}\} \leq \{h, h^{-1}\}$ if either $L(g) < L(h)$ or $L(g) = L(h)$ and the leading half of g strictly precedes that of h, or $L(g) = L(h)$, the leading halves of g and h coincide, and the leading half of g^{-1} precedes that of h^{-1}.

Thus if $\{g, g^{-1}\} < \{h, h^{-1}\}$ and $\{h, h^{-1}\} < \{g, g^{-1}\}$ then g and h differ only in the kernel; since this can occur with $g \neq h$ then $<$ is only a pre-order.

For $g \in G$ let the leading half of $g^{\epsilon(g)}$, $\epsilon(g) = \pm 1$, precede that of $g^{-\epsilon(g)}$. A finite system $\{g_1, \dots, g_m\}$ in G is called *shorter* than a system $\{h_1, \dots, h_m\}$ if $\{g_i^{\epsilon(g_i)}, g_i^{-\epsilon(g_i)}\} < \{h_i^{\epsilon(h_i)}, h_i^{-\epsilon(h_i)}\}$ holds for all $i \in \{1, \dots, m\}$ and for at least one $i \in \{1, \dots, m\}$, $\{h_i^{\epsilon(h_i)}, h_i^{-\epsilon(h_i)}\} < \{g_i^{\epsilon(g_i)}, g_i^{-\epsilon(g_i)}\}$ fails to hold.

A finite system $\{g_1, \dots, g_m\}$ in G is said to be *Nielsen reduced* or *minimal* with respect to $<$ if $\{g_1, \dots, g_m\}$ cannot be carried by a Nielsen transformation into a system $\{h_1, \dots, h_m\}$ with $h_i = 1$ for some $i \in \{1, \dots, m\}$ and there is no system Nielsen equivalent to $\{g_1, \dots, g_m\}$ which is shorter. If G is countable then every finite system can be carried by a Nielsen transformation into a minimal system. In general, as already mentioned, for a given finite system, a suitable order can always be chosen such that this finite system can be carried by a Nielsen transformation into a minimal system.

The Nielsen reduction method in G now refers to Nielsen transformations from given systems to shorter systems and the resulting investigation of minimal systems.

An analysis of the result of H. Zieschang [257] for G {see also Rosenberger [220]} produces the following result.

Theorem 2.6.2. *Let $G = H_1 *_A H_2$. If $\{x_1, \dots, x_m\}$ is a finite system of elements in G then there is a Nielsen transformation from $\{x_1, \dots, x_m\}$ to a system $\{y_1, \dots, y_m\}$ for which one of the following cases hold:*

(i) $y_i = 1$ *for some* $i \in \{1, \dots, m\}$

(1) *Each* $w \in \langle y_1, \dots, y_m \rangle$ *can be written as* $w = \prod_{i=1}^{q} y_{\nu_i}^{\epsilon_i}, \epsilon_i = \pm 1, \epsilon_i = \epsilon_{i+1}$ *if* $\nu_i = \nu_{i+1}$ *with* $L(y_{\nu_i}) \leq L(w)$ *for* $i = 1, \dots, q$;

(ii) *There is a product* $a = \prod_{i=1}^{q} y_{\nu_i}^{\epsilon_i}, a \neq 1$ *with* $y_{\nu_i} \in A\ (i = 1, \dots, q)$ *and in one of the factors* H_j *there is an element* $x \notin A$ *with* $x^{-1}ax \in A$;

(iii) *There is a* $g \in G$ *such that for some* $i \in \{1, \dots, m\}$ *we have* $y_i \notin gAg^{-1}$, *but for a suitable natural number k we have* $y_i^k \in gAg^{-1}$;

(iv) *Of the* y_i *there are* $p \geq 1$ *contained in a subgroup of G conjugate to* H_1 *or* H_2 *and a certain product of them is conjugate to a non-trivial element of A.*

The Nielsen transformation can be chosen so that $\{y_1, \dots, y_m\}$ *is shorter (with respect to the length and a suitable order) than* $\{x_1, \dots, x_m\}$ *or the lengths of the elements of* $\{x_1, \dots, x_m\}$ *are preserved. Further if* $\{x_1, \dots, x_m\}$ *is a generating system of G then in case (v) we find* $p \geq 2$ *for in this case conjugations determine a Nielsen transformation.*

If we are interested in the combinatorial description of $\langle x_1, \dots, x_m\rangle$ *in terms of generators and relations we find again that* $p \geq 2$ *in case (v), possibly after suitable conjugations.*

There is a further theory of Nielsen reduction in HNN groups. This theory was developed by N. Peczynski and W. Reiwer [210] as a generalization of the theory over free products with amalgamation.

Let

$$K = \langle B, t; \text{rel } B, t^{-1}K_1t = K_{-1}\rangle$$

be an HNN group with *base group B*, *stable letter* or *free part t* and *associated subgroups* K_1, K_{-1}. (see [176] for additional information on the HNN construction). For an element $x \in K$ a representation

$$x = h_1t^{\epsilon_1}h_2t^{\epsilon_2}\dots h_nt^{\epsilon_n}h_{n+1}$$

with $\epsilon_i = \pm 1, h_i \in B$ is said to be *reduced* if $\epsilon_{i+1} = -\epsilon_i$ implies that $h_{i+1} \notin K_{\epsilon_{i+1}}$.

Choose left transversals R_1 of K_1 and R_{-1} of K_{-1} in B where K_1 and K_{-1} are represented by 1. Then each $x \in K$ may be uniquely represented as

$$x = l_1t^{\epsilon_1}l_2t^{\epsilon_2}\dots l_nt^{\epsilon_n}b$$

with $\epsilon_i = \pm 1, b \in B, l_i \in R_{\epsilon_i}$ and $\epsilon_i = \epsilon_{i+1}$ whenever $l_{i+1} = 1$. The *length* $L(x)$ of x is then defined to be n.

As right transversals of K_1, K_{-1}, we take the inverses R_1^{-1}, R_{-1}^{-1}. Then each $x \in K$ has a reduced representation

$$x = l_1t^{\epsilon_1}\dots l_mt^{\epsilon_m}k_xt^{\nu_m}r_m\dots t^{\nu_1}r_1$$

with $m \geq 0$, $\epsilon_i, \nu_i \in \{\pm 1\}, l_i \in R_{\epsilon_i}, r_i \in R_{-\nu_i}^{-1}$ and $k_x = h_1t^{\epsilon}h_2, h_1, h_2 \in B, \epsilon = \pm 1$, if $L(x)$ is odd or $k_x \in B$ if $L(x)$ is even. In this representation $l_1t^{\epsilon_1}\dots l_mt^{\epsilon_m}$ is called the *leading half*, $t^{\nu_m}r_m\dots t^{\nu_1}r_1$ the *rear half* and k_x the *kernel* of x respectively. The above reduced representation is then called a *symmetric form* for x.

We now introduce, as in the amalgamated free product situation, an ordering on K. We may assume that the groups are countable. This is no restriction if one considers a given finite system in K, for given a finite system a suitable order can always be chosen so that this system can be carried by a Nielsen transformation into what we will call a Nielsen reduced system. Choose a total order of the transversals R_1, R_{-1}, and order products $l_1 t^{\epsilon_1} \dots l_m t^{\epsilon_m}$ by using the lexicographic order on the sequences $(l_1, \dots, l_m)$. Next we extend this order to the set of pairs $\{g, g^{-1}\}, g \in K$, where the notation is chosen such that the leading half of g precedes that of g^{-1} with respect to the above ordering. Let $\{g, g^{-1}\} < \{h, h^{-1}\}$ if either $L(g) < L(h)$ or $L(g) = L(h)$ and the leading half of g strictly precedes that of h or $L(g) = L(h)$ and the leading halves of g and h coincide while the leading half of g^{-1} precedes that of h^{-1}. Hence if $\{g, g^{-1}\} < \{h, h^{-1}\}$ and $\{h, h^{-1}\} < \{g, g^{-1}\}$ then at most the kernels of g and h may be different.

For $g \in K$ let the leading half of $g^{\epsilon(g)}, \epsilon(g) = \pm 1$, precede that of $g^{-\epsilon(g)}$.

A finite system $\{g_1, \dots, g_m\}$ in K is called *shorter* than a system $\{h_1, \dots, h_m\}$ if $\{g_i^{\epsilon(g_i)}, g_i^{-\epsilon(g_i)}\} < \{h_i^{\epsilon(h_i)}, h_i^{-\epsilon(h_i)}\}$ holds for all $i \in \{1, \dots, m\}$ and at least for one $i \in \{1, \dots, m\}$, $\{h_i^{\epsilon(h_i)}, h_i^{-\epsilon(h_i)}\} < \{g_i^{\epsilon(g_i)}, g_i^{-\epsilon(g_i)}\}$ fails to hold.

A system $\{g_1, \dots, g_m\}$ in K is said to be *Nielsen reduced* or *minimal* with respect to $<$ if $\{g_1, \dots, g_m\}$ cannot be carried into a system $\{h_1, \dots, h_m\}$ with $h_i = 1$ for some $i \in \{1, \dots, m\}$ and there is no system Nielsen equivalent to $\{g_1, \dots, g_m\}$ which is shorter. If the group K is countable, then each finite system, as in the case of a free product with amalgamation, can be carried by a Nielsen transformation into a minimal system. In general for a given finite system a suitable order can always be chosen so that this finite system can be carried by a Nielsen transformation into a minimal system.

The following theorem is a slightly refined summary of the results of Peczynski and Reiwer [210].

Theorem 2.6.3. *Let $K = \langle t, B; rel.B, t^{-1}K_1 t = K_{-1}\rangle$ be an HNN group. If $\{x_1, \dots, x_m\}$ is a finite system of elements in K then there is a Nielsen transformation from $\{x_1, \dots, x_m\}$ to a system $\{y_1, \dots, y_m\}$ for which one of the following cases holds:*
(i) *$y_i = 1$ for some $i \in \{1, \dots, m\}$*
(ii) *Each $w \in \langle y_1, \dots, y_m\rangle$ can be written as*

$$w = \prod_{i=1}^{q} y_{\nu_i}^{\epsilon_i}$$

with $\epsilon_i = \pm 1$, and $\epsilon_i = \epsilon_{i+1}$ if $\nu_i = \nu_{i+1}$ with $L(y_{\nu_i}) \le L(w)$ for $i = 1, \dots, q$.
(iii) *Some subgroup of B contains $p, p \ge 1$, of the y_i and some product of these y_i is conjugate to a non-trivial element of K_1 or K_{-1}.*

The Nielsen transformation can be chosen in finitely many steps such that $\{y_1, \dots, y_m\}$ is shorter than $\{x_1, \dots, x_m\}$ or the lengths of the elements of $\{x_1, \dots, x_m\}$ are preserved.

We mention that the theory over HNN groups has been used extensively in studies by Fine, Roehl and Rosenberger [100], [101] and Fine, Gaglione, Spellman and Rosenberger [85] on freeness properties of subgroups of HNN groups.

2.7 Bass–Serre theory

A very powerful geometric technique for handling group amalgams was developed by Bass and Serre. It involves the action of a group G on a tree X. By analyzing this action the amalgam structure of G can be deduced. The theory also recovers, in a relatively easy manner, the main theorems of group amalgams – Nielsen–Schreier, Kurosh, Karrass–Solitar. This recovery involves a case by case analysis by amalgam type. The Bass–Serre technqiue is related to the cell complex approach discussed in Section 2.4.

In this section we give a very brief description of Bass–Serre Theory. A complete treatment is found in the book by Serre [238]. We consider a group G acting on a tree X, that is a graph without circuits. We assume further that G *acts without inversions* that is $gy \neq y^{-1}$ for all $g \in G$ and y and edge in X and y^{-1} the inversely oriented edge to y.

We first consider the connection between this action and free groups. We say that *G acts freely on the graph X* if it acts without inversion and fixes no vertex. We then get.

Theorem 2.7.1. *A group G is a free group if and only if G acts freely on a tree X.*

From the above result we obtain perhaps the simplest proof of the Nielsen–Schreier theorem. Let H be a subgroup of a free group G. From the above theorem there exists a tree X on which G acts freely. H must clearly also act freely on this tree and therefore H is free. The explicit form of the Nielsen–Schreier theorem giving the form for the generators of H can also be obtained. We again refer the reader to [238].

Theorem 2.7.1 establishes an equivalence between free groups and free actions on trees. We now establish such an equivalence between amalgams and other actions on trees.

If G acts on a tree X then a *fundamental domain for X mod G* is a subgraph T of X such that $T \to G/X$ is an isomorphism (G/X is the quotient graph). For groups acting on trees a fundamental domain exists if and only if G/X is also a tree. A segment is a graph of the form

$$P \overset{y}{\longleftrightarrow} Q$$

We then get.

Theorem 2.7.2.

(1) *Let G act on a tree X and suppose that*

$$P \overset{y}{\longleftrightarrow} Q$$

is a segment of X. Suppose that this segment is a fundamental domain for X mod G. Then

$$G = G_P *_{G_y} G_Q$$

where G_P, G_Q and G_y are the stabilizers in G of the vertices P, Q and the edge y respectively.

(2) *Conversely suppose that $G = G_1 *_A G_2$. Then there exists a tree X such that G acts on X with fundamental domain a segment*

$$P \overset{y}{\longleftrightarrow} Q$$

and such that $G_P = G_1, G_Q = G_2$ and $G_y = A$ for the respective stabilizers.

This theorem establishes an equivalence between free products with amalgamation and groups acting on trees with a segment as a fundamental domain. The next result establishes a similar equivalence between HNN groups and group actions on trees with loops as the quotient graph. A *loop* is a graph of the form

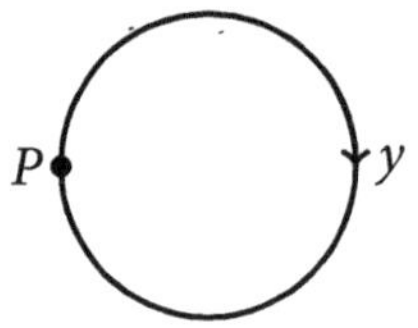

Theorem 2.7.3.

(1) *Let G act on tree X with a loop {as above} as the quotient graph for X mod G. Then $G = G_P$ is an HNN group with base G_y.*

(2) *Conversely an HNN group can be made to act on a tree so that it has a loop as a quotient graph.*

The ideas of Bass–Serre Theory have been extended to a theory of groups acting on more general tree-like structures. This has evolved, along with the notion of a hyperbolic group and an automatic group into *geometric group theory*. We introduce these ideas formally in the next chapter.

3 Geometric group theory

3.1 Geometric group theory

It is a classical idea to study a geometric object by looking at its group of isometries. Gromov [122] somewhat reversed this notion by making a group itself a geometric object. This was done by considering properties of groups whose Cayley graph (see Section 3) satisfies certain geometric properties. In particlar he introduced *hyperbolic groups* whose Cayley graph satisfies a geometric property of hyperbolic geometry. This ushered in a whole new branch of combinatorial group theory called *geometric group theory*. Gromov's original stated purpose was twofold: to generalize small cancellation theory (see Section 3.2) and to extend to hyperbolic three manifold groups the group theoretic techniques used with Kleinian groups (see [122]).

The scope of geometric group theory has been expanded beyond solely hyperbolic groups to also include the theory of groups acting on more general trees and the theory of automatic groups. What ties these all together is the relationship between the group structure and the geometric structure of its Cayley graph. Geometric group theory is used extensively in the elementary theory of groups so in this chapter we introduce the fundamental ideas: Cayley graphs, small cancellation theory, hyperbolic groups, group actions on trees and automatic groups.

3.2 The Cayley graph

Much of geometric group theory is motivated by the relationship between a group given by a presentation and a graphical representation of this presentation called its Cayley graph. If we fill in cells and make this a 2-dimensional complex we get what is called the Cayley complex. Cayley and then Dehn introduced for each group G, a graph, now called its *Cayley graph* as a tool to apply complexes to the study of G. The Cayley graph is actually tied to a presentation and not to the group itself. Gromov reversed the procedure and showed that by considering the geometry of the Cayley graph one could get information about the group. This led to the development of the theory of hyperbolic groups which we will discuss in Section 3.4.

Definition 3.2.1. Let $G = \langle X; R \rangle$ be a group presentation. We form a graph $\Gamma(G, X)$ in the following way. Let $A = X \cup X^{-1}$. For the vertex set of $\Gamma(G, X)$ we take the elements of G, that is $V(\Gamma) = \{g\colon g \in G\}$. The edges of Γ are given by the set $\{(g, x)\colon g \in G, x \in A\}$. We call g the initial point and gx is the terminal point. That is two points g, g_1 in the vertex set are connected by an edge if $g_1 = gx$ for some $x \in A$. We have $(g, x)^{-1} = (gx, x^{-1})$. This gives an oriented graph called the *Cayley graph* of G on the generating set X.

Call x the *label* on the edge (g, x). Given a $g \in G$ then g is represented by at least one word W in A. This represents a path in the Cayley graph. The length of the word W is the length of the path. This is equivalent to making each edge have length one. If we take the distance between 2 points as the minimum path length we make the Cayley graph a metric space. This metric is called the *word metric*. If we extend this metric to all pairs of points in the Cayley graph in the obvious way (making each edge a unit real interval) then the Cayley graph becomes a geodesic metric space. Each closed path in the Cayley graph represents a relator.

By left multiplication the group G acts on the Cayley graph as a group of isometries. Further the action of G on the Cayley graph is without inversion, that is $ge \neq e^{-1}$ if e is an edge.

If we sew in a 2-cell for each closed path in the Cayley graph we get a simply connected 2-complex called the *Cayley complex*.

In Figure 3.1 we present the Cayley graph for the symmetric group S_3 on 3 symbols. Recall this has the presentation

$$\langle a, b; a^2 = b^3 = (ab)^2 = 1\rangle .$$

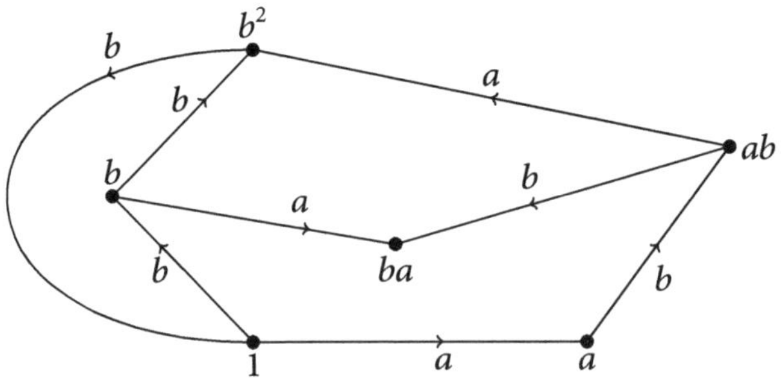

Fig. 3.1. Cayley Graph for S_3.

3.3 Dehn algorithms and small cancellation theory

Max Dehn in his pioneering work on combinatorial group theory [64] introduced the following three fundamental *group decision problems*.

(1) Word Problem: Suppose G is a group given by a finite presentation. Does there exist an algorithm to determine if an arbitrary word w in the generators of G defines the identity element of G?

(2) Conjugacy Problem: Suppose G is a group given by a finite presentation. Does there exist an algorithm to determine if an arbitary pair of words u, v in the generators of G define conjugate elements of G?

(3) Isomorphism Problem: Does there exist an algorithm to determine given two arbitary finite presentations whether the groups they present are isomorphic or not?

All three of these problems have negative answers in general (see [176]) but attempts for solutions and for solutions in restricted cases have been of central importance in the field. For this reason combinatorial group theory has always searched for and studied classes of groups in which these decision problems are solvable.

For finitely generated free groups there are simple and elegant solutions to all three problems. If F is a free group on $x_1, \ldots, x_n$ and w is a freely reduced word in $x_1, \ldots, x_n$ then $w \neq 1$ if and only if $L(w) \geq 1$. Since freely reducing any word to a freely reduced word is algorithmic this provides a solution to the word problem. Further a freely reduced word $w = x_{v_1}^{e_1} x_{v_2}^{e_2} \ldots x_{v_n}^{e_n}$ is *cyclically reduced* if $v_1 \neq v_n$ or if $v_1 = v_n$ then $e_1 \neq -e_n$. Clearly then every element of a free group is conjugate to an element given by a cyclically reduced word called a cyclic reduction. This leads to a solution to the conjugacy problem. Suppose that v and w are two words in the generators of F and $\overline{v}, \overline{w}$ are respective cyclic reductions. Then v is conjugate to w if and only if $\overline{v}$ is a cyclic permutation of $\overline{w}$. Finally two finitely generated free groups are isomorphic if and only if they have the same rank.

Dehn in 1912 [65] provided a solution to the word problem for a finitely generated orientable surface group. Dehn proved that in a surface group S_g with $g \geq 2$ any non-empty word w in the generators which represents the identity, must contain at least half of the original relator (see Section 3.5), $R = [a_1, b_1] \ldots [a_g, b_g]$ where $a_1, b_1, \ldots, a_g, b_g$ are the generators, that is if $w = 1$ in S_g, then $w = bcd$ where for some cyclic permutation R' of R, $R' = ct$ with $L(t) < L(c)$ where L represents free group length. It follows then that $w = bt^{-1}d$ in S_g and this word representation of w has shorter length than the original. Given an arbitrary w in S_g we can apply this type of reduction process to obtain shorter words. After a finite number of steps we will either arrive at 1 showing that $w = 1$ or at a word that cannot be shortened in which case $w \neq 1$. This procedure solves the word problem for S_g and is known as *Dehn's Algorithm* for a surface group. Dehn's original approach was geometric and relied on an analysis of the tessellation of the hyperbolic plane provided by a surface group.

The idea of a Dehn algorithm can be generalized in the following manner. Suppose G has a finite presentation $\langle X; R \rangle$ (R here is a set of words in X). Let F be the free group on X and N the normal closure in F of R, $N = N_F(R)$ so that $G = F/N$. G, or more precisely the finite presentation $\langle X; R \rangle$, has a *Dehn Algorithm*, if there exists a finite set of words $D \subset N$ such that any non-empty word w in N can be shortened by applying a relator in D. That is, given any non-empty w in N, w has a factorization $w = ubv$ where there is an element of the form bc in D with $L(c) < L(b)$. Then applying bc to

w we have $w = uc^{-1}v$ in G where $L(uc^{-1}v) < L(ubv)$. By the same argument as in the surface group case the existence of a Dehn Algorithm leads to a solution of the word problem. Further Dehn also presented an algorithm based on the word problem algorithm to solve the conjugacy problem in surface groups.

Here we just mention that Lyndon placed the study of small cancellation theory in a geometric context and this is the way it is most often looked at.

The general idea of a Dehn algorithm is clearly that there is "not much cancellation possible in multiplying relators". Although Dehn's approach was geometric, the ideas can be phrased purely algebraically. This is the basic notion of *small cancellation theory*. This theory was initiated in 1947 by Tartakovskii [246] who showed, using purely algebraic methods, that certain groups, besides one relator groups, also satisfy a Dehn Algorithm. His conditions were, that in these groups again there is not much cancellation in multiplying relators. Greendlinger [119], Schiek [229] and Britton [37] introduced other *small cancellation conditions* and also obtained Dehn Algorithms, and thus greatly expanded the class of groups with solvable word problem. Lyndon [169, 170, 171, 172] in the mid 1960s placed the whole theory in a geometric context and thus returned to Dehn's original approach. Lyndon used this geometric approach to reprove the Freiheitssatz of Magnus concerning one-relator groups (see [90]). The geometric constructions used by Lyndon, now called Lyndon–Van Kampen diagrams, have been extended and modified for use in proving the Freiheitssatz in many different contexts (see [90]). Essentially Lyndon's geometric techniques lead to planar geometric tesselations and the solutions to the various decision problems occur when these tesselations are non-Euclidean. This led Gromov to define negatively curved or hyperbolic groups. We will see this in Section 3.4.

A complete and readable account of small cancellation theory can be found in Chapter 5 of Lyndon and Schupp's book [176]. The proofs, both algebraic and geometric, are quite complex. What we will do in this section is define the small cancellation conditions for free groups and then introduce the geometric constructions that go along with them.

Suppose F is free on a set of generators X. Let R be a *symmetrized* set of words in F. By this we mean that all elements of R are cyclically reduced and for each r in R all cyclically reduced conjugates of both r and r^{-1} are in R. If r_1 and r_2 are distinct elements of R with $r_1 = bc_1$ and $r_2 = bc_2$, then b is called a *piece*. Pieces represent those subwords of elements of R which can be cancelled by multiplying two non-inverse elements of R. The small cancellation hypotheses state that pieces must be relatively small parts of elements of R. We define three small cancellation conditions, the most common being the first which is a metric condition.

The first small cancellation condition is a metric condition denoted $T'(\lambda)$ where λ is a positive real number. This condition asserts that if r is an element of R with $r = bc$ and b a piece, then $L(b) < \lambda L(c)$. If G is a group with a presentation $\langle X; R\rangle$ where R is symmetrized and satisfies $T'(\lambda)$, then G is called a *λ-group*. So for example, if $\lambda = 1/6$, G is a *sixth group*, etc.

If p is a natural number, the second small cancellation condition is a non-metric one denoted $T(p)$. This asserts that no element of R is a product of fewer than p pieces. Notice that $T'(\lambda)$ implies $T(p)$ for $\lambda \geq 1/(p-1)$.

The final small cancellation condition is also a non-metric condition denoted $T(q)$ for q a natural number. This asserts the following: Suppose $r_1, \ldots, r_h$ with $3 \leq h < q$ are elements of R with no successive pair inverses. Then at least one of the products $r_1r_2, \ldots, r_{h-1}r_h, r_hr_1$ is reduced without cancellation. $T(q)$ is dual to $T(p)$ for $(1/p) + (1/q) = 1/2$ in a suitable geometric context [176].

Greendlinger [119] proved purely algebraically that sixth-groups satisfy a Dehn Algorithm while Schiek showed the same for fourth groups also satisfying T(4) [229]. Lyndon [169] placed the study of small cancellation theory in a geometric context and this is the way it is most often looked at. We briefly describe how this geometric interpretation arises.

Suppose that the group G has a finite presentation $\langle X; R\rangle$ with R a symmetrized set of words. As before let F be free on X and let N be the normal closure in F of R so that $G = F/N$. If w is a word in G, then $w = 1$ if and only if w as a word in F is a product of conjugates of elements of R, that is, $w = u_1r_1u_1^{-1} \ldots u_mr_mu_m^{-1}$ where the u_i are words in F and the r_i are elements of R. The sequence $u_1r_1u_1^{-1}, \ldots, u_mr_mu_m^{-1}$ is called an *R-sequence of length m* for w. A *minimal R-sequence* for w is an R-sequence of minimum length. We will associate to any R-sequence for w a connected, simply-connected *diagram* in the Euclidean plane called an *R-diagram*. The small cancellation hypotheses are examined by analyzing these diagrams.

A *Lyndon–Van Kampen diagram* for a group F consists of a collection M of pairwise distinct vertices, oriented edges and regions in the Euclidean plane together with a labelling function f assigning to each oriented edge e an element $f(e)$ of F. This labelling function must satisfy $f(e^{-1}) = f(e)^{-1}$ where e^{-1} is the oppositely oriented edge of e. Further if α is a path in M with $\alpha = e_1 \ldots e_n$, then $f(\alpha)$ is defined as $f(e_1) \ldots f(e_n)$. If D is a region in M, a label of D is an element $f(\alpha)$ for any boundary cycle of D. We have the following result which summarizes many of the geometric properties and existence of these diagrams:

Theorem 3.3.1 ([LS]). *Let F be a free group and $c_1, \ldots, c_m$ a sequence of non-trivial elements of F. Then there exists a diagram $M = M(c_1, \ldots, c_m)$ over F satisfying the following properties.*

(i) *If e is an edge of M, $f(e) \neq 1$.*
(ii) *M is connected and simply connected with a distinguished vertex 0 on the boundary of M. There is a boundary cycle $e_1, \ldots, e_n$ of M (a cycle in M of minimal length which contains all the edges in the boundary of M) beginning at 0 such that the product $f(e_1) \ldots f(e_n)$ is reduced without cancellation and $f(e_1) \ldots f(e_n) = c_1 \ldots c_m$.*
(iii) *If D is a region of M and $e_1, \ldots, e_j$ is a boundary cycle of D, then $f(e_1) \ldots f(e_j)$ is reduced without cancellation and is a cyclically reduced conjugate of some c_i.*

The next theorem provides a converse to the above result and further allows us to relate this to R-sequences.

Theorem 3.3.2 ([LS]). *Let M be a connected, simply connected diagram over a group F with regions $D_1, \dots, D_m$. Let α be a boundary cycle of M beginning at a vertex v_0 on the boundary of M and let $w = f(\alpha)$. Then there exists boundary labels r_i of D_i and elements u_i of F, $1 \leq i \leq m$, such that $w = u_1 r_1 u_1^{-1} \dots u_m r_m u_m^{-1}$.*

Now suppose that R is a symmetrized subset of words in a free group F. An *R-diagram* is a diagram M over F such that if ∂ is any boundary cycle of any region D of M, then $f(\partial)$ is in R. If $G = F/N$ as before, then from the two theorems we obtain the following fact. A word w in F is in N if and only if there exists a connected, simply connected R-diagram M such that the label on the boundary of M is w. Thus connected, simply connected diagrams provide an adequate tool for studying membership in normal subgroups of free groups. The analysis of the small cancellation conditions lies in analyzing the structure of R-diagrams under these conditions.

Historically, Van Kampen discovered these diagrams in the 1930s [251], but they were apparently not used to any great extent until Lyndon [168] and Weinbaum [254] applied them to small cancellation theory. In doing this Lyndon provided many combinatorial generalizations of regular tessellations of the plane (see also [176]), again returning to the ideas of Dehn. In particular Lyndon used Van Kampen diagrams to translate Magnus' original proof of the Freiheitssatz into combinatorial geometry. The main tool employed in this was the *maximum-minimum modulus principle* for these diagrams, which can be described in the following manner. Consider a presentation $G = \langle X; R \rangle$, and assume that X is a disjoint union of subsets X_α. The generators in each of the partitioning subsets are said to be of the same *type*.

Assume further that there is an integer-valued function assigning a subscript to each generator in X. If w is a word on X and α is a type, then $\max(\alpha 0(w))$ will denote the generator of type α with maximum subscript which occurs in w (when w involves a generator of type α). Similarly we define $\min \alpha(w)$. The presentation $\langle X; R \rangle$ is *staggered* if every relator in R contains at least one generator of each type and the following condition holds for every type α: If $i < j$, then $\max \alpha(r_i) < \max \alpha(r_j)$ and $\min \alpha(r_i) < \min \alpha(r_j)$. ($x_i < x_j$ if x_i and x_j are generators of the same type with $i < j$.) Let $\langle X; R \rangle$ be a staggered presentation and $R*$ the symmetrized set generated by R. Let M be a connected, simply connected reduced $R*$-diagram (see below). For any type α, $\max \alpha(M)$ (respectively $\min \alpha(M)$) will denote the generator of type α with maximum (resp. minimum) subscript which occurs as the label on an edge of M. The maximum-minimum modulus principle then says that for each type α, there are edges in the boundary ∂M labelled by $\max \alpha(M)$ and $\min \alpha(M)$. Lyndon's proof of the Freiheitssatz not only worked for one-relator groups but also for one-relator products where the factors are subgroups of the additive group of the reals.

A diagram over F is a *reduced diagram* if it satisfies the following additional property. Let D_1 and D_2 be regions of M with an edge $e \subseteq \partial(D_1) \cup \partial(D_2)$. Let $e\delta_1$ and

$\delta_2 e^{-1}$ be boundary cycles of D_1 and D_2 respectively. Let $f(\delta_1) = f_1$ and $f(\delta_2) = f_2$. Then one never has $f_2 = f_1^{-1}$. If M is the diagram of a minimal R-sequence, then M is reduced.

Although the original work on small cancellation theory involved using these conditions to solve the word problem, the small cancellation hypotheses have also been instrumental in handling the conjugacy problem. We mention one result in this direction (see [176]).

Theorem 3.3.3. *Let R be a finite non-empty symmetrized subset of a finitely generated free group F and and let N be the normal subgroup of F generated by R. If R satisfies one of the conditions $C(6)$, $C(4)$ and $T(4)$ or $C(3)$ and $T(6)$, then the conjugacy problem for $G = F/N$ is solvable.*

3.4 Hyperbolic groups

In hyperbolic geometry there is a universal constant A such that triangles are *A-thin*. By this we mean that if $\triangle$ is any geodesic triangle then any point on one side is at a distance less than A from some point on one of the other two sides. Now suppose G is a finitely generated group with fixed finite generating set X. Let Γ be the Cayley graph of G relative to this generating set X equipped with the word metric. A *geodesic* in the Cayley graph is a path between two points with minimal length relative to the word metric. A geodesic triangle is a triangle with geodesic sides. A geodesic triangle in Γ is *δ-thin* if any point on one side is at a distance less than δ from some point on one of the other two sides. Γ is *δ-hyperbolic* if every geodesic triangle is δ-thin. Finally G is *word-hyperbolic* or just *hyperbolic* if G is δ-hyperbolic with respect to some generating set X with $1 \notin X$ and some fixed $\delta \geq 0$. Gromov further showed that being hyperbolic is independent of the generating set although the δ may change, that is if G is hyperbolic with respect to one finite generating set it is hyperbolic with respect to all finite generating sets. For a full account of hyperbolic groups see the original paper of Gromov [122], the book [42] or the notes edited by H. Short [242].

Further suppose G is a finitely generated group with a finite presentation $\langle X; R\rangle$. If w is a freely reduced word in the finitely generated free group $F(X)$ on X of length $L(w)$ and $\overline{w} = 1$ in G then there are words $P_i \in F(X)$ and relators $R_i \in R$ such that

$$w = \prod_{i=1}^{N} P_i R_i^{\epsilon_i} P_i^{-1} \text{ in } F(X)$$

where $\epsilon_i = \pm 1$ for each i. G then satisfies a *linear isoperimetric inequality* if there exists a constant K such that for all words w we have

$$N < KL(w).$$

Gromov, and independently Lysenok and Shapiro, proved that hyperbolic groups are intimately tied to small cancellation theory in that a finitely presented group is hyperbolic if and only if it satisfies a Dehn algorithm. A summary of results of Gromov and independently of Lysenok and Shapiro (see [242]) ties all these ideas together

Theorem 3.4.1 ([242]). *The following conditions on a finitely presented group are equivalent:*
(1) *G is hyperbolic;*
(2) *G satisfies a linear isoperimetric inequality;*
(3) *G has a Dehn algorithm.*

If G is hyperbolic then it must have a Dehn algorithm. Further for a group G being finitely presented and having a Dehn algorithm implies that G is hyperbolic. It follows that all the orientable surface groups of genus greater than or equal to 2 and the nonorientable surface groups of genus greater than or equal to 4 are hyperbolic. Further Gromov showed that all fundamental groups of closed compact hyperbolic manifolds are hyperbolic. From the the existence of the Dehn algorithm we obtain the following corollaries.

Corollary 3.4.1. *A hyperbolic group G is finitely presented.*

In fact if G is δ-hyperbolic with generating set X and if we define

$$R = \{w \in F(X) \colon L(w) \leq 8\delta \text{ and } \overline{w} = 1 \text{ in } G\}$$

then $\langle X; R\rangle$ is a Dehn presentation for G. By a *Dehn presentation* we mean a group presentation with a Dehn algorithm. The Dehn algorithm provides as it does in small cancellation theory a solution to the word problem, We then have:

Corollary 3.4.2. *A hyperbolic group G has a solvable word problem.*

Gromov [122] further has proved that hyperbolic groups have solvable conjugacy problem while Sela [232] has shown that the isomorphism problem is solvable for the class of torsion-free hyperbolic groups and Dahmiani and Guiradel [62] for all hyperbolic groups.

Theorem 3.4.2. *The conjugacy problem is solvable for hyperbolic groups. The isomorphism problem is solvable for the class of hyperbolic groups.*

Corollary 3.4.3. *The isomorphism problem for one-relator groups with torsion is solvable.*

Hence, relative to the decision problems, the class of hyperbolic groups behaves like the class of finitely generated free groups. Further hyperbolic groups, especially torsion-free hyperbolic groups, share many additional properties with free groups. We mention some of them.

Theorem 3.4.3. *A hyperbolic group can contain no subgroup isomorphic to $\mathbb{Z} \times \mathbb{Z}$. In particular an Abelian subgroup of a hyperbolic group which contains an element of infinite order is cyclic by finite.*

Corollary 3.4.4. *An Abelian subgroup of a torsion-free hyperbolic group is cyclic.*

Free groups are torsion-free. Hyperbolic groups can have elements of finite order. However the existence of a Dehn algorithm assures that there are only finitely many conjugacy classes of torsion elements.

Theorem 3.4.4. *In a hyperbolic group there are only finitely many conjugacy classes of elements of finite order.*

Since subgroups of finitely presented groups need not be finitely generated it follows that subgroups of hyperbolic groups need not be hyperbolic. However there are certain geometrically defined subgroups which must themselves be hyperbolic.

A *geodesic metric space* is a metric space where between any two points is a geodesic or minimum length path. If X is a geodesic metric space then a subset A is *quasiconvex* if there is a constant ϵ such that for any geodesic $\overline{ab}$ with endpoints $a, b \in A$ then $\overline{ab}$ is within an ϵ neighborhood of A. A subgroup of a finitely generated group is quasiconvex if the vertices in the subgroup form a quasiconvex set in the Cayley graph.

Theorem 3.4.5. *A quasiconvex subgroup of a hyperbolic group is hyperbolic.*

Thinking of the Cayley graph of a free group as a tree it is clear that a finitely generated subgroup of a finitely generated free group is quasiconvex. Further an infinitely generated subgroup is not hyperbolic and hence not quasiconvex. Therefore we have.

Corollary 3.4.5. *A subgroup of a finitely generated free group is quasiconvex if and only if it is finitely generated.*

Gromov [122] states that if H is a quasiconvex subgroup of a hyperbolic group G and $g \notin H$ then there is a large enough power n such that the subgroup generated by H and g^n is their free product. This is fairly clear for finitely generated free groups. Fine and Rosenberger [95] and independently Cohen and Lustig [58] studied groups with a restricted version of this property. Specifically a group G is a *restricted Gromov group* or *RG group* if given any non-trivial elements $x, y \in G$ then the subgroup $\langle x, y\rangle$ is cyclic or there exists a positive integer t with $g^t \neq 1, h^t \neq 1$ and $\langle g^t, h^t\rangle = \langle g^t\rangle * \langle h^t\rangle$.

Bestvina and Feign [28] have shown that an amalgam of two hyperbolic groups over a cyclic subgroup is still hyperbolic. Kharlamapovich and Myasnikov [151] have a more general result that the amalgam of two hyperbolic groups is again hyperbolic whenever one of the amalgamated subgroups is quasiconvex and malnormal in its respective factor. More general results in a different direction were proved by Juhasz and Rosenberger [142].

Theorem 3.4.6. *If H_1, H_2 are hyperbolic and $H_1 \cap H_2 = H$ is a quasiconvex subgroup, malnormal in either H_1 or H_2, then the amalgamated product $H_1 *_H H_2$ is hyperbolic. In particular if W_1, W_2 are elements of infinite order in H_1 and H_2 respectively and neither is a proper power then*

$$H_1 \underset{W_1=W_2}{*} H_2$$

is hyperbolic. Since finitely generated subgroups of free groups are quasiconvex it follows that if A and B are free groups and $A \cap B = H$ is a finitely generated subgroup malnormal in either A or B then the amalgamated product $A *_H B$ is hyperbolic.

Both Kharlamapovich and Myasnikov [151] and Bestvinna and Feign [28] have unrelated results concerning the hyperbolicity of HNN extensions. If A and B are subgroups of a group G then we say that A and B are *conjugately separated* if $g^{-1}Ag \cap B = \{1\}$ for any $g \in G$. An HNN group of the form $\langle G, t : t^{-1}At = B\rangle$ where A and B are conjugately separated subgroups of G is called a *separated HNN extension.*

Theorem 3.4.7 ([151]). *A separated HNN extension*

$$\langle G, t; t^{-1}At = B\rangle$$

of a hyperbolic group G is hyperbolic if the associated subgroups A and B are quasiconvex in G and at least one is malnormal.

Theorem 3.4.8 ([28]). *Let F be a finitely generated free group with basis $a_1, \ldots, a_n$. Let $f \in \mathrm{Aut}(F)$. Then the mapping torus*

$$M = \langle a_1, \ldots, a_n, t; t^{-1}a_it = f(a_i), i = 1, \ldots, n\rangle$$

is hyperbolic if and only if the automorphism f has no non-trivial periodic conjugacy classes.

Recall that a free group acts freely on a tree and in fact this tree can be taken as its Cayley graph. Rips constructed for hyperbolic groups a simplicial complex, now called the Rips complex, on which the group acts. In particular Rips proved the following (see [232]).

Theorem 3.4.9. *A hyperbolic group G acts simplicially on a simplicial complex P satisfying*
(1) *P is contractible, locally finite and finite dimensional*
(2) *on the vertices of P, G acts freely and transitively*
(3) *the quotient complex P/G is compact*

We close the section by mentioning a conjecture due to Gersten.

Conjecture (Gersten Conjecture). *A torsion-free one-relator group is hyperbolic if and only if it contains no subgroup isomorphic to a Baumslag–Solitar group*

$$B_{m,n} = \langle x, y; yx^my^{-1} = x^n, mn \neq 0\rangle .$$

3.5 Free actions on trees: arboreal group theory

In Bass–Serre Theory (see Section 2.7) the amalgam structure of a group is determined by its action as a group of isometries on a *tree*. Here a tree is a connected graph or one-dimensional simplicial complex without loops. Particularly striking is the result for free groups which completely classifies free groups as those groups which admit free actions on trees. The use of group actions on trees has been extended to the study of group actions on more general tree-like structures. In this wider context a standard tree as above is called a *simplicial tree* or a $\mathbb{Z}$-tree. The name *arboreal group theory* has been adopted to cover all the studies of groups acting on trees.

Chiswell [47] and independently Tits (see [53] introduced the construction of an $\mathbb{R}$-tree. An $\mathbb{R}$-tree is a connected metric space which is "tree-like". A precise definition will be given below. The theory of $\mathbb{R}$-trees became prominent because of work of Morgan and Shalen [188]. Their work concerned studying a finitely generated group G by considering the space of discrete faithful representations of G as a group of orientation preserving isometries of a hyperbolic space. This space factored out by equivalent representations had a compactification. The ideal points are obtained from certain actions of G on $\mathbb{R}$-trees. The basic question on these $\mathbb{R}$-trees became whether there was a classification of groups admitting free actions, in analogy with the Bass–Serre results. The answer is yes, in the finitely generated case, (Rips Theorem below).

An $\mathbb{R}$-tree T is a non-empty metric space with metric d such that there is no subspace homeomorphic to a circle and such that if $u, v \in T$ with $r = d(u, v)$ then there exists a unique isometry $\alpha\colon [0, r] \to T$ with $\alpha(0) = u$ and $\alpha(r) = v$.

A *segment* in a $\mathbb{R}$-tree is the image of an isometry $\alpha\colon [0, r] \to T$. $\alpha(0), \alpha(r)$ are the endpoints of the segment. To see that such a structure is "tree-like" it is clear that the above definition is equivalent to the following, which says that there is a type of branching at every point. A non-empty metric space is an $\mathbb{R}$-tree if

(1) Given $u, v \in T$ there is a segment with endpoints u, v.
(2) The intersection of two segments with common endpoint is a segment.
(3) If two segments intersect in a single point which is an endpoint of both then their union is a segment.

Now suppose a group G acts on an $\mathbb{R}$-tree T. We say $g \in G$ is an *inversion* if g leaves a segment invariant but g has no fixed points. As before G acts freely on an $\mathbb{R}$-tree if there are no fixed points. An $\mathbb{R}$-*free group* is a group which acts freely and without inversions on an $\mathbb{R}$-tree. Clearly free groups are $\mathbb{R}$-free. Further free Abelian groups and all orientable surface groups of any genus as well as all non-orientable surface groups of genus ≥ 4 also act freely on $\mathbb{R}$-trees. In fact, in a sense these are the only finitely generated examples. The following result, given by Rips, (see [30] or [107]) gives the classification of finitely generated groups acting on $\mathbb{R}$-trees. The book by Chiswell [47] contains a proof.

Theorem 3.5.1 (Rips Theorem). *A finitely generated $\mathbb{R}$-free group is a free product of finitely many finitely generated free Abelian groups and surface groups.*

Morgan and Shalen [188] further extended the concept of an $\mathbb{R}$-tree to a Λ-tree where Λ is an arbitrary ordered Abelian group. This concept arose in their paper from an example of a valuation on a field where the ordered Abelian group is the valuation group. In Serre's original work on groups acting on trees [238], certain trees were constructed from discrete valuations and Morgan and Shalen's construction can be considered a generalization of this. A complete discussion of Λ-trees can be found in the book by Chiswell [47] as well as in the survey articles by Morgan [187], Shalen [239], [240], the work of Bass [6] and Alperin and Bass [4] or the papers of Chiswell [47, 48, 49, 50, 51].

Let Λ be an ordered Abelian group written additively. If for each $a, b \in \Lambda$ with $a > 0$ there exists a positive integer n such that $b < na$ then Λ is an archimedean ordered Abelian group. In particular all additive subgroups of the reals $\mathbb{R}$ are archimedean. If Λ_1 and Λ_2 are ordered Abelian groups then the direct sum $\Lambda_1 \oplus \Lambda_2$ can also be made into an ordered Abelian group with the lexicographic ordering. However the archimedean property is not necessarily preserved under this construction. For example under the lexicographic ordering $\mathbb{Z} \oplus \mathbb{Z}$ is a non-archimedan ordered Abelian group. In particular if $a = (1, 0)$ and $b = (0, 1)$ then $b < na$ for all positive integers n.

If Λ is an ordered Abelian group and X is a set then an Λ-*metric* on X is a Λ-valued function $d\colon X \times X \to \Lambda$ satisfying the ususal metric space properties; (i) $d(x, y) \geq 0$ and $d(x, y) = 0$ iff $x = y$ (ii) $d(x, y) = d(y, x)$ and (iii) $d(x, y) \leq d(x, z) + d(z, y)$. The pair (X, d) is then an Λ-metric space. In this context an ordinary metric space is just an $\mathbb{R}$-metric space.

The construction of a Λ-tree now mimics the definition of an $\mathbb{R}$-tree. A closed interval in Λ is defined as for $\mathbb{R}$: $[a, b]_\Lambda = \{x \in \Lambda; a \leq x \leq b\}$ for $a, b \in \Lambda$ with $a \leq b$ and $[b, a]_\Lambda = [a, b]_\Lambda$. As for $\mathbb{R}$ a *segment* in a Λ-metric space is the image of an isometry $\alpha\colon [a, b]_\Lambda \to X$ for some $a, b \in \Lambda$ with $a \leq b$. The endpoints of the segment are $\alpha(a), \alpha(b)$ and $a = b$ is allowed. A non-empty Λ-metric space (X, d) is an Λ-tree if

(1) (X, d) is *geodesic*. This means that if $x, y \in X$ then there is a segment in X with endpoints x, y.
(2) The intersection of two segments with common endpoint is a segment.
(3) If two segments intersect in a single point which is an endpoint of both then their union is a segment.

Chiswell proves that actions on Λ-trees are closely tied to Λ-valued Lyndon length functions. If G is a group and Λ an ordered Abelian group then a Λ-valued *Lyndon length function* is a map $L\colon G \to \Lambda$ satisfying

(1) $L(1) = 0$
(2) For all $g \in G$, $L(g) = L(g^{-1})$
(3) For all $g, h, k \in G$, $c(g, h) \geq \min(c(h, k), c(k, g))$ where $c(g, h) = 1/2(L(g) + L(h) - L(g^{-1}h))$
(4) For all $g, h \in G$, $c(g, h) \in \Lambda$.

Suppose the group G acts by isometries on the Λ-tree (X, d). Let $x \in X$. Define for each $g \in G$, $L_x(g) = d(x, gx)$. This will then define a Λ-valued Lyndon length function of G. In fact Chiswell's theorem [47] actually shows that all Λ-valued Lyndon length functions arise from such actions of the group G on Λ-trees.

Theorem 3.5.2 ([47]). *Let G be a group and $L\colon G \to \Lambda$ a Λ-valued Lyndon length function. Then there is a Λ-tree (X, d), an action of G on X and a point $x \in X$ such that $L = L_x$.*

As with simplicial trees and $\mathbb{R}$-trees the first paramount question is to examine the structure of groups which admit free actions on Λ-trees. The definitions again mimic those for $\mathbb{R}$-trees. We suppose that G acts as a group of isometries on a Λ-tree (X, d). An element $g \in G$ is an *inversion* if g leaves a segment invariant but g has no fixed points. As before G acts freely on a Λ-tree if there are no fixed points. A *Λ-free group* is a group which acts freely and without inversions on an Λ-tree. More generally a group is *tree-free* if it is Λ-free for some ordered Abelian group Λ.

It is easy to see that subgroups of Λ are Λ-free since they act on Λ by translations. Harrison [129] using Lyndon length functions proved that free products of Λ-free groups are again Λ-free. However, at present there is no general structure theorem for groups acting freely on arbitrary Λ-trees. In Chapter 4 we will discuss the connection between tree-free groups and certain questions in logic. Then we will also present certain examples due to Alperin and Bass [4] of tree-free groups.

We close this section by giving one general result on tree-free groups. Harrison [129] had proved (with a correction by Chiswell ([48]) that an $\mathbb{R}$-free group has the property that any two-generator subgroup is either free or Abelian. This is also a very easy consequence in the finitely generated case of Rips Theorem. It was conjectured in [100] that this was true in general for tree-free groups. That is, in a tree-free group any two generator subgroup is either free or Abelian. This general result was proved by Chiswell [48] and independently by Urbanski and Zamboni [250].

Theorem 3.5.3. *Let G be a tree-free group. Then any two generator subgroup of G is either free or Abelian.*

3.6 Automatic groups

We next introduce *automatic groups*. These grew out of a study of the interplay between group theory, computer science and geometric tesselations. Recall that elements of free groups are represented by unique normal forms. Further free reduction and Nielsen reduction are effective rewriting methods to transform single words or groups of words into these unique forms. This is precisely what is necessary in many applications in computer science where computer programs can be thought of as manipulations of lists of words from some given alphabet. Hence there has been a great deal of attention paid to the connections between group theory and computer science.

This has developed into the theory of *automatic groups*. A complete description of the theory can be found in the book by Epstein, Cannon, Holt, Patterson and Thurston [68] and results in this section which are not attributed can be found there. In this section we mention the basic definitions and results. Automatic groups were invented by [68] although they were hinted at earlier by Cannon [44].

There are some preliminary notions that must be defined, specifically *regular language, finite state automata, normal forms* and *rewrite rules*. Let A be an alphabet and A^* the free monoid on A. A *language* is a set of words from A^*. A *finite state automaton* M is a computing device which reads an input string or word from a language A^* from a tape and then eventually either accepts or rejects the input string. We can also think of accepting as recognizing the string and rejecting as not recognizing. M reads in only one direction (no backing up) and can write only in its fixed internal storage. (A formal definition of a finite state automaton can be found in [68].) The set of words L in A^* which M accepts is called the language recognized by M. A *regular language* is a language accepted by some finite state automaton. A *context-free language* is a language accepted by a push-down automaton, which is a more complicated nondeterministic finite state automaton with a push-down stack storage device. A *two tape automaton* is a finite state automaton which can read two tapes simultaneously. If it reads the tapes at the same rate it is a *synchronous two tape automaton* and if it can read the two tapes at different rates it is an *asynchronous two tape automaton*. Two tape automata can be used to compare two words from L although it may be necessary to pad the shorter word by a symbol outside of A^* so that they have the same length.

Suppose now that G is a group with a generating set X, with $1 \notin X$, and let Γ be the free monoid on $X \cup X^{-1}$. Let $\mu: \Gamma \to G$ so that for a word $W \in \Gamma$, $\mu(W)$ is the element of G represented by W. A *set of normal forms* for G is a subset T of Γ such that $\mu_{|T}$ is a bijection. If T is closed under the operation of taking subwords then T is a *hereditary set of normal forms*. Equivalently a hereditary set of normal forms is a two-sided Schreier transversal for a normal subgroup N of the free group F on X. Clearly every group has a hereditary set of normal forms although if X is finite, T constructed as a two-sided Schreier transversal will be recursive if and only if G has a solvable word problem. The problem arises how to effectively transform words into normal forms. A *rewrite rule* is an ordered pair (u, v) of words of Γ such that $\mu(u) = \mu(v)$. An *application of the rewrite rule* is then to replace a subword of the form u in a word w by v to obtain a new word w'. Note that this need not shorten the word and in fact may lengthen it. Let Λ be a collection of rewrite rules. A word w is Λ-reduced if it is impossible to apply any of the rewrite rules in Λ to w, that is no subword of w is the left hand side of a rewrite rule in Λ. The set Λ is a *complete rewriting system* if it satisfies

(1) the set T of Λ-reduced words is a hereditary set of normal forms
(2) there are no infinite chains of applications of rewrite rules from Λ. Free reduction is clearly a complete rewriting system for a finitely generated free group. If Λ is a complete rewriting system then the following is true.

Lemma 3.6.1. *If Λ is a complete rewriting system for a group G then*
(1) *Starting with any word w the rewrite rules of Λ can be applied successively to obtain a unique normal form for w.*
(2) *If $\mu(w_1) = \mu(w_2)$ then applying rewrite rules in any order to w_1, w_2 will eventually lead to the same word. This property is called confluence.*
(3) *The collection of equations $u = v$ with $(u, v) \in \Lambda$ gives a complete set of relations for G.*

From a computer science point of view what is of interest is the class of finitely generated groups with complete rewriting systems. Free groups and finite groups are clearly in this class. It has also been shown that all polycyclic-by-finite groups as well as finitely generated free extensions of finitely generated free groups are in this class.

Theorem 3.6.1. *Let G be a finitely generated group with a complete rewriting system Λ. Then G is finitely presented and the set of Λ-reduced words forms a regular language. Further the word problem is solvable.*

Rewrite rules need not shorten the length of a word. However we do get the following interesting tie with both free groups and hyperbolic groups.

Theorem 3.6.2. *Let G be a finitely generated group with a finite, length reducing, complete rewriting system. Then this system gives a Dehn algorithm and hence G must be hyperbolic. Further G is virtually free, that is has a free subgroup of finite index.*

Another remarkable result along these lines was obtained by Muller and Schupp [189]. Let G be a finitely generated group with generating system X. The word problem of G is the formal language of words over the generators which are equal to the identity in the group. If this language happens to be context-free, then the group is called *context-free*. Finitely generated virtually free groups are context-free. Muller and Schupp proved the converse result that a context-free group is virtually free.

Theorem 3.6.3 ([189]). *A finitely generated group G is context-free if and only if it is virtually free.*

Now let G be a group with generating set X and let Γ and μ be as above. A *synchronously automatic structure* for G relative to X is a regular language $L \subset \Gamma$ such that $\mu(L) = G$ together with a synchronous two tape finite state automaton M which accepts the set of pairs of elements of L which represent elements of G lying at most one unit apart in the Cayley graph relative to X. That is the automaton accepts the set of pairs

$$\{(U, V) : U, V \in L, \mu(U) = \mu(Vx) \quad \text{for some } x \in X \cup X^{-1} \cup \{1\}\}.$$

If in addition there is a finite state automaton M' which accepts pairs of words that start one unit part, that is accepts the set of pairs

$$\{(U, V) : U, V \in L, \mu(U) = \mu(xV) \quad \text{for some } x \in X \cup X^{-1} \cup \{1\}\}$$

then the triple (L, M, M') is a *biautomatic structure*. Further, if instead, M is an asynchronous automaton then (L, M) is an *asyncronous automatic structure*. A group G is an *automatic group* if it has an automatic structure. Similarly it is a *biautomatic group* if it has a biautomatic structure and an *asynchronously automatic group* if it has an asynchronous automatic structure. Biautomatic groups are clearly automatic but it is not known whether the two classes coincide. Further the class of asynchronously automatic groups is much larger than the class of automatic groups.

Automatic groups are finitely presented and all the classes of automatic groups have solvable word problems while the conjugacy problem is recursively solvable for biautomatic groups.

Theorem 3.6.4. *Asynchronously automatic groups have solvable word problem. The conjugacy problem is recursively solvable for biautomatic groups but there exist asynchronously automatic groups with unsolvable conjugacy problem. Further the isomorphism problem for asynchronously automatic groups is recursively unsolvable.*

Free groups are automatic. However, in addition to the ties to free groups and computer science, automatic groups are of interest because many other commonly occurring finitely presented groups are automatic. In particular all the hyperbolic groups are automatic [68]. Further the class of automatic groups is closed under finite free products, finite direct products and finite extensions. It is not closed under finitely generated subgroups. Baumslag, Gersten, Shapiro and Short [13] have further examined closure properties of the class of automatic groups under group amalgams – free products with amalgamation and HNN extension. There has also been additional work on the automaticity of graph groups and Coxeter groups. We close by summarizing some of these results (see [68] and [13]).

Theorem 3.6.5. *The following groups are all automatic*
(1) *finite groups,*
(2) *finitely generated free groups,*
(3) *finitely generated Abelian groups,*
(4) *hyperbolic groups – in fact even biautomatic. Here if G is a hyperbolic group and Δ is a Dehn algorithm for G then the Δ-reduced words form a regular language. In addition the set of geodesic paths in the Cayley graph starting at the identity also forms a regular language,*
(5) *all non-metric small cancellation groups are automatic,*
(6) *finitely generated graph groups are automatic.*

Theorem 3.6.6. *The class of automatic groups is closed under finite free products, finite direct products and finite extensions.*

Theorem 3.6.7. *Let G be the amalgamated free product*

$$G = A *_C B.$$

Then

(1) *if* A, B *are finitely generated Abelian groups,* G *is automatic,*
(2) *if* A, B *are hyperbolic and* C *is cyclic,* G *is automatic,*
(3) *if* A, B *are finitely generated free groups and* C *is finitely generated,* G *is asynchronously automatic,*
(4) *if* A, B *are asynchronously automatic and* C *has finite index in both factors,* G *is asynchronously automatic.*

We close by mentioning that Gilman [117] classified hyperbolic groups in terms of defining language property. This is a beautiful extension of the result of Miller and Schupp.

3.7 Stallings foldings and subgroups of free groups

As we have seen, the Cayley graph provides a very useful geometric way of studying finitely generated groups. For free groups F, the group can be identified with the fundamental group of this graph and hence from covering space theory each subgroup corresponds to a covering map of another graph to the original graph. Using this viewpoint J. Stallings [244] introduced the concept of foldings of graphs now called *Stallings foldings*. These have been used in many different applications (see [145]) and especially in the study of automorphisms of free groups. In [145], Kapovich and Myasnikov, recast Stallings foldings in a more combinatorial framework, and applied them systematically to the study of subgroups of free groups. In this section we introduce this fundamental idea. In [183], Markus–Epstein, extended these ideas to the study of free products of finite groups.

Let F be a free group with free basis A. The basic idea in the use of Stallings foldings in the study of the subgroup structure of free groups is to associate to each subgroup H of the free group F a directed graph $\Gamma(H)$, whose edges are labeled by the elements of $A^{\pm 1}$, and which carries all the essential information about the subgroup H. Further, if H is finitely generated then $\Gamma(H)$ is finite and easy to construct. A finite labeled directed graph can be identified with a finite state automaton and thus for each finitely genertated subgroup H the constructed $\Gamma(H)$ is a finite state automaton which accepts precisely the reduced words in $A^{\pm 1}$ which belong to H. This automaton approach provides an extremely useful method to handle computational and algorithmic problems in free groups. Considering $\Gamma(H)$ in this automaton mode often speeds up several algorithms in free groups (see [145]).

Topologically a free group F on A can be identified as the fundamental group of a wedge of circles of cardinality $|A|$. The graph automaton $\Gamma(H)$ represents the topological core of the covering space, relative to H, of this wedge of circles. Algebraically $\Gamma(H)$ can be viewed as the essential part of the relative coset Cayely graph of F/H with respect to A. This last approach was used effectively by C. Simms [243]. Finally $\Gamma(H)$

can be viewed in the context of Bass–Serre theory. The free group F acts on its standard Cayley graph which is a regular tree (see Chapter 3). There is a unique H-invariant subtree $T(H)$ that is minimal among H-invariant subtrees containing the identity. The graph automaton $\Gamma(H)$ can be identified with the quotient graph $T(H)/H$.

Now let $F = F(A)$ be a free group with free basis $A = \{a_1, \ldots, a_m\}$ and let H be a subgroup of F generated by elements $h_1, \ldots, h_n \in F$. We show how to construct a finite automaton $\Gamma(H)$ which accepts precisely the reduced words in $A^{\pm 1}$ which belong to H. To start, we look a bit more deeply at finite state automata in terms of directed graphs.

Let S be a finite set and Γ a directed S-labeled graph with vertex set $V(\Gamma)$, that is the edges of Γ are labeled by S-elements. Γ has the *one way reading property*, denoted OR, if for each vertex $v \in V(\Gamma)$, and each label $s \in S$, there exists at most one edge in Γ starting at v with label s.

The tuple (Γ, S, V_0, Y, μ) defines a finite state automation $\mathcal{A}$ where Γ is a directed S-graph with property OR, $V_0 \in V(\Gamma)$ is a distinguished vertex, $Y \subset V(\Gamma)$ and μ is a transition function (defined below). The automaton $\mathcal{A}$ has the set of states $Q = V(\Gamma) \cup \oplus$ where $\oplus$ is the fail state, the alphabet S or $S^{\pm 1}$, the starting state V_0, the accepting states Y and the transition function $\mu\colon Q \times S \to Q$ defined by

$$\begin{aligned} \mu(v, s) &= w \text{ if } v \in V(\Gamma) \text{ and there is an edge from } v \text{ to } w \text{ with label } s \\ &= \oplus \text{ otherwise}. \end{aligned}$$

Let S^* denote the set of all words in the alphabet S. Then a subset $L \subset S^*$ is *regular* if there is a finite state automaton which has L as its accepting language. If L is regular then the set of words $\overline{L}$ that consists of reduced words from L is also regular.

Example 3.7.1. Let Γ be the following directed labeled S-graph where $S = \{a^{\pm 1}\}$.

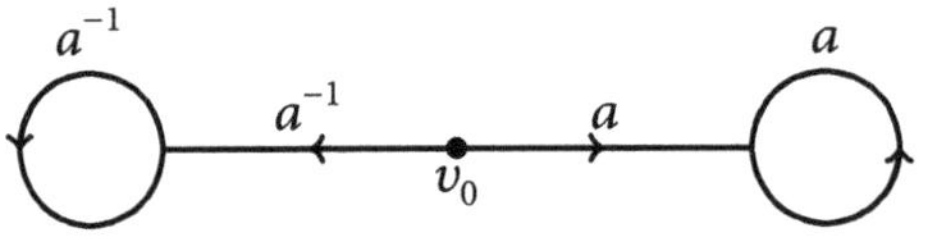

Fig. 3.2. Directed S-Graph.

Let $Y = V(\Gamma)$. Then Γ corresponds to a finite automaton $\mathcal{A}$ which accepts all words of the form $a^n, n \in \mathbb{Z}$. Hence the language $L(\mathcal{A})$ of $\mathcal{A}$ is precisely the infinite cyclic group on a. In this case we say that $\mathcal{A}$ accepts the infinite cyclic group generated by a.

Notice that for any group G with generating system S the Cayley graph $\Gamma(G, S)$ satisfies the one way reading property. If we choose $V_0 = \{1\}$ and $Y = \{1\}$ then the Cayley graph will lead to an automaton $\mathcal{A}$ which is infinite if G is infinite and which accepts a word w in the alphabet $S^{\pm 1}$ if and only if $w = 1$ in G.

To start the construction of $\Gamma(H)$ we begin with a finite non-oriented labeled graph Γ with labels from $A^{\pm 1}$, a distinguished vertex 1 and n simple loops $p_1, \ldots, p_n$ with labels $h_1, \ldots, h_n$. In a labeled graph if x is a label of an edge e then x^{-1} is the label of the edge $\bar{e}$. In the graph Γ we place an orientation by making the positive edges those with labels $a_1, \ldots, a_m$. As an example consider the free group F on two generators a, b and let H be the subgroup generated by the words

$$h_1 = aba^{-1}b^{-1}a \text{ and } h_2 = ab^{-1} .$$

The corresponding graph Γ is then pictured in Figure 3.3.

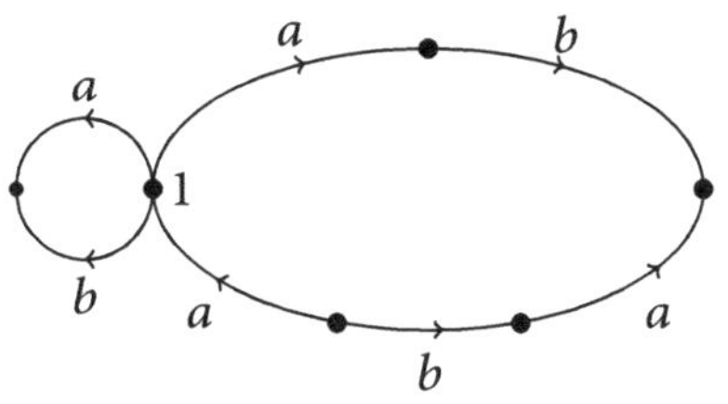

Fig. 3.3. Graph Γ.

Notice that every loop p in Γ at the vertex 1 is a product of finitely many simple loops $p_1^{\pm 1}, \ldots, p_n^{\pm 1}$. It follows that the label of p is a finite product of the generators $h_1^{\pm 1}, \ldots, h_n^{\pm 1}$ of H and hence p defines an element of H. Conversely if a word $w \in F$ is in H there must be a loop at 1 with label w. Hence for any subgroup $H = \langle h_1, \ldots, h_n \rangle$ the graph Γ has the following property:

Property L An element $f \in F$ belongs to H if and only if there exists a loop in Γ at the vertex 1 with label w where the word w defines $f \in F$.

Notice that in general the graph Γ does not satisfy the one way reading property OR and therefore does not give rise immediately to an automaton. The problem, as illustrated by the example above, is that there may be two different edges from 1 with the same label. To correct this problem we *fold* the graph.

Suppose that in a labeled graph Γ there are two different edges starting at the same vertex v and labeled by the same label a as in the diagram below. We transform the graph Γ into a graph Γ' by identifying the two edges and their terminal vertices. This is called a *folding* at v. If there are no such vertices then we say that the graph is *completely folded*. It is clear that a finite graph can be folded in a finite number of steps to a completely folded graph.

What is crucial is the following which is straightforward.

Lemma 3.7.1. *If a graph Γ satisfies Property L and Γ' is a folding of Γ then Γ' also satisfies Property L.*

Now let Γ be the graph constructed from the subgroup H. The finite automaton that goes with H is $\Gamma(H)$, the completely folded graph from Γ. Now $\Gamma(H)$ satisfies both

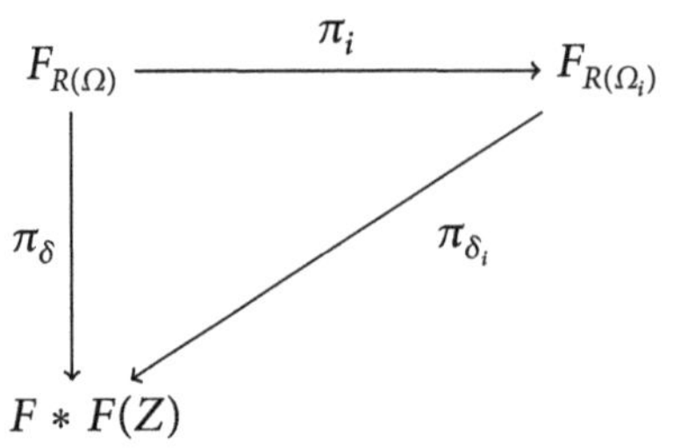

Fig. 3.4. Folded Graph.

the one way reading property and satisfies Property L from the Lemma above. Hence it forms an automaton. Then if we put $v_0 = 1$ and $Y = \{1\}$ if follows that $\Gamma(H)$ is a finite automaton which accepts H, that is $\Gamma(H)$ accepts a reduced word in $A^{\pm 1}$ if and only if this word defines an element of H. Further this construction is algorithmically effective and hence we have.

Theorem 3.7.1. *Let $H = \langle h_1, \ldots, h_n\rangle$ be a finitely generated subgroup of a free group $F = F(a_1, \ldots, a_m)$. Then one can effectively construct a finite automaton $\Gamma(H)$ which accpets H.*

There are many classical results on free groups that can be proved much more easily using this automaton. We refer the reader to the paper of Kapovich and Myasnikov [145] for most of these but we mention several further items here.

There are at most

$$cm(|h_1| + \cdots + |h_n|)^2$$

steps necessary to construct $\Gamma(H)$ where c is a constant. This provides a bound for the solution of the membership problem in a free group.

Corollary 3.7.1. *Let F be a free group of finite rank. Then the membership problem in F is decidable in quadratic time complexity. That is given words $w, h_1, \ldots, h_n \in F$ one can effectively decide whether or not $w \in \langle h_1, \ldots, h_n\rangle$ and it takes at most*

$$C(|h_1| + \cdots + |h_n|)^2$$

steps, where C is a constant which depends linearly on the rank of F.

This method also provides another proof of the Nielsen–Schreier theorem (in the finitely generated case) (see [145]).

The graph automaton $\Gamma(H)$ also provides a method to determine if the subgroup H is of finite index. In particular we say that an A-labled graph Γ is regular if for any vertex $v \in \Gamma$ and any label $a \in A$ there exists an edge in Γ starting at v and with label a.

Theorem 3.7.2. *A finitely generated subgroup H of F has finite index if and only if $\Gamma(H)$ is a regular $A^{\pm 1}$-labeled graph.*

Proof. Suppose $\Gamma(H)$ is not a regular A-labeled graph. Then there exists a vertex $v \in \Gamma(H)$ and a label $a \in A^{\pm 1}$ such that there is no edge in $\Gamma(H)$ starting at v and labeled by a. Let p be a reduced path from 1 to v in $\Gamma(H)$ and let g be the label of the path p. Choose a letter $b \in A$ not equal to a or a^{-1} and define a word w as follows

$$w = ga \quad \text{if } g \text{ does not begin with } a^{-1}$$
$$w = gab \quad \text{otherwise}\,.$$

Clearly w is a cyclically reduced word and hence any power w^k is reduced as written. Further $w^k \notin H$ since we cannot read fa in $\Gamma(H)$. It follows that

$$Hw^k = Hw^l \implies w^{k-l} \in H \implies k = l.$$

Therefore in this case H must be of infinite index in F.

Conversely suppose that $\Gamma(H)$ is a regular $A^{\pm 1}$-labeled graph. Let T be a maximal subtree of $\Gamma(H)$. For any vertex $v \in \Gamma(H)$ denote by $\gamma(v)$ the path from 1 to v inside T. If w is a reduced word in $A^{\pm 1}$ then we can read off w inside $\Gamma(H)$ starting at 1. Let v_w be the vertex where we finish reading w inside $\Gamma(H)$.

Then $w\gamma(v_w)^{-1}$ is the label of a loop in $\Gamma(H)$ at 1 and hence $w\gamma(v_w) \in H$. Therefore

$$Hw = H\gamma(vw)$$

and $\gamma(v_w)$ is a representative of w modulo H. Put

$$R_T = \{\gamma(v) \colon v \in \Gamma(H)\}$$

then R_T is a transversal of H in F. Since $\Gamma(H)$ is a finite graph it follows that $|R_T|$ is finite and H is of finite index in F. □

We close this section and this chapter by briefly explaining the relationship between $\Gamma(H)$ and the Cayley graph $\Gamma(F, A)$.

Let G be a group, A a generating system with $1 \notin A$ and Γ the Cayley graph of G with respect to A. The group G acts on $\Gamma(G, A)$ by left multiplication

$$v \xrightarrow{g} gv, (v, a) \xrightarrow{g} (gv, a)\,;$$

where here v is a vertex in Γ, (v, a) is an edge in Γ and $g \in G$. This map is an isomorphism of Γ.

Let H be a subgroup of G. Then H also acts on Γ. Denote by Γ_H the quotient graph under this action

$$\Gamma_H = \Gamma / H\,.$$

The vertices of Γ_H are the orbits of the action of H on Γ and hence they are just the right cosets of H in G,

$$Hf = \{g \in G \colon g = hf \text{ for some } h \in H\}\,.$$

There is an edge with label $a \in A$ from Hf into Hg if and only if $Hfa = Hg$. This turns Γ_H into an A-labeled graph. Notice that every loop in Γ_H at the vertex H defines an element of H. Moreover Γ_H has the one way reading property. Hence if we put $v_0 = H, Y = \{H\}$ then Γ_H gives rise to an automaton that accepts H. If G is a free group than $\Gamma(H)$ is embeddable into Γ_H and any loop at H in Γ_H is a product of loops in $\Gamma(H)$.

4 First order languages and model theory

4.1 First order language for group theory

The elementary theory of groups is tied to first-order logic and to model theory. In this chapter, we discuss the necessary material from these areas. We start with a first-order language appropriate for group theory. This language which we denote by L_0 is the first-order language with equality (always interpreted as the identity relation) containing a binary operation symbol $\cdot$ (often suppressed in favor of juxtaposition), unary operation symbol $^{-1}$, and a constant symbol, 1. In particular, this is what it means for L_0 to be *appropriate* for group theory. A *formula* in this language is a logical expression containing a string of variables $\overline{x} = (x_1, \ldots, x_n)$, the logical connectives $\vee, \wedge, \sim$, and the quantifiers $\forall, \exists$. Here $\vee$ stands for the conjunction of two propositions, $\wedge$ for the disjunction of two propositions and $\sim$ for for the negation.

A variable in a formula is called *bound* (or occurs bound) if it is restricted by a quantifier $(\forall, \exists)$. Otherwise, the variable is called *free* (or occurs free). A *sentence* is a formula of L_0 in which all variables are bound, or in other words there are no free occurrences of any variable. A *universal sentence* of L_0 is one of the form $\forall\overline{x}(\varphi(\overline{x}))$ where $\overline{x} = (x_1, \ldots, x_n)$ is a tuple of distinct variables, $\varphi(\overline{x})$ is a formula of L_0 containing no quantifiers and containing at most the variables of $\overline{x}$. Similarly an *existential sentence* is one of the form $\exists\overline{x}(\varphi(\overline{x}))$ where $\overline{x}$ and $\varphi(\overline{x})$ are as above. A *universal-existential sentence* is one of the form $\forall\overline{x}\exists\overline{y}(\varphi(\overline{x}, \overline{y}))$. Similarly defined is an *existential-universal sentence*. It is known that every sentence of L_0 is logically equivalent to one of the form

$$Q_1x_1 \ldots Q_nx_n(\varphi(\overline{x}))$$

where $\overline{x} = (x_1, \ldots, x_n)$ is a tuple of distinct variables, each Q_i for $i = 1, \ldots, n$ is a quantifier, either $\forall$ or $\exists$, and $\varphi(\overline{x})$ is a formula of L_0 containing no quantifiers and containing at most the variables of $\overline{x}$. This is called the *Prenex Normal Form* of a sentence (See [183].) Further vacuous quantifications are permitted. Finally a *positive sentence* is one logically equivalent to a sentence constructed using (at most) the connectives $\vee, \wedge, \forall, \exists$, i.e., no negations are involved.

If G is a group then the *universal theory* of G consists of the set of all universal sentences of L_0 true in G. We denote the universal theory of a group G by $Th_\forall(G)$. Since any universal sentence is equivalent to the negation of an existential sentence it follows that two groups have the same universal theory if and only if they have the same *existential theory*. The set of all sentences of L_0 true in G is called the *first-order theory* or the *elementary theory* of G. We denote this by $Th(G)$. We note that being *first-order* or *elementary* means that in the intended interpretation of any formula or sentence all of the variables (free or bound) are assumed to take on as values only individual group elements – never, for example, subsets of, nor functions on, the group in which they are interpreted.

Two very important concepts in the elementary theory of groups, are *completeness* and *decidability*. Given a non-empty class of groups $\mathcal{X}$ closed under isomorphism then we say its first-order theory is *complete* if given a sentence ϕ of L_0 then either ϕ is true in every group in $\mathcal{X}$ or ϕ is false in every group in $\mathcal{X}$. The first-order theory of $\mathcal{X}$ is *decidable* if there exists a recursive algorithm which, given a sentence ϕ of L_0 decides whether or not ϕ is true in every group in $\mathcal{X}$. We will see these types of questions relative to the class of non-Abelian free groups in regard to the Tarski problems.

4.2 Elementary equivalence

Extremely important for the solution of the Tarski problem is the notion of elementary equivalence of groups.

Definition 4.2.1. Two groups G and H are *elementarily equivalent* (symbolically $G \equiv H$) if they have the same first-order theory, that is $Th(G) = Th(H)$.

Group monomorphisms which preserve first-order formulas are called elementary embeddings. Formally:

Definition 4.2.2. If H and G are groups and $f : H \to G$ is a monomorphism then f is an *elementary embedding* provided whenever $\phi(x_0, \ldots, x_n)$ is a formula of L_0 containing free at most the distinct variables $x_0, \ldots, x_n$ and $(h_0, \ldots, h_n) \in H^{n+1}$ then $\phi(h_0, \ldots, h_n)$ is true in H if and only if the formula $\phi(f(h_0), \ldots, f(h_n))$ is true in G. If H is a subgroup of G and the inclusion map $i : H \to G$ is an elementary embedding then we say that G is an *elementary extension* of H.

We note that the existence of an elementary embedding $f : H \to G$ is a sufficient condition for H and G to be elementarily equivalent. Further an isomorphism from G onto H is an elementary embedding. Another example of an elementary embedding is given by the following. we will discuss ultrafilters and ultrapowers in Section 4.5.

Example 4.2.1. Let G be a group and let I be a nonempty set. Let $\delta : G^I \to G$ be the diagonal map, i.e., $\delta(g)(i) = g$ for all $i \in I$. Let D be an ultrafilter on I. Then the map d from G into the *ultrapower* G^I/D given by $g \longmapsto [\delta(g)]_D$ is an elementary embedding (see Section 4.5).

We note that two groups can be elementarily equivalent without one being an elementary extension of the other.

Example 4.2.2. Let F be a free group of countably infinite rank ω with basis $A = \{a_n : n < \omega\}$. Let F' be the commutator subgroup of F. Then F and F' are elementarily equivalent since they are isomorphic. None the less F is not an elementary extension of F'. For example, the formula

$$\exists x\,([[a_0, , a_1], x] = [[a_0, a_1], a_2])$$

is true in F but false in F'. Here $[x, y]$ is the commutator $[x, y] = x^{-1}y^{-1}xy$.

A sufficient condition to prove that an inclusion map is an elementary embedding is provided by the next theorem

Theorem 4.2.1. *Suppose H_0 is a subgroup of H and suppose that to every finite subset $\{a_1, \ldots, a_n\}$ of H_0 and every element $b \in H$ there exists an automorphism σ of H fixing $a_1, \ldots, a_n$ and mapping b into H_0. Then the inclusion map from H_0 into H is an elementary embedding.*

Proof. Suppose that $\Phi(x_1, \ldots, x_n, x_{n+1})$ is a formula of L_0 containing free at most the distinct variables $x_1, \ldots, x_n, x_{n+1}$. Suppose that $((a_1, \ldots, a_n), b) \in (H_0)^n \times H$ is such that $\Phi(a_1, \ldots, a_n, b)$ holds in H.

By the hypothesis of the theorem there is an automorphism σ of H fixing each of $a_1, \ldots, a_n$ and mapping b into H_0. Since σ is an automorphism it is an elementary embedding of H into itself, so, $\Phi(a_1, \ldots, a_n, \sigma(b))$ holds in H. Therefore the condition of Corollary 1.9 on Page 76 of [25] is satisfied which implies that H is an elementary extension of H_0. □

We will see this used as a criterion in showing that two infinite rank free groups are elementarily equivalent.

In general, given two groups G and H, it is difficult to determine whether or not G and H are elementarily equivalent. Sometimes the following general criterion, which is due to Keisler and Shelah (see [40]) is useful: two groups are elementary equivalent if and only if their ultrapowers (with respect to some non-principal ultrafilter) are isomorphic. Keisler first proved the result assuming the Generalized Continuum Hypothesis and subsequently Shelah gave a proof avoiding this assumption (see [40]). Another general test for elementary equivalence is based on games (increasing chains of sets of finite partial isomorphisms) and is due to Ehrenfeucht [70].

Szmielew [245] has completely characterized elementary equivalence of Abelian groups. Recall that a group G has *finite exponent* if there is a positive integer n such that $x^n = 1$ for all $x \in G$ and G has *infinite exponent* otherwise. Szmielew distinguishes between two types of linear independence in an Abelian group A (which we shall write additively). If m is a positive integer and $(a_i)_{i \in I}$ is a sequence of elements of A containing only finitely many nonzero terms, then $(a_i)_{i \in I}$ is *linearly independent modulo m* provided that

$$\sum_{i \in I} n_i a_i = 0 \Rightarrow n_i \equiv 0 (\operatorname{mod} m)$$

for all $i \in I$.

The sequence $(a_i)_{i \in I}$ is *linearly independent modulo m in the stronger sense* provided that

$$\sum_{i \in I} n_i a_i \in mA \Rightarrow n_i \equiv 0 (\operatorname{mod} m)$$

for all $i \in I$.

Szmielew then defines, for each prime p and each positive integer k, three quantities $\rho^{(i)}[p,k](A)$, $i = 1,2,3$, each of which is either a nonnegative integer or the symbol ∞, as follows:

(1) $\rho^{(1)}[p,k](A)$ is the maximal number (if it exists) of elements of order p^k and linearly independent modulo p^k.
(2) $\rho^{(2)}[p,k](A)$ is the maximal number (if it exists) of elements linearly independent modulo p^k in the stronger sense.
(3) $\rho^{(3)}[p,k](A)$ is the maximal number (if it exists) of elements of order p^k and linearly independent modulo p^k in the stronger sense.

Theorem 4.2.2 ([245]). *Let A and B be Abelian groups. Then A and B are elementarily equivalent if and only if the following two conditions are satisfied:*
(1) *Either A and B both have the same finite exponent or they both have infinite exponent.*
(2) *For all primes p and positive integers k, one has*

$$\rho^{(i)}[p,k](A) = \rho^{(i)}[p,k](B)$$

for $i = 1,2,3$.

It is easy to see that the elementary theory of a given finitely generated Abelian group is decidable. This follows directly from the decidability of the elementary theory of Presburger arithmetic (non-negative integers with addition) (see [245]). A general description of Abelian groups with decidable elementary theory is given also by Szmielew [245]:

Theorem 4.2.3 ([245]). (1) *The elementary theory of the class of all Abelian groups is decidable.*
(2) *The elementary theory of an Abelian group A is decidable if and only if the invariants $\rho^{(i)}[p,k](A)$ are computable (i.e., the functions $(p,k) \to \rho^{(i)}[p,k](A)$ are computable for $i = 1,2,3$).*

The proofs of Smielew's results are quite long and involved. We refer the reader to the original papers [245].

There have also been other results concerning the problems of elementary equivalence and decidability of elementary theories for specific classes of groups. A. I. Malcev solved the problems for the classical linear groups [181]. It turns out that in this case everything depends on the field of coefficients. Let H and K be commutative fields. In analogy to the group theory case we say that $K \equiv F$ if K and F satisfy the same elementary sentences in a language appropriate for field theory. Mal'cev's main result is the following.

Theorem 4.2.4 ([181]). *Let H, K be commutative fields and suppose that $n, m \geq 2$. Then $GL(n,K) \equiv GL(m,F)$ if and only if $n = m$ and $K \equiv F$, that is K and F are ele-*

mentarily equivalent in a language appropriate for field theory. Further $Th(GL(n, F))$ is decidable if and only if $Th(F)$ is decidable.

The problem of decidability of the theory finitely generated solvable groups was solved in three steps: by Y. Ershov (nilpotent case) [69], by N. Romanovskii (polycyclic case) [218], and by G. Noskov [204] (general case). A summary of the results is:

Theorem 4.2.5. *A finitely generated virtually solvable group G has decidable elementary theory if and only if it is virtually Abelian.*

The proofs of all the cases reduce the problem of decidability of $Th(G)$ to the corresponding problem for the ring of integers, which is, undecidable.

The problem of elementary equivalence of finitely generated nilpotent groups and unipotent groups (as well as the problem of decidability of their elementary theories) was solved in a series of papers by A. Myasnikov [191], A. Myasnikov and V. Remeslennikov [192, 193, 194], and F. Oger [206]. The question of elementary equivalence for finitely generated solvable groups remains open. We will discuss these Tarski-like problems in the next chapter.

Before moving on to discuss ideas from model theory we mention two other results concerning elementary equivalence.

Theorem 4.2.6 (Elementary Chain Theorem). *If*

$$G_0 \to G_1 \to \cdots \to G_\zeta \to \ldots$$

is a chain of elementary embeddings then the direct union G is an elementary extension of each group in the chain.

Theorem 4.2.7 (Downward Lowenheim–Skolem Theorem for Groups). *If S is an infinite subset of a group H then there is a subgroup H_0 of H containing S such that $|H_0| = |S|$, that is H_0 and S have the same cardinality, and the inclusion map $H_0 \to H$ is an elementary embedding.*

4.3 Models and model theory

We now introduce some basic ideas from model theory.

Definition 4.3.1. If Φ is a consistent set of sentences of L_0, then the class of all groups G satisfying every sentence φ in Φ is the *model class* of Φ. We denote the model class of Φ by $\mathbb{M}(\Phi)$.

For the remainder of this chapter we tacitly assume that our sets of sentences of Φ contain the group axioms.

We note that any model class is nonempty and is closed under isomorphism. Further the model class operator $\mathbb{M}$ reverses inclusions. That is, if Φ and Ψ are consistent sets of sentences of L_0 and $\Phi \subseteq \Psi$, then $\mathbb{M}(\Psi) \subseteq \mathbb{M}(\Phi)$.

A class of groups which is the model class of some set of first-order sentences is said to be axiomatic. Formally:

Definition 4.3.2. If $\mathcal{X}$ is a nonempty class of groups closed under isomorphism, then $\mathcal{X}$ is *axiomatic* provided that there is at least one set Φ of sentences of L_0 such that $\mathcal{X} = \mathbb{M}(\Phi)$.

An axiomatic class is *finitely axiomatizable* if there exists a finite set of axioms.

We will need the following results.

Lemma 4.3.1. *Let G and H be groups. Then every universal sentence of L_0 true in G is also true in H if and only if H is embeddable in an elementary extension $^*G$ of G.*

Lemma 4.3.2. *Let $\mathcal{X}$ be an axiomatic class of groups. Then $\mathcal{X}$ is universally axiomatizable (i.e., has at least one set of universal axioms) if and only if $\mathcal{X}$ is closed under subgroups.*

Lemma 4.3.3. *Let $\mathcal{X}$ be an axiomatic class of groups. If $\mathcal{X}$ is closed under direct products of two factors, then $\mathcal{X}$ is closed under direct products of an arbitrary number (finite or infinite) of factors.*

For a proof of Lemma 4.3.1 see either Gratzer [118] Theorem 3 in Chapter 7 or Bell and Slomson [25] Lemma 3.8 in Chapter 9. For Lemma 4.3.2 a proof is in Chang and Keisler [40] Chapter 5 Theorem 5.2.4 while a proof of Lemma 4.3.3 is in Gratzer [118] Chapter 7, Corollary 2.

4.4 Varieties and quasivarieties

The model theory of groups is also closely tied to the theory of group varieties. Recall that if F is a free group on $x_1, \dots, x_n$ and $w = w(x_1, \dots, x_n)$ is a word in the generators then we say that the group G satisfies the *law* $w(x_1, \dots, x_n) = 1$ if the substitution of any elements $g_1, \dots, g_n \in G$ for $x_1, \dots, x_n$ gives the identity in G. In the logical language we introduced in Section 4.1, this says that G satisfies the universal sentence

$$\forall x_1, \dots, x_n (w(x_1, \dots, x_n) = 1) .$$

A *group variety* is the class of all groups satisfying each one of a given set of laws. Again in the language of model theory this can be described as the model class of the given set of laws. By its definition a group variety is axiomatic. The theory of group varieties is examined in great detail in the book of Hanna Neumann [202]. Group varieties can be described purely group theoretically as follows:

Theorem 4.4.1 ([202]). *A nonempty class of groups $\mathcal{X}$ closed under isomorphism is a group variety if and only it is closed under subgroups, factor groups and unrestricted direct products.*

Some commonly studied varieties are the variety of all Abelian groups, the variety of all Abelian groups of exponent n, the variety of all nilpotent groups of class at most some fixed class c and the variety of solvable groups of length at most some fixed solvability length.

If G is a group there is a minimal variety containing G. We call this the variety generated by G and denote it $var(G)$. Similarly if $\mathcal{Y}$ is any class of groups there is a minimal variety containing $\mathcal{Y}$. We denote this minimal variety by $var(\mathcal{Y})$. If V represents a set of laws defining the group variety $\mathcal{V}$ and F is a rank n free group then $F/V(F)$ is the *free group of rank n in the variety* $\mathcal{V}$. Here $V(F)$ is the verbal subgroup of F generated by the free group words in the variables of the set of laws V. Every group in a variety $\mathcal{V}$ is a homomorphic image of some free group in the variety.

Certain extensions of the varietal concept also play a role in the elementary theory of groups.

Definition 4.4.1. Let $\mathcal{X}$ be a nonempty class of groups closed under isomorphism. Then $\mathcal{X}$ is a *prevariety* if $\mathcal{X}$ is closed under subgroups and $\mathcal{X}$ is closed under arbitrary direct products.

Observe that since the trivial group 1 is a subgroup of any group G, every prevariety $\mathcal{X}$ must contain at least 1. Note also that the intersection of any family of prevarieties is again a prevariety; so, if $\mathcal{Y}$ is any class of groups there is a least prevariety $pvar(\mathcal{Y})$ containing $\mathcal{Y}$. This is the prevariety *generated* by $\mathcal{Y}$. In the case that $\mathcal{Y} = G$ is a singleton, we write $pvar(G)$ for the prevariety *generated by* G.

Prevarieties can also be characterized as a *separating family of groups.*

Definition 4.4.2. Let $\mathcal{X}$ be a nonempty class of groups and let H be a group. Then $\mathcal{X}$ *separates* H provided that for every nontrivial element $h \in H$ there is a $G_h \in \mathcal{X}$ and a homomorphism $\varphi_h\colon H \to G_h$ such that $\varphi_h(h) \neq 1$. The family $\mathcal{X}$ is *separating* if any group that $\mathcal{X}$ separates lies in $\mathcal{X}$.

Theorem 4.4.2. *Let $\mathcal{X}$ be a class of groups. Then $\mathcal{X}$ is a prevariety of groups if and only if $\mathcal{X}$ is a separating family of groups.*

Proof. Suppose first that $\mathcal{X}$ is a prevariety of groups. Let the group H be separated by $\mathcal{X}$. For each $h \neq 1$ in H there is a group $G_h \in \mathcal{X}$ and a homomorphism $\varphi_h\colon H \to G_h$ such that $\varphi_h(h) \neq 1$. Then H embeds into $\prod_{h\in H\setminus\{1\}} G_h$; hence, H is, up to isomorphism, a subgroup of a direct product of groups in $\mathcal{X}$. It follows that H lies in $\mathcal{X}$ whenever H is separated by $\mathcal{X}$. We have thus proven that every prevariety is separating.

Now let $\mathcal{X}$ be a separating family of groups. Suppose $H \in \mathcal{X}$ and G is a subgroup of H. Given any $g \neq 1$ in G, the inclusion map $\iota\colon G \to H$ does not annihilate g. Thus G is separated by $\mathcal{X}$; hence, G lies in $\mathcal{X}$. Thus $\mathcal{X}$ is closed under subgroups. Now let

$(G_i)_{i\in I}$ be an indexed family from $\mathcal{X}$. Let $g = (g_i) \neq 1$ lie in $\prod_{i\in I} G_i$. Choose $i_0 \in I$ such that $g_{i_0} \neq 1$. If $\pi\colon \prod_{i\in I} G_i \to G_{i_0}$ is projection onto the i_0-th coordinate, then $\pi(g) = g_{i_0} \neq 1$. Hence $\mathcal{X}$ separates $\prod_{i\in I} G_i$; therefore, $\prod_{i\in I} G_i$ lies in $\mathcal{X}$. Thus $\mathcal{X}$ is closed under direct products. We have therefore shown that every separating family of groups is a prevariety of groups. □

We now continue on to quasivarieties. A universal sentence of the form

$$\forall \overline{x}\Big(\bigwedge_i (u_i(\overline{x}) = 1) \to (w(\overline{x}) = 1)\Big)$$

is called a *quasilaw* or *quasi-identity*. Note that every identity $\forall \overline{x}(w(\overline{x}) = 1)$ is equivalent to a quasilaw

$$\forall \overline{x}, y \left((y \cdot y^{-1} = 1) \to (w(\overline{x}) = 1)\right) .$$

A *quasivariety* is the model class of a set of quasilaws. The quasivariety $\mathrm{qvar}(\mathcal{X})$ *generated by* $\mathcal{X}$ is the model class of those quasilaws φ true in every group in $\mathcal{X}$. If $\mathcal{X}$ is a nonempty class of groups then the *universal closure* $\mathrm{ucl}(\mathcal{X})$ of $\mathcal{X}$ is the model class of the set of all universal sentences φ true in every group G in $\mathcal{X}$. If $\mathcal{X} = \{G\}$ is a singleton, then we write $\mathrm{ucl}(G)$ for the universal closure of G and $\mathrm{qvar}(G)$ for the quasivariety generated by G.

It is clear that both $\mathrm{ucl}(\mathcal{X})$ and $\mathrm{qvar}(\mathcal{X})$ are axiomatic. Moreover $\mathrm{ucl}(\mathcal{X})$ is the least universally axiomatizable class containing $\mathcal{X}$ and $\mathrm{qvar}(\mathcal{X})$ is the least quasivariety containing $\mathcal{X}$. If Φ and Ψ are consistent sets of sentences of L_0 and $\Phi \subseteq \Psi$, then $\mathbb{M}(\Psi) \subseteq \mathbb{M}(\Phi)$. It follows from this that $\mathrm{ucl}(\mathcal{X}) \subseteq \mathrm{qvar}(\mathcal{X})$. Further it is straightforward to verify that every quasivariety contains the trivial group 1, is closed under subgroups and is closed under direct products. Mal'cev [181] gives the following:

Lemma 4.4.1. *Let $\mathcal{X}$ be a universally axiomatizable class of groups. Then $\mathcal{X}$ is a quasivariety if and only if the trivial group 1 lies in $\mathcal{X}$, (equivalently: $\mathcal{X}$ is nonempty), and $\mathcal{X}$ is closed under direct products.*

Corollary 4.4.1 (Mal'cev). *A prevariety is a quasivariety if and only if it is axiomatic.*

It follows that every axiomatic prevariety is a quasivariety. However not every prevariety need be axiomatic and hence need not be a quasivariety. The following example shows this.

Example 4.4.1. Call an Abelian group *reduced* provided it contains no nontrivial divisible subgroup. It is straightforward to verify that the class of all reduced groups is a prevariety. Here we use additive notation since we are dealing with Abelian groups. Indeed, this class is clearly closed under taking subgroups. Consider a direct sum $\sum_\lambda G_\lambda$ of reduced Abelian groups G_λ. Recall that the homormorphic image of a divisible group is divisible. (As a reference for all facts about divisible groups see [161], Volume I. Section 23). So if $H \subseteq \sum_\lambda G_\lambda$ is divisible, then so is its projection into the

λ^{th} component, i.e., $\pi_\lambda(H) \subseteq G_\lambda$. But since G_λ is reduced, this forces $\pi_\lambda(H) = 1$. But this must be true for all λ. Thus the components of H in $\sum_\lambda G_\lambda$ are trivial for all λ. This means $H = 1$. So $\sum_\lambda G_\lambda$ is reduced. This verifies our claim that the class of reduced groups is a prevarity.

One can construct an ultrapower (see next section) of the integers, $\mathbb{Z}$, which contains a copy of the rationals, $\mathbb{Q}$. It follows that the $pvar(\mathbb{Z})$ is not axiomatic, hence $pvar(\mathbb{Z})$ is not a quasivariety. To produce such an ultrapower of $\mathbb{Z}$, we proceed as follows. Let $\mathbb{N}$ be the set of positive integers and let D be a nonprincipal ultrafilter on $\mathbb{N}$. (See next section.) Let ${}^*\mathbb{Z}$ be the ultrapower $\mathbb{Z}^{\mathbb{N}}/D$. Then ${}^*\mathbb{Z}$ is an additively written nontrivial torsion free Abelian group. For each $k \in \mathbb{N}$, let $a_k \colon \mathbb{N} \to \mathbb{Z}$ be defined by

$$\begin{aligned} a_k(n) &= 0 \quad \text{if } n < k \\ &= \frac{n!}{k} \quad \text{if } n \geq k \end{aligned}$$

and let $\alpha_k = [a_k]_D$ be the element of ${}^*\mathbb{Z}$ determined by a_k (See next section.). Then letting H be subgroup of ${}^*\mathbb{Z}$ generated by $\{\alpha_k \colon k \in \mathbb{N}\}$, it is easy to see that H is divisible and hence by the theory of divisible groups H contains a copy of $\mathbb{Q}$. (As a matter of fact, $H \cong \mathbb{Q}$, but all that we need is that ${}^*\mathbb{Z}$ contains a copy of $\mathbb{Q}$ – so it is not reduced.) To see that H is divisible, observe that

$$\{n \in \mathbb{N} \colon ka_k(n) \neq a_1(n)\} \subseteq \{1, 2, \ldots, k-1\}$$

and so is finite. Thus its complement

$$\{n \in \mathbb{N} \colon ka_k(n) = a_1(n)\}$$

is cofinite. But any nonprincipal ultrafilter must contain all of the cofinite sets (see next section).Thus $\{n \in \mathbb{N} \colon ka_k(n) = a_1(n)\} \in D$. This means that $k\alpha_k = \alpha_1$ (see next section). Moreover, $\alpha_1 = [(1!, 2!, \ldots, n!, \ldots)]_D \neq 0$. Next note that it is easy to show that

$$d\alpha_{dm} = \alpha_m$$

Now consider an arbitrary element of H

$$\alpha = n_1\alpha_1 + \cdots + n_m\alpha_m \in H\,.$$

Let k be any positive integer. We need to show that α is the kth multiple of an element of H. But this is true. For let

$$N = kLCM(1, \ldots, m)\,.$$

Then using $(\star)$ above, we see that

$$\alpha = \sum_{j=1}^{m} n_j\alpha_j = \left(\sum_{j=1}^{m} n_j k\frac{N}{jk}\right)\alpha_N = k\left(\sum_{j=1}^{m} n_j\frac{N}{jk}\right)\alpha_N\,.$$

Thus $H \subseteq^* \mathbb{Z}$ is divisible.

4.5 Filters and ultraproducts

Let I be a nonempty set. Let $P(I)$ be the power set of I. If $J \in P(I)$ let J' be is complement, $I \backslash J$. If a statement about $P(I)$ involves the logical symbols $\sim, \cup, \cap, \subset, \emptyset, I$, then its *dual* is obtained by simply replacing these, respectively, by $\sim, \cap, \cup, \supset, I, \emptyset$. A subset $M_0 \subset P(I)$ will be called an *ideal* in the ring $P(I)$ if $\emptyset \in M_0$, M_0 is closed under finite unions and M_0 is closed under the formation of subsets. M_0 will be a *proper ideal* in $P(I)$ if is an ideal in $P(I)$ and $I \notin M_0$. The dual of a proper ideal is called a proper filter. Specifically, a subset $D_0 \subset P(I)$ is a *proper filter* on I provided
(1) $I \in D_0$;
(2) $A, B \in D_0 \implies A \cap B \in D_0$;
(3) $A \in D_0$ and $A \subset B \subset I \implies B \in D_0$;
(4) $\emptyset \notin D_0$.

We will usually suppress the adjective proper and so call a proper filter simply a filter. Intuitively a filter on I is a collection of subsets of I which contains the *big* subsets of I. The trivial filter on I is $\{I\}$, The power set $P(I)$ is called an *improper filter*. A subset $S \subseteq P(I)$ satisfies the *finite intersection property*, denoted (fip), provided whenever $A_1, \ldots, A_n$ are finitely many members of S the intersection $A_1 \cap \ldots \cap A_n \neq \emptyset$. Clearly in view of (2) and (4) above, a necessary condition for a set $S \subset P(I)$ to lie in a filter is that it satisfy fip. It turns out that fip is also sufficient, For suppose that $S \subset P(I)$ satisfies fip. Then let D_0 be the family of all elements $B \in P(I)$ such that B contains some finite intersection $A_1 \cap \ldots \cap A_n$ of elements of S. Then it is easy to see that D_0 is a filter on I containing S.

As a matter of fact, D_0, is clearly the smallest filter on I containing the elements of S. For if F is any filter and $S \subset F$, then $D_0 \subseteq F$. D_0 as constructed above is called the filter on I *generated by* S. A filter generated by a single nonempty subset $S = \{A\}, \emptyset \subsetneqq A \subseteq I$ is called a *principal filter* on I. A *nonprincipal filter* on I is one which is not principal. We note that a principal filter is the dual of a (proper) principal ideal.

A filter D on I is an *ultrafilter* on I provided it is maximal in the class of filters on I, i.e., $D \subset E$ and E a filter on I implies that $D = E$. Now if D_0 is any filter on I and $A \subseteq I$, then we can't have both $A \in D_0$ and $A' \in D_0$. Thus for every such A at most one of $A \in D_0$ and $A' \in D_0$ holds. We claim that if D is a filter, then D is an ultrafilter if and only if for every element $A \in P(I)$ exactly one of $A \in D$ and $A' \in D$ holds. First suppose that neither one of $A \in D$ nor $A' \in D$ holds. Then let E be the collection of all supersets B such that $C \cap A \subseteq B \subseteq I$ as C varies over all elements of D. Then it is easy to show that E is a filter on I. Also since $C \cap A \subseteq C$ and $C \cap A \subseteq A$, we see that $D \subset E$ properly. Thus in this case D is not an ultrafilter. Now suppose that exactly one of $A \in D_0$ or $A' \in D_0$ holds and $D \subseteq E$ where E is a filter on I. If there were a $B \in E \setminus D$, then $B \notin D$. By hypothesis then $B' \in D \subseteq E$. But this forces B and B' to lie in E, a contradiction. Therefore $E \setminus D = \emptyset$. So $E = D$ and D is an ultrafilter.

For every $i_0 \in I$ the set

$$D = \{A \in P(I)\colon i_0 \in A\}$$

is a principal ultrafilter on I; moreover, we claim that these exhaust the principal ultrafilters on I. To see this, we first note that D is clearly an ultrafilter on I. Now suppose that E is any principal ultrafilter on I. Then E is the smallest filter on I containing some fixed generator $A \subseteq I$. Now $\emptyset \notin E$, so $A \neq \emptyset$. Let $i_0 \in A$ be fixed. But since $i_0 \in A$, $A \in D$. Thus $D \subseteq E$. But since D is an ultrafilter, $D = E$. Now suppose that D is an ultrafilter on I generated by $X = \{A\}$ where $\emptyset \subsetneqq A \subseteq I$, We claim $A = \{i_0\}$ for some $i_0 \in I$, i.e., A is a singleton. Since $A \neq \emptyset$, there is an $i_0 \in A$. But then the filter E_0 generated by $Y_0 = \{\{i_0\}\}$ contains D. Thus by maximality $E_0 = D$. Furthermore, if $i_1 \in A \setminus \{i_0\}$, then $\{i_0\} \cap \{i_1\} = \emptyset$. So the filter E_1 on I generated by $Y_1 = \{\{i_1\}\}$ cannot contain $\{i_0\}$. But this contradicts the fact that $E_1 = D = E_0$ again by maximality of D. Thus we have shown that the principal ultrafiliters on I are precisely those of the form

$$\{X \in P(I)\colon i_0 \in X\}$$

for each fixed $i_0 \in I$.

Now consider the complement of $\{i_0\}$, i.e., $\{i_0\}'$. It is *cofinite* (i.e., it has a finite complement $\{i_0\}$). But

$$\{i_0\}' \notin \{X \in P(I)\colon i_0 \in X\}\ .$$

Thus no principal ultrafilter contains all the cofinite subsets of I. Taking the contrapositive, we see that if an ultrafilter contains all of the cofinite subsets, then it must be nonprincipal. The converse is also true as we shall soon see.

Now suppose that I is finite. We claim that any ultrafilter on I is principal. For suppose that D is a nonprincipal ultrafilter on I, Suppose also that $i_0 \in A \in D$. Now A is finite. But since D is an utrafilter either $\{i_0) \in D$ or $\{i_0\}' \in D$. But if $\{i_0\} \in D$, then $E = \{B \subseteq I\colon i_0 \in B\} \subseteq D$. But E is the principal ultrafilter generated by $\{i_0\}$. (That E is maximal follows from the fact that for every subset $C \subseteq I$ exactly one of $i_0 \in C$ or $i_0 \notin C$ holds.) Thus $E = D$ by maximality and D is principal – contrary to hypothesis. Thus suppose that $I \setminus \{i_0\} = \{i_0\}' \in D$ for every $i_0 \in A$. Now let $B = \{i_1, \ldots, i_n\}$. Then $I \setminus \{i_1\}, \ldots, I \setminus \{i_n\} \in D$. Thus

$$\bigcap_{j=1}^{n} (I \setminus \{i_j\}) \in D\,.$$

By DeMorgan's Laws,

$$I \setminus \bigcup_{j=1}^{n} (\{i_j\}) \in D$$

i.e., $I \setminus B \in D$. This shows that no finite subset can be in D But if I is finite then so is $I \setminus B$. This contradiction shows that D must be principal.

We note that the above argument also shows that if I is infinite, then every nonprincipal ultrafilter on I must contain all the cofinite subsets of I. Also if I is infinite, then we may appeal to Zorn's Lemma to prove that every filter D_0 on I extends to an ultrafilter D on I. Thus we assume

Lemma 4.5.1 (Zorn's Lemma). *If P is a nonempty, partially ordered set such that each chain C contained in P has an upper bound $u \in P$, then P has a maximal element $m \in P$.*

Lemma 4.5.2. *If $E \subseteq P(I)$ and E satisfies fip, then there exists an ultrafilter D on I such that $E \subset D$.*

Proof. Let F be the filter generated by E. So $\emptyset \notin F$ whence F is a proper filter. Now if C is any nonempty chain of proper filters on I, then it is easy to show that $\cup C$ is a proper filter on I. Moreover, if $D \in C$ and $E \subset D$, then $E \subset \cup C$, Thus by Zorn's Lemma the class of all proper filters on I containing E has a maximal element, say D. This D is an ultrafilter on I and $E \subset D$. □

Theorem 4.5.1 (Prime Ideal Theorem). *Any proper filter on I can be extended to an ultrafilter on I.*

Proof. Any proper filter on I satisfies fip, so the above Lemma 4.5.1 applies to it. □

Assuming the consistency of the Zermelo–Frankel Axioms (see [25] for an explanation), the Prime Ideal Theorem is provably weaker than the Axiom of Choice (see for instance [128]).

Based on filters we define the important constructions of reduced products and ultraproducts. Let I be a nonempty set and D a proper filter on I and let $\{G_i\}_{i\in I}$ be a family of groups indexed by I. Then it is straightforward to show that the relation on the direct product $\prod_{i\in I} G_i$ defined by $f \equiv_D g$

$$\{i \in I; f(i) = g(i)\} \in D$$

is a congruence relation. A relation on a group G is a congruence relation provided it is an equivalence relation that preserves the group operation. It follows that the subset K of the direct product given by

$$K = \left\{ f \in \prod_{i\in I} G_i; f \equiv_D 1 \right\}$$

actually defines a normal subgroup.

Definition 4.5.1. Let I be a nonempty set and D a proper filter on I and let $\{G_i\}_{i\in I}$ be a family of groups indexed by I. Then the *reduced product* $\prod_{i\in I} G_i/D$ is the quotient of the direct product $\prod_{i\in I} G_i$ modulo the normal subgroup $K = \{f \in \prod_{i\in I} G_i; f \equiv_D 1\}$. We write $[f]_D$ for the equivalence class Kf of f modulo the proper filter D on I. If D

is an ultrafilter on I then the reduced product is the *ultraproduct* of the family $\{G_i\}_{i\in I}$ modulo the ultrafilter D on I. If the family

$$\{G_i\}_{i\in I}$$

consists of a single group G so that the direct product is G^I then the ultraproduct is an *ultrapower*.

If D is the trivial filter on I then $\prod_{i\in I} G_i/D \cong \prod_{i\in I} G_i$ and therefore every direct product is up to isomorphism a reduced product.

A family $\phi_i\colon H_i \to G_i$ of group homomorphisms determines in a natural way a homomorphism $\phi\colon \prod_{i\in I} H_i \to \prod_{i\in I} G_i$. If D is a proper filter on I then ϕ induces a well-defined map

$$\phi^*\colon \prod_{i\in I} H_i/D \to \prod_{i\in I} G_i/D\,.$$

Further ϕ^* will be monic if each ϕ_i is monic.

Lemma 4.5.3. *Let I be a nonempty set and D a proper filter on I and let $\{G_i\}_{i\in I}$ be a family of groups indexed by I. Then*

$$[(f,g)]_D \to ([f]_D, [g]_D)$$

determines a well-defined isomorphism from $\prod_{i\in I}(G_i \times G_i)/D$ onto $(\prod_{i\in I} G_i/D) \times (\prod_{i\in I} G_i/D)$

The proof is straightforward.

Ultraproducts and ultrapowers play a major role in determining when a class of groups is axiomatic and are closely tied to elementary equivalence.

Lemma 4.5.4. *Let $\mathcal{X}$ be a class of groups. Then $\mathcal{X}$ is axiomatic if and only if $\mathcal{X}$ is closed under both ultraproducts and elementary equivalence.*

Lemma 4.5.5. *Reduced products preserve elementary equivalence, i.e., if I is a nonempty set, and G_i is elementarily equivalent to H_i for all $i \in I$ and D is a proper filter on I, then $\prod_{i\in I} G_i/D$ is elementarily equivalent to $\prod_{i\in I} H_i/D$.*

For a proof of Lemma 4.5.4 see Gr Chapter 6, Theorem 3 and for Lemma 4.5.5 see [40] Chapter 6, Theorem 6.3.4.

Corollary 4.5.1. *Direct products and ultraproducts preserve elementary equivalence.*

Proof. Since direct products and ultraproducts are special cases of reduced products Lemma 4.5.5 establishes the result. □

Finally we mention again the result of Keisler and Shelah introduced in Section 4.2.

Theorem 4.5.2. *Two groups are elementary equivalent if and only if their ultrapowers (with respect to some non-principal ultrafilter) are isomorphic.*

Also important in this direction is Los's Theorem which implies that a group G and any ultrapower share the same elementary theory (see [25]).

Theorem 4.5.3 ([177] (Relativized for groups)). *Let $(G_i)_{i\in I}$ be a family of groups indexed by a nonempty set. Let D be an ultrafilter on I and let $U = \prod_{i\in I} G_i/D$ be the corresponding ultraproduct of the family. Let $\varphi(\overline{x})$ be a formula of L_0 and $\overline{[f]_D}$ be a tuple of elements of U. Then*

$$U \models \varphi\left(\overline{[f]_D}\right) \quad \text{if and only if } \left\{i \in I\colon G_i \models \varphi(\overline{f(i)})\right\} \in D\,.$$

Before proving Los' Theorem, we observe that it has the following corollary:

Corollary 4.5.2. *Let G be a group and $^*G$ an ultrapower of G. Then G and $^*G$ have the same elementary theory.*

Proof of Corollary 4.5.2. Let I be a nonempty index set and D an ultrafilter on I. If G is a group and $^*G = G^I/D$ is an ultrapower and

$$d\colon G \to^* G$$

by

$$d(g) = [\delta(g)]_D$$

where $\delta(g)(i) = g$ for all $i \in I$ is the canonical embedding. Then it follows from Los' Theorem that d is an elementary embedding and hence $G \equiv^* G$. To see this, we note that $*G \models \varphi(\overline{d(g)})$ if and only if

$$\{i \in I\colon G \models \varphi(\overline{g})\} = \left\{i \in I\colon G \models \varphi(\overline{\delta(g)(i)})\right\} \in D\,.$$

But

$$\begin{aligned}\{i \in I\colon G \models \varphi(\overline{g})\} &= \phi \text{ if } G \not\models \varphi(\overline{g})\\ &= I \text{ if } G \models \varphi(\overline{g})\end{aligned}$$

so $\{i \in I\colon G \models \varphi(\overline{g})\} \in D$ if and only if $G \models \varphi(\overline{g})$. That is, $*G \models \varphi(\overline{d(g)})$ if and only if $G \models \varphi(\overline{g})$ so d is elementary as claimed. □

Proof. (Los' Theorem) We may assume that L_0 contains the logical symbols $\sim$, $\wedge$, and $\exists$ and the others are defined in terms of these. We induce on the complexity of $\varphi(\overline{x})$ as measured by the number of logical symbols it contains.

Suppose first that $\varphi(\overline{x})$ is atomic. Then it has the form

$$s(x_1, \dots, x_n) = t(x_1, \dots, x_n)$$

where s and t are terms of L_0. Now $U \models s([f_1]_D, \dots, [f_n]_D) = t([f_1]_D, \dots, [f_n]_D)$ if and only if

$$\{i \in I\colon s(f_1(i), \dots, f_n(i)) = t(f_1(i), \dots, f_n(i))\} \in D$$

by definition of the ultraproduct. That starts the induction.

$$U \models\sim \psi([f_1]_D, \ldots, [f_n]_D) \quad \text{if and only if } U \not\models \psi([f_1]_D, \ldots, [f_n]_D)\,.$$

But, by inductive hypothesis,

$$U \models \psi([f_1]_D, \ldots, [f_n]_D) \quad \text{if and only if } \{i \in I : G_i \models \psi(f_1(i), \ldots, f_n(i))\} \in D\,.$$

It follows that

$$U \models\sim \psi([f_1]_D, \ldots, [f_n]_D) \quad \text{if and only if } \{i \in I : G_i \models \psi(f_1(i), \ldots, f_n(i))\} \notin D$$

if and only if

$$I \setminus \{i \in I : G_i \models \psi(f_1(i), \ldots, f_n(i))\} \in D$$

(since D is an ultrafilter) if and only if

$$\{i \in I : G_i \not\models \psi(f_1(i), \ldots, f_n(i))\} \in D$$

if and only if

$$\{i \in I : G_i \models\sim \psi(f_1(i), \ldots, f_n(i))\} \in D\,.$$

Now $U \models \psi([f_1]_D, \ldots, [f_n]_D) \wedge \pi([f_1]_D, \ldots, [f_n]_D)$ if and only if

$$U \models \psi([f_1]_D, \ldots, [f_n]_D) \text{ and } U \models \pi([f_1]_D, \ldots, [f_n]_D)\,.$$

By inductive hypothesis,

$$U \models \psi([f_1]_D, \ldots, [f_n]_D) \quad \text{if and only if } \{i \in I : G_i \models \psi(f_1(i), \ldots, f_n(i))\} \in D$$

and

$$U \models \pi([f_1]_D, \ldots, [f_n]_D) \quad \text{if and only if } \{i \in I : G_i \models \pi(f_1(i), \ldots, f_n(i))\} \in D\,.$$

But

$$\begin{aligned}&\{i \in I : G_i \models \psi(f_1(i), \ldots, f_n(i)) \wedge \pi(f_1(i), \ldots, f_n(i))\} = \\ &\qquad \{i \in I : G_i \models \psi(f_1(i), \ldots, f_n(i))\} \cap \{i \in I : G_i \models \pi(f_1(i), \ldots, f_n(i))\}\,.\end{aligned}$$

Note that, since $A \cap B \subseteq A$ and $A \cap B \subseteq B$, $A \cap B \in D$ if and only if $A \in D$ and $B \in D$. It follows that

$$U \models \psi([f_1]_D, \ldots, [f_n]_D) \wedge \pi([f_1]_D, \ldots, [f_n]_D)$$

if and only if

$$\{i \in I : G_i \models \psi(f_1(i), \ldots, f_n(i)) \wedge \pi(f_1(i), \ldots, f_n(i))\} \in D\,.$$

Suppose $U \models \exists x(\psi(x, [f_1]_D, \ldots, [f_n]_D))$. Then $U \models \psi([f]_D, [f_1]_D, \ldots, [f_n]_D)$ for some $f \in \prod_{i\in I} G_i$. By inductive hypothesis,

$$A = \{i \in I : G_i \models \psi(f(i), f_1(i), \ldots, f_n(i))\} \in D\,.$$

But

$$\{i \in I : G_i \models \exists x(\psi(x, , f_1(i), \ldots, f_n(i))\} \supseteq A$$

so

$$\{i \in I : G_i \models \exists x(\psi(x, , f_1(i), \ldots, f_n(i))\} \in D\,.$$

Hence, $U \models \exists x(\psi(x, [f_1]_D, \ldots, [f_n]_D))$ implies

$$\{i \in I : G_i \models \exists x(\psi(x, , f_1(i), \ldots, f_n(i))\} \in D\,.$$

Now suppose that

$$B = \{i \in I : G_i \models \exists x(\psi(x, , f_1(i), \ldots, f_n(i))\} \in D\,.$$

For each $i \in I$ choose an element $f(i) \in G_i$ such that $G_i \models \psi(f(i), f_1(i), \ldots, f_n(i))$. Define $f(i) = 1$ for all $i \in I \setminus B$. Then

$$\{i \in I : G_i \models \exists x(\psi(f(i), , f_1(i), \ldots, f_n(i))\} \supseteq B$$

so $\{i \in I : G_i \models \exists x(\psi(f(i), , f_1(i), \ldots, f_n(i))\} \in D$. By inductive hypothesis,

$$U \models \psi([f]_D, [f_1]_D, \ldots, [f_n]_D).$$

Therefore $U \models \exists x(\psi(x, [f_1]_D, \ldots, [f_n]_D))$. Thus,

$$\{i \in I : G_i \models \exists x(\psi(x, f_1(i), \ldots, f_n(i)))\} \in D$$

implies

$$U \models \exists x(\psi(x, [f_1]_D, \ldots, [f_n]_D)).$$

Since we have shown implications in both directions,

$$U \models \exists x(\psi(x, [f_1]_D, \ldots, [f_n]_D))$$

if and only if

$$\{i \in I : G_i \models \exists x(\psi(x, f_1(i), \ldots, f_n(i)))\} \in D\,.$$

That completes the induction and proves the theorem. □

One final concept that will be needed in Chapter 11 is that of an *ultralimit*. Given any sequnce of real or complex numbers (x_n) and an ultrafilter U on the natural numbers $\mathbb{N}$ then a value x is an *ultralimit* of the sequence with respect to U if for each $\epsilon > 0$ the sets $\{n : |x_n - x| \leq \epsilon\} \in U$. The concept can be extended to sequences over any metric space and even to sequneces of spaces themsleves.

5 The Tarski problems

5.1 The Tarski problems

From a group theoretical point of view, interest in elementary theory has been motivated in large part by the *Tarski problems* or *Tarski conjectures* formulated by A. Tarski around 1945. In a 1988 peper surveying combinatorial group theory [175] Roger Lyndon called the Tarski problems, which he described as folklore, among the outstanding open problems (at that time) in the field. Basically Tarski conjectured that all non-Abelian free groups have exactly the same elementary theory. We'll state this in a more formal manner below. The genesis of this conjecture seems to be the following straightforward observations on free groups which show immediately that all non-Abelian free groups have the same universal theory.

Let α, β be ordinal numbers with $\alpha < \beta$. Let F_α be the free group of rank α, and F_β the free group of rank β. Then there is a natural embedding of F_α into F_β as a proper free factor. Furthermore if ω is the first limit ordinal then there are monomorphisms embedding F_ω into F_2, the free group of rank 2, and hence into F_n for all $2 \leq n < \omega$. However F_ω cannot be a free factor in F_n for any finite n. It follows that if F and G are any two countable non-Abelian free groups, then each is embeddable in the other. Hence the elementary theory of any countable non-Abelian free group must strongly resemble the elementary theory of any other countable non-Abelian free group.

Notice that if $\forall \overline{x} \phi(\overline{x})$ is a universal sentence true in a group G, then the same sentence must also be true in any subgroup of G. Hence every universal sentence true on a group is true in every subgroup. However clearly there may be universal sentences true in a subgroup that are not true in the overgroup. It follows that if H is embedded in G then the universal theory of G is embedded in the universal theory of H, that is

$$Th_\forall(G) \subset Th_\forall(H) .$$

Since any two countable non-Abelian free groups are embeddable in each other this argument together with the Lowenheim–Skolem Theorem (see Chapter 4) shows that any two non-Abelian free groups have exactly the same universal theory. By a theorem of Vaught (see [118] Chapter 6) every free group of rank $r \geq \omega$ is elementarily equivalent to F_ω and hence must be universally equivalent to F_ω. Combining these observations we obtain.

Theorem 5.1.1. *Any two non-Abelian free groups have exactly the same universal theory and hence exactly the same existential theory.*

On the other hand the universal theory of free groups is not complete in that a group not elementarily equivalent to any non-Abelian free group can have the same universal theory as the class of non-Abelian free groups. We call a group *universally free* if it has the same universal theory as the class of non-Abelian free groups. The above statement on completeness means that there exist nonfree, universally free groups which can be distinguished from free groups by first-order properties. In the next chapter we discuss this further and give a very beautiful group theoretical description of universally free groups.

Observations of this type were the basis for the Tarski conjectures which we now make precise.

Conjecture (Tarski Conjecture 1). *Any two non-Abelian free groups are elementarily equivalent. That is any two non-Abelian free groups satisfy exactly the same first-order theory.*

Conjecture (Tarski Conjecture 2). *If the non-Abelian free group H is a free factor in the free group G then the inclusion map $H \to G$ is an elementary embedding.*

Clearly the second conjecture is stronger than the first and implies the first. If true then the theory of the non-Abelian free groups would be complete, that is given a sentence ϕ of L_0 then either ϕ is true in every non-Abelian free group or ϕ is false in every non-Abelian free group.

In addition to the completeness of the theory of the non-Abelian free groups the question of its decidability also arises. Recall that this means the question of whether there exists a recursive algorithm which, given a sentence ϕ of L_0 decides whether or not ϕ is true in every non-Abelian free group. Tarski further conjectured that the theory of the non-Abelian free groups is decidable.

Conjecture (Tarski Conjecture 3). *The elementary theory of the countable non-Abelian free groups is decidable.*

Kharlampovich and Myasnikov [152, 153, 154, 155, 156] and Sela [233, 234, 235, 236, 237] have recently proved that the first two Tarski conjectures are indeed true, while Kharlampovich and Myasnikov also prove the third on decidability. The proof by Kharlampovich and Myasnikov grew out of work on equations over non-Abelian free groups, residually free groups and on the algebraic geometry of groups. Sela calls his approach Diophantine Geometry over Groups. In Chapter 9 we will briefly outline the Kharlampovich–Myasnikov version of the proof. A separate version of the proof is being written up by Bestvinna and Feign [29].

Theorem 5.1.2. *If the non-Abelian free group H is a free factor in the free group G then the inclusion map $H \to G$ is an elementary embedding. Hence any two non-Abelian free groups are elementarily equivalent. Moreover the theory of the non-Abelian free groups is decidable.*

5.2 Initial work on the Tarski problems

In this section we review some of the inital work on the Tarski questions.

The first progress was due to Vaught who showed that the Tarski conjectures 1,2 are true if G and H are both free groups of infinite rank.

Theorem 5.2.1. *If the infinite rank free group H is a free factor in the free group G then the inclusion map is an elementary embedding. In particular if G and H are free groups of infinite rank then $Th(G) = Th(H)$.*

Proof. The basic idea in Vaught's proof is to use the criteria for elementary embeddings given in Theorem 4.2.1. Recall that this says that if H_0 is a subgroup of H and that to every finite subset $\{a_1, \ldots, a_n\}$ of H_0 and every element $b \in H$ there exists an automorphism σ of H fixing $a_1, \ldots, a_n$ and mapping b into H_0, then the inclusion map from H_0 into H is an elementary embedding. Applying this to free groups of infinite rank suppose that F is free on an infinite subset S and that G is free on an infinite subset S_0 of S. Then permutations of S will induce enough automorphisms to guarantee that the inclusion map of G into F is an elementary embedding We note that Vaught's proof extends to infinite rank free algebras in any variety of algebras of any type. We now give the details of the proof.

Let H be an infinite rank free factor in the free group F. Let $h_1, \ldots, h_n$ be finitely many elements of H and let $f \in F$. We show that there is an automorphism of F which preserves any fixed finite set in H and also maps any fixed element $f \in F$ into H.

Suppose that F is the free product $H * K$ and let A be a free basis for H and B be a free basis for K. Then, necessarily:

(1) A and B are disjoint;
(2) $A \cup B$ is a free basis for F;
(3) A is infinite.

Now the elements $h_1, \ldots, h_n$ and f, when expressed as words on $A \cup B$, can involve only finitely many letters from A; moreover, in addition to the letters from A, f can involve only finitely many letters from B. Say $A_0 = \{a_1, \ldots, a_p\}$ is the set of letters from A in $h_1, \ldots, h_n$ and f and $B_0 = \{b_1, \ldots, b_q\}$ is the set of letters from B involved in f. Say $h_i = u_i(a_1, \ldots, a_p), i = 1, \ldots, n$ and $f = w(a_1, \ldots, a_p, b_1, \ldots, b_q)$. Since A is infinite we can find a subset A_1 of A, disjoint from A_0, with $|A_1| = |B_0|$. Say, $A_1 = \{a_{p+1}, \ldots, a_{p+q}\}$. Now consider the permutation π of $A \cup B$ which is the identity on the complement of $A_1 \cup B_0$ and which is the following product of disjoint transpositions on $A_1 \cup B_0$: $\pi = (a_{p+1}, b_1)(a_{p+2}, b_2) \ldots (a_{p+q}, b_q)$. Let σ be the automorphism of F determined by the permutation π of its free basis $A \cup B$. Then

$$\begin{aligned}\sigma(h_i) &= \sigma(u_i(a_1, \ldots, a_p)) = u_i(\sigma(a_1), \ldots, \sigma(a_p)) \\ &= u_i(\pi(a_1), \ldots, \pi(a_p)) = u_i(a_1, \ldots, a_p) = h_i,\end{aligned}$$

for $i = 1, \dots, n$ and

$$\sigma(f) = \sigma(w(a_1, \dots, a_p, b_1, \dots, b_q)) = w(\sigma(a_1), \dots, \sigma(a_p), \sigma(b_1), \dots, \sigma(b_q))$$
$$= w(\pi(a_1), \dots, \pi(a_p), \pi(b_1), \dots, \pi(b_q)) = w(a_1, \dots, a_p, a_{p+1}, \dots, a_{p+q}) \in H . \square$$

The elementary chain theorem (Theorem 4.2.6) gives that the direct union of elementary embeddings is elementary. Therefore Vaught's proof that the Tarski conjectures 1,2 are true for free groups of infinite rank reduced the conjectures to free groups of finite rank.

The next significant progress was due to Merzljakov. Recall that a *positive sentence* is a first-order sentence which is logically equivalent to a sentence constructed using (at most) the connectives $\vee, \wedge, \forall, \exists$. The *positive theory* of a group G consists of all the positive sentences true in G. Merzljakov [185] showed that the non-Abelian free groups have the same positive theory.

Theorem 5.2.2. *Two non-Abelian free groups have the same positive theory.*

A proof of Merzljakov [185] can be given involving generalized equations and a quantifier elimination process. This was a precursor to the methods used in the eventual solution of the overall Tarski problems.

Since positive sentences are preserved by homomorphic images and since every group is a homomorphic image of a free group this theorem has the following curious corollary. If the positive sentence ϕ is true in some non-Abelian free group and hence in all non-Abelian free groups then it is true in any group whatsoever.

Further work following Merzljakov centered on restricted theories of free groups. We have seen that any two non-Abelian free groups satisfy the same universal theory. Sacerdote [226] proved that this could be extended to *universal-existential sentences* (see Section 1).

Theorem 5.2.3. *Any two non-Abelian free groups satisfy the same universal-existential theory, that is they satisfy exactly the same universal-existential sentences.*

In the next chapter we give further results about universal and existential freeness. Advances in a different direction were given by Makanin and Razborov. Makanin [179] proved that there exists an algorithm to determine, given a finite system of equations in a free group, whether the system possesses at least one solution. In other words, the Diophantine problem is decidable for free groups (compare this to undecidability of the Diophantine problems for integers). Razborov [213]. working with the Makanin algorithm determined an algorithm to effectively describe the solution sets of a finite system of equations in a free group. Kharlampovich and Myasnikov further developed the Makanin–Razborov method. Their technique allows one to transform arbitrary finite systems of equations in free groups to some "canonical forms" and describe precisely the irreducible components of algebraic sets in free groups. These canonical forms consist of finitely many quadratic equations in a triangular form (see Definition 6.3 and Theorem 6.2). Observe that quadratic equations over free groups (Comerford and Edmonds [60]), and hyperbolic groups (see Grigorchuk and Kurchanov [120]) are well-studied, and there are good methods for solving them.

The following result is a corollary of the decidability of the Diophantine problem

Theorem 5.2.4 ([180]). (1) *The existential (and hence the universal) theory of a free group is decidable.*
(2) *The positive theory of a free group is decidable*

In addition to the Tarski problems themselves, the quest for a solution has inspired many other results in group theory. This is especially true concerning the theory of solutions of equations within groups. In 1959 Vaught asked the question whether the sentence

$$\forall x \forall y \forall z (x^2 y^2 z^2 = 1 \to xy = yx \wedge xz = xz \wedge yz = zy)$$

holds in all free groups. R. Lyndon [172] then proved that for each solution of $x^2 y^2 z^2 = 1$ in a free group the elements commute pairwise. This result launched the theory of equations over free groups. The first general results in this direction are were due to Lyndon [172], Lorenc [166] and Appel [5] where they described the solution set of an arbitrary one variable equation over a free group. In 1966 Malcev described the solution set of the equation $[x, y] = [a, b]$ over the free group $F(a, b)$ a problem considered earlier by Nielsen [203]. A version of this was also solved by Zieschang (see [259]) and by Csorgo, Fine and Rosenberger [61]. Malcev's solution has the following nontrivial implication for the elementary theory of a free group of rank 2; the set of all free bases of $F(a, b)$ can be defined by a first order formula in the language of group theory. That is the elements $u, v \in F(a, b)$ are a free basis if and only if they satisfy the following formula with constants a, b;

$$\exists z([u, v] = z^{-1}[a, b]z \vee [u, v] = z^{-1}[b, a]z) .$$

The focus of study eventually turned to quadratic equations over free groups that is equations in which every variable occurs exactly twice. Quadratic group relators have always been essential in combinatorial group theory due to their close connection with surface group relators (see Chapters 2 and 3). Comerford and Edmonds [60] and Grigorchuk and Kurchanov [120] described the solution sets of quadratic equations over arbitrary free groups. Further work of Hoare, Karrass and Solitar shows that every quadratic equation over a free group is automorphically equiavlent to a standard one.

Makhanin in 1982 [179, 180] proved that if a given equation over a free group F has a solution in F then this equation has a solution of bounded length and this bound can be effectively computed from the equation itself. Makhanin's work allowed Razborov [213] to describe the solution set of a system of equations over F. Makhanin further extended his results [180] proving that the universal theory of a non-Abelian free group F is algorithmically decidable.

5.3 The positive solution to the Tarski problems

The proofs of the Tarski problems both by Kharlampovich and Myasnikov and by Sela are very long and very involved. In Chapter 9 we will try to outline the Kharlampovich–Myasnikov proof. Even this outline requires us to describe the theory of fully residually free groups which we will do in the next two chapters and then introduce algebraic geometry over groups which we will do in Chapter 8. In this section we describe the fundamental ingredients that must be developed for the proof of the Tarski problems. There are four basic ingredients necessary, each of which will be touched on in subsequent chapters. These are: algebraic geometry over groups, the theory of fully residually free groups including the relationship to free exponential groups, an implicit function theorem (over free groups) and a description of irreducible algebraic varieties (over free groups) in terms of triangular quasi-quadratic systems.

It was clear from the beginning that to deal with the Tarski problems it was necessary to give a precise definition of solution sets of equations and inequations over free groups. In analogy with the classical theory of equations over number fields, algebraic geometry over groups was developed by G. Baumslag, A. Myasnikov and V. Remeslennikov (see [16], [195]). The theory of algebraic geometry over groups translated the basic notions of the classical algebraic geometry: algebraic sets, the Zariski topology, Noetherian domains, irreducible varieties, radicals and coordinate groups to the setting of equations over groups. This provided the necessary machinery to transcribe important geometric ideas into pure group theory. Some of these ideas go back to R. Bryant [38], V. Guba [123], B. Plotkin [211] and E. Rips [217]. We will introduce the ideas of algebraic geometry over groups in Chapter 8.

The next basic ingredient in the approach to the Tarski problems is the idea of a group being discriminated by a free group. This led to a systematic study of fully residually free groups. A group G is *fully residually free* provided to every finite set $S \subset G \setminus \{1\}$ of non-trivial elements of G there is a free group F_S and an epimorphism $h_S\colon G \to F_S$ such that $h_S(g) \neq 1$ for all $g \in S$. In other language we say that the group G is *discriminated* by a free group. Remeslennikov [214] and independently Gaglione and Spellman [108] established a surprising connection between fully residually free groups and the universal theory of free groups. In particular a finitely generated non-Abelian group G is fully residually free if and only if it is universally free, that is has the same universal theory as the non-Abelian free groups. Since elementary free groups must be universally free, fully residually free groups must play a prominent role in the study of the elementary theory. We discuss the extensive research on fully residually free groups in the next two chapters. We note that in Sela's work, finitely generated fully residually free groups are called limit groups and appear as limiting groups of sets of homomorphisms into free groups. In Chapter 7 we will make Sela's approach more precise.

R. Lyndon [173] introduced the concept of an exponential group, that is a group which allows parametric exponents in an associative unitary ring A. In particular he studied the free exponential group $F^{\mathbb{Z}[t]}$ where exponents are allowed from the polynomial ring $\mathbb{Z}[t]$ over the integers $\mathbb{Z}$ (see Chapter 6). Lyndon established that the free exponential group $F^{\mathbb{Z}[t]}$, and hence any finitely generated subgroup of it, is fully residually free and hence, if it is non-Abelian, universally free. Kharlampovich and Myasnikov (see [156, 158] established the converse; therefore a finitely generated group is fully residually free if and only if it is embeddable in $F^{\mathbb{Z}[t]}$.

Lyndon's original motivation for introducing exponential groups was from the solution sets of equations over free groups. In [174] he found that the solution set of any equation with one variable over a free group F can be obtained from finitely many paramtetric words by specializing their parameters in the integers. A parametric word with parameters in $\mathbb{Z}[t_1, \ldots, t_n]$ is a formal expression obtained from a basis for F by finitely many concatenations and exponentiations from $\mathbb{Z}[t_1, \ldots, t_n]$. If one specializes the parameters $t_1, \ldots, t_n$ in $\mathbb{Z}$ one obtains an element of F. Lyndon proved that for any equation with one variable over a free group F one can effectively find a finite set of parametric words with parameters form the ring $\mathbb{Z}[t_1, \ldots, t_n]$ such that any solution of this equation can be obtained from some specialization of these words. K. Appel refined Lyndon's result and showed that the solution set of a one variable equation over a free group can be parametrized by a finite set of words of the form $fg^t h$ where $f, g, h \in F$ and $t \in \mathbb{Z}[t]$.

A further detailed study of the structure of exponential groups was carried out by A. Myasnikov and V. Remeslennikov [196, 197]. They proved that the group $F^{\mathbb{Z}[t]}$ can be obtained starting from F by an infinite chain of a special type of HNN extensions called *extensions of centralizers*. If G is a group and C is the centralizer of a nontrivial element in G then

$$G(C, s) = \langle G, s; s^{-1}cs = c, c \in C\rangle$$

is a free extension of the centralizer C by s. From the work of Myasnikov and Remeslennikov, to construct $F^{\mathbb{Z}[t]}$ one needs to extend each centralizer sufficiently many times so that every proper centralizer is isomorphic to a free Abelian group of infinite rank – the additive group of $\mathbb{Z}[t]$. This further implies that any finitely generated subgroup of $F^{\mathbb{Z}[t]}$ and hence any fully residually free group, is actually a subgroup of a group that can be obtained from F by finitely many extensions of centralizers. For such groups Bass–Serre theory (see Chapter 2) can be used to determine the structure.

The free exponential groups and their ties to the fully residually free groups will be discussed in the next chapter.

The final ingredients of the proof of the Tarski problems is a reduction of systems of equations over free groups to certain normal forms or systems of quadratic equations and then an implicit function theorem. We will describe these in detail in Chapter 9 after we introduce the necessary material on fully residually free groups and algebraic geometry.

5.4 Tarski-like problems

Although Tarski phrased his original questions in terms of free non-Abelian groups the basic questions of elementary equivalence and decidability can be asked about any class of algebraic objects. We'll concentrate here, as in the rest of this monograph, on groups. However the questions are equally valid for classes of semigroups or classes of rings or classes of algebras.

The first basic question concerns elementary equivalence. Let $\mathcal{K}$ be a class of groups. This could be a variety such as the Abelian groups or nilpotent groups of a particular class, or a union of varieties such as the nilpotent groups in general or just a well-defined class of groups such as the hyperbolic groups. The elementary equivalence problem asks for a description of the elementarily equivalent groups within the class $\mathcal{K}$.

Elementary Equivalence Problem Let $\mathcal{K}$ be a class of groups. Describe the elementarily equivalent groups within $\mathcal{K}$.

Related to this, as with the Tarski problem for free groups, is the *decidability question*, that is to describe within the class $\mathcal{K}$ those groups with decidable elementary theory.

Decidability of Theories of $\mathcal{K}$ Let $\mathcal{K}$ be a class of groups. Describe within the class $\mathcal{K}$ those groups with decidable elementary theory.

These types of questions have been considered for many classes of groups beyond the non-Abelian free group. In general these Tarski-like problems and their solutions provide a compact way to describe a host of purely algebraic questions about equations, algebraic structure, automorphisms, homomorphic images, residual properties and algorithmic problems in the class $\mathcal{K}$. It is for these reasons that the Tarski-like problems are of interest to algebraists as well as logicians.

Tarski himself provided the first solution of these Tarski-like problems applied to the class of finitely generated Abelian groups.

Theorem 5.4.1 (Tarski). *Two finitely generated Abelian groups are elementarily equivalent if and only if they are isomorphic.*

Somewhat later the decidability of the elementary theory of the finitely generated Abelian groups was determined by Szmielew [245]. We'll see more of Szmielew's results in Chapter Ten when we discuss discriminating groups.

Theorem 5.4.2 (Szmielew). *The elementary theories of the finitely generated Abelian groups are decidable.*

In terms of free Abelian groups of finite rank these can be expressed as

Theorem 5.4.3. *Let $\mathbb{Z}^n$ denote the free Abelian group of rank n. Then $Th(\mathbb{Z}^n) = Th(\mathbb{Z}^m)$ if and only if $n = m$. Further the theory $Th(\mathbb{Z}^n)$ is decidable.*

Oger (see [206]) handled elementary equivalence for the case of finitely generated non-Abelian nilpotent groups (see aslo Myasnikov and Remeslennikov [192]. Earlier Malcev [181] and Ershov [69] showed that the theories of the finitely generated non-Abelian nilpotent groups is undecidable.

Theorem 5.4.4 (Myasnikov, Oger). *Finitely generated non-Abelian nilpotent groups G and H are elementarily equivalent if and only if $G \times \mathbb{Z} \cong H \times \mathbb{Z}$.*

Theorem 5.4.5 (Mal'cev, Ershov). *The elementary theories of the finitely generated non-Abelian nilpotent groups are undecidable.*

In terms of free nilpotent groups this can be phrased as:

Theorem 5.4.6. *Let N_c stand for the class of nilpotent groups of class c and $F_{N_c}(n)$ denote the free nilpotent group of class c and rank n. Then $Th(F_{N_c}(n)) = Th(F_{N_c'}(n'))$ if and only if $n = n'$ and $c = c'$ for $c \geq 2, n \geq 2$. Further the theory $Th(F_{N_c}(n))$ is undecidable.*

For finitely generated non-Abelian free solvable groups Malcev showed that they are elementarily equivalent if and only if they are isomorphic and further their theories are undecidable.

For free solvable groups the results are analogous.

Theorem 5.4.7 (Mal'cev). *Finitely generated non-Abelian free solvable groups G and H are elementarily equivalent if and only if they are isomorphic. In particular if S_c denotes the class of solvable groups of class c and $F_{S_c}(n)$ the free group in S_c of rank n, then $Th(F_{S_c}(n)) = Th(F_{S_c'}(n'))$ if and only if $n = n'$ and $c = c'$ for $c \geq 2, n \geq 2$. Further their elementary theories are undecidable.*

What is called the Novosibirsk Theorem (Mal'cev, Ershov, Romanovskii and Novikov) handled the decidability of the theory of finitely generated solvable groups.

Theorem 5.4.8. *Let G be a finitely generated solvable group. Then $Th(G)$ is decidable if and only if G is virtually Abelian.*

For non-Abelian free groups the proofs of Kharlampovich and Myasnikov and the proof of Sela shows that if $X \subset Y$ with $|X| \geq 2$ and F_X is the free group on X then the embedding F_X into F_Y is an elementary embedding. Although the total Tarski results are different, the same result holds for free nilpotent and free solvable groups of infinite rank.

Theorem 5.4.9. *Let X be an infinite set and $X \subset Y$. Then $F_{N_c}(X)$ embeds elementarily into $F_{N_c}(Y)$ and $F_{S_c}(X)$ embeds elemntarily into $F_{S_c}(Y)$.*

Recall that a *polycyclic group* G is a group with finite normal series with cyclic factors. A *metabelian group* is a group G whose derived group G' is Abelian. Both the elementary equivalence question and the decidability question are open for each of these classes. There are some results on equations in metAbelian groups by Remeslennikov, Timoshenko and Stoehr (see [216]).

Sela applied the methods of his proof that all non-Abelian free groups have the same elementary theory to the case of torsion-free hyperbolic groups. He first proved the following; any finitely generated group elementarily equivalent to a torsion-free hyperbolic group must be a torsion-free hyperbolic group.

Theorem 5.4.10 ([232]). *Let Γ be a torsion-free hyperbolic group. If a finitley generated group G is elementary equivalent to Γ, then G is a torsion-free hyperbolic group.*

Proceeding from this, Sela is able to present the full classification of hyperbolic groups relative to elementary equivalence. A *rigid hyeperbolic group Γ* is a hyperbolic group which is freely indecomposable and does not admit any nontrial cyclic splitting (see Chapter 7).

Theorem 5.4.11 ([232]). *Let Γ_1, Γ_2 be non-elementary torsion-free rigid hyperbolic groups Then Γ_1 is elementary equivalent to Γ_2 if and only if they are isomorphic.*

Sela associates with every (torsion-free) hyperbolic group Γ a subgroup of it, called the *elementary core* of Γ, denoted $EC(\Gamma)$. See [232] for a precise definition. The elementary core is a retract of the hyperbolic group Γ, and although it is not canonical, its isomorphism type is an invariant of the hyperbolic group. The elementary cores provide the basis for the elementary classification of torsion-free hyperbolic groups.

Theorem 5.4.12. *Let Γ_1, Γ_2 be two non-elementary torsion-free hyperbolic groups. Then Γ_1 and Γ_2 are elementary equivalent if and only if their elementary cores $EC(\Gamma_1)$ and $EC(\Gamma_2)$ are isomorphic.*

This result show that the elementary class of a torsion-free hyperbolic group is determined by the isomorphism class of its elementary core. Hence, in order to be able to decide whether two torsion-free hyperbolic groups are elementary equivalent one needs to compute their elementary core, and to decide if the two elementary cores are isomorphic. Both can be done using Sela's solution to the isomorphism problem for torsion-free hyperbolic groups. Then:

Theorem 5.4.13 ([232]). *Let Γ_1, Γ_2 be two torsion-free hyperbolic groups. Then it is decidable if Γ_1 is elementary equivalent to Γ_2.*

On the other hand the decidability question for theory of the class of hyperbolic groups is still open.

A straightforward group theory question is whether elementary equivalence and decidability is closed under various types of group products and amalgams. For direct products the solution is known.

Theorem 5.4.14. *Suppose that A_1, B_1, A_2, B_2 are groups. Then if $A_1 \equiv B_1$ and $A_2 \equiv B_2$ it follows that $A_1 \times A_2 \equiv B_1 \times B_2$.*

Theorem 5.4.15. *If A and B are groups and $Th(A)$ and $Th(B)$ are both decidable then $Th(A \times B)$ is also decidable.*

The closure of elementary equivalence under free products is still an open question. That is if $A_1 \equiv B_1$ and $A_2 \equiv B_2$ it is not known whether $A_1 * A_2 \equiv B_1 * B_2$.

We close this chapter with a final set of questions. A *graph group* or *partially commutative group* which we abbreviate PC group is a group $G = G(\Gamma)$ formed in the following manner.

Let Γ be a finite graph and G a group with a generator x_i for each vertex in Γ and relations of the form $[x_i, x_j] = 1$ if the vertices corresponding to x_i and x_j are adjacent. Notice if Γ is a fully connected graph the $G(\Gamma)$ is just the free Abelian group on the vertices. This construction can be extended so that at each vertex there is a group G_i with $\Gamma(G)$ the group generated by the union of the G_i and additional relations given $[G_i, G_j] = 1$. The group $F \times F$ where F is a free group is of this form. This group is quite interesting because it has many bad properties such as an unsolvable conjugacy problem. The basic questions here are the following.

Question (PC 1). *Describe in the language of graphs when two PC groups $G(\Gamma_1)$ and $G(\Gamma_2)$ are elementarily equivalent.*

Question (PC 2). *Does every PC group have decidable elementary theory?*

The group $F \times F$ is known to have decidable elementary theory. Further there are examples of finitely generated groups elementarily equivalent to $F \times F$. However the decidability of the elementary theory of $F \times F$ is relative to base language L_0. If constants from $F \times F$ were allowed to extend the langauge L_0 to $L_0[F \times F]$ the decidability of this extended language would imply the decidability in $F \times F$ of the sentence $\exists x(b = x^{-1}ax)$. This implies that the conjugacy problem would be solvable (see Chapter 2) whereas we know that it is not solvable in $F \times F$.

6 Fully residually free groups I

6.1 Residually free and fully residually free groups

Crucial to the study of the elementary theory of groups in general, and the solution of the Tarski problems in particular, is the class of *fully residually free groups*. In the next two chapters we present a detailed overview of the extensive research on this class of groups and its impact on the elementary theory. In Sela's approach finitely generated fully residually free groups are known as *limit groups*. In this guise they were studied by Sela (see [233, 234, 235, 236, 237] and [29]) in terms of studying homomorphisms of general groups into free groups.

Residual properties have always played a role in the study of infinite groups. If $\mathcal{P}$ is a group property inherited by subgroups, and G is a group, then we say that G is *residually-*$\mathcal{P}$ if for all $g \in G$ with $g \neq 1$ there exists a group H_g having property $\mathcal{P}$ and an epimorphism

$$\phi_g : G \to H_g$$

such that $\phi_g(g) \neq 1$.

Equivalently G is residually-$\mathcal{P}$ if the intersection of all normal subgroups N such that G/N has property $\mathcal{P}$ is just the identity. Residual finiteness is perhaps the most widely known residual property. We concentrate on residual freeness.

Definition 6.1.1. A group G is *residually free* if for each non-trivial $g \in G$ there is a free group F_g and an epimorphism $h_g : G \to F_g$ such that $h_g(g) \neq 1$. Equivalently for each $g \in G$ there is a normal subgroup N_g such that G/N_g is free and $g \notin N_g$.

We extend this concept to allow mappings into free groups that preserve finite sets of elements.

Definition 6.1.2. A group G is *n-residually free* for a natural number n provided to every ordered n-tuple $(g_1, \dots, g_n) \in (G \setminus \{1\})^n$ of non-identity elements of G there is a free group F and an epimorphism $h : G \to F$ such that $h(g_i) \neq 1$ for all $i = 1, \dots, n$. G is *fully residually free* or *ω-residually free* provided it is n-residually free for every $n \in \mathbb{N}$. Hence G is *fully residually free* provided to every finite set $S \subset G \setminus \{1\}$ of non-trivial elements of G there is a free group F_S and an epimorphism $h_S : G \to F_S$ such that $h_S(g) \neq 1$ for all $g \in S$.

If a non-Abelian group G is residually free we say that G is *separated* by a non-Abelian free group while if G is fully residually free we say that G is *discriminated* by a non-Abelian free group. Clearly fully residually free implies residually free. Notice further that since free groups are residually finite it follows that residually free groups are also residually finite.

As simple examples of residually free groups we have free Abelian groups and the direct product $F \times F$ where F is a non-Abelian free group. The following theorem provides an extremely important nontrivial example whose proof we will look at later in the chapter.

Recall that a *surface group* is the fundamental group of a compact orientable or non-orientable surface. If the genus of the surface is g then we say that the corresponding surface group also has genus g. It can be shown (see [90] and the references there) that an orientable surface group S_g of genus $g \geq 1$ has a one-relator presentation of the form

$$S_g = \langle a_1, b_1, \ldots, a_g, b_g; [a_1, b_1] \ldots [a_g, b_g] = 1 \rangle \tag{6.1.1}$$

while a non-orientable surface group T_g of genus $g \geq 2$ also has a one-relator presentation, now of the form

$$T_g = \langle a_1, a_2, \ldots, a_g; a_1^2 a_2^2 \ldots a_g^2 = 1 \rangle. \tag{6.1.2}$$

Theorem 6.1.1. *An orientable surface group of genus $g \geq 1$ is a residually free group.*

The result is clear if $g = 1$ since then S_g is a free Abelian group of rank 2. For $g \geq 2$ this theorem was originally proved by G.Baumslag [7] answering a question of Magnus. However the actual results were subsequently shown to be much stronger. Orientable surface groups of genus $g \geq 2$ and nonorientable surface groups of genus $g \geq 4$ are actually fully residually free and elementary free. We will see this later in the chapter. Surface groups have always played a central role in combinatorial group theory. Subsequently they have been shown to be of central importance in the solution of the Tarski problem. The paper by Ackermann, Fine and Rosenberger [1] exhibits the tremendous importance of surface groups in combinatorial group theory.

It is not a difficult exercise to show that residual finiteness implies full residual finiteness. Full residual finiteness for a group G means that given any finite set $g_1, \ldots, g_n$ of nontrivial elements of G there is a homomorphism ϕ from G onto a finite group such that $\phi(g_i) \neq 1$ for all g_i.

Lemma 6.1.1. *If a group G is residually finite then it is fully residually finite.*

Proof. Let $g_1, \ldots, g_n$ be a finite set of nontrivial elements of the residually finite group G. Then for each $i = 1, \ldots, n$ there exists a finite group H_i and a homomorphism $\phi_i \colon G \to H_i$ such that $\phi_i(g_i) \neq 1$. Let ϕ be the homomorphism form G to $H_1 \times \cdots \times H_n$ given by

$$\phi(g) = (\phi_1(g), \phi_2(g), \ldots, \phi_n(g)) \quad \text{for} \quad g \in G.$$

Then $\phi(g_i) \neq 1$ for all $i = 1, \ldots, n$ proving the lemma. □

For the property of being a free group, in distinction to finiteness, it is not true that residual freeness implies full residual freeness. To show that these are not equivalent concepts and to continue the study we need to introduce the following concept.

Definition 6.1.3. A group G is *commutative transitive*, abbreviated CT, provided the relation of commutativity is transitive on the non-identity elements of G.

This property holds in all free groups as well as in all Fuchsian groups and is preserved under certain amalgam operations. Further discussions of commutative transitivity can be found in [86, 256, 164, 96, 2] and [90]. The following immediate result due to Harrison [129] gives equivalent formulations of the CT property.

Lemma 6.1.2. *Let G be a group. The following three statements are pairwise equivalent.*
(i) *G is commutative transitive.*
(ii) *The centralizer $C(x)$ of every nontrivial element in $x \in G$ is Abelian.*
(iii) *Every pair of distinct maximal Abelian subgroups in G has trivial intersection.*

In free groups two elements commute only if they are powers of a common element. Using this fact the following result is relatively easy.

Lemma 6.1.3. *Any fully residually free group G is CT.*

Proof. Suppose that G is fully residually free and a, b, c are nontrivial elements of G satisfying $[a, b] = 1$ and $[b, c] = 1$. Suppose that $[a, c] \neq 1$. Then there exists a homomorphism $\phi\colon G \to F$ with F free such that $\phi(a) = \overline{a} \neq 1$, $\phi(b) = \overline{b} \neq 1$, $\phi(c) = \overline{c} \neq 1$ and $\phi([a, c]) = [\overline{a}, \overline{c}] \neq 1$. Since ϕ is a homomorphism $[\overline{a}, \overline{b}] = 1$ and $[\overline{b}, \overline{c}] = 1$. Further F is a free group so $[\overline{a}, \overline{c}] \neq 1$ giving a contradiction. Therefore $[a, c] = 1$ and G is CT. □

Commutative transitivity is given by the universal sentence

$$\forall x, y, z((y \neq 1) \wedge (xy = yx) \wedge (yz = zy)) \to (xz = zx)) \,.$$

This universal sentence holds in any free group and hence in any universally free group. Therefore universally free groups are also CT.

Lemma 6.1.4. *Any universally free group G is CT.*

Now using the CT property one can show that the class of fully residually free groups is a proper subclass of the class of residually free groups. Indeed, if F is a free non-Abelian group, then the direct square $F \times F$ is residually free, but not commutative-transitive, hence not fully residually free.

Lemma 6.1.5. *The class of fully residually free groups is a proper subclass of the class of residually free groups.*

The type of argument in Lemma 6.1.4 using commutative transitivity is extremely powerful and used often in the study of fully residually free groups.

Before giving a beautiful theorem tying together fully residually free and universally free groups we make some observations whose proofs are immediate.

Lemma 6.1.6.
(1) *Residually free groups are torsion free.*
(2) *If G is residually free and $g, h \in G$ then $\langle g, h\rangle$ is either free or Abelian.*
(3) *A subdirect product and in particular a direct product of residually free groups is residually free.*

The situation for free products of residually free groups is more complicated. First of all the free product of residually free groups need not be residually free. Consider

$$A = \langle a_0, a_1, a_2; [a_0, a_1] = [a_0, a_2] = 1\rangle \quad \text{and} \quad B = \langle b_1, b_2; \ \rangle .$$

The group A is the direct product of a free group of rank 1 and a free group of rank 2 and hence is residually free while B is free and so residually free. Let $G = A * B$ be their free product and consider the element

$$w = [[[b_1, a_0], [b_2, a_0]], [a_2, a_1]] .$$

If G were residually free there would be a free group F_w and a map $\phi_w\colon G \to F_w$ which would not annihilate w. From $\phi_w(w) \neq 1$ we can easily deduce that $\phi_w(a_0)$ and $[\phi_w(a_2), \phi_w(a_1)]$ must be nontrivial in F_w. However, since $[a_0, a_1] = [a_0, a_2] = 1$, it must follow that $\phi_w(a_0)$ must commute with both of $\phi_w(a_1)$ and $\phi_w(a_2)$. These must then lie in the cyclic subgroup of F_w containing the nontrivial element $\phi_w(a_0)$. This forces $\phi_w(a_1)$ and $\phi_w(a_2)$ to commute contrary to $[\phi_w(a_2), \phi_w(a_1)] \neq 1$. Therefore G cannot be residually free.

However if both factors are fully residually free then the free product is residually free. Suppose H and K are fully residually free groups and $w = h_1k_1h_2k_2 \ldots h_nk_n$ is a nontrivial element of $G = H * K$ where h_i, k_i with $i = 1, \ldots, n$ are nontrivial elements of H and K respectively. Then there exist maps $\phi_H\colon H \to F_H$ and $\phi_K\colon K \to F_K$ with F_H, F_K free which do not annihilate any of the syllables in w. This can then be extended to a map $\phi_w\colon G \to F_H * F_K$ which does not annihilate w, showing that G is residually free. Presently we shall show that the class of fully residually free groups is actually closed under taking free products so that G is actually fully residually free. The complete solution to the free product question was given by B. Baumslag [22] who showed that

Theorem 6.1.2. *Let H and K be non-trivial residually free groups. Then their free product $H * K$ is residually free if and only if H and K are each fully residually free. In this case the free product is also fully residually free.*

Proof. In the next section we show that a residually free group, that is also commutative transitive, must be fully residually free. We assume this fact for the proof of Theorem 6.1.2.

Suppose that H and K are non-trivial fully residually free groups. Then the argument given before shows that the free product $H * K$ is residually free. Further a free product of CT groups must be CT so $H * K$ is CT. Since it is both CT and residually free it must be fully residually free.

Conversely suppose that H and K are non-trivial residually free groups and their free product $H * K$ is residually free. We show that H must be fully residually free. If H were residually free and not fully residually free then H cannot be CT from the result mentioned above. Hence there exist elements $a, b, c \in H$ with $a \neq 1$ such that

$[a, b] = 1, [a, c] = 1$ but $[b, c] \neq 1$. Since H is residually free and free groups are CT, it follows that for any homomorphism f from H onto a free group, if $f([b, c]) \neq 1$ then $f(a) = 1$. Let $k_1, k_2 \in K$ with $k_1 \neq k_2$ and, as in the example of a free product of two residually free groups that is not residually free, let $w = [[[k_1, a], [k_2, a]], [b, c]]$. If $H * K$ were residually free there would be a free group F_w and a map $\phi_w : H * K \to F_w$ which would not annihilate w. From $\phi_w(w) \neq 1$ we can easily deduce that $\phi_w(a)$ and $[\phi_w(b), \phi_w(c)] = \phi_w([b, c])$ must be nontrivial in F_w contradicting the statements above. Hence H must be fully residually free. An analogous argument works for K. □

The question of when an amalgamated free product of fully residually free groups is again residually free is still an open question (see Section 8).

A fully residually free group is 1-free (i.e. torsion free) but not necessarily 2-free. Indeed, every residually free group is torsion free. Suppose G is residually free and $g \in G \setminus \{1\}$. Then there is a free group F and an epimorphism $h: G \to F, x \to \overline{x}$ such that $\overline{g} \neq 1$. If $n > 1$ is an integer, then $\overline{g}^n \neq 1$. It follows that $g^n \neq 1$ and G is torsion free as claimed. $\mathbb{Z} \times \mathbb{Z}$ is an example of a fully residually free group which is not 2-free. Of course, every free group is both fully residually free and n-free for every positive integer n. In [73] it was proved that every 2-free, fully residually free group must also be 3-free. This result is best possible since, for example, the non-orientable surface group $\langle a_1, a_2, a_3, a_4; a_1^2 a_2^2 a_3^2 a_4^2 = 1\rangle$ is a 3-free, fully residually free group which is not 4-free. We will present a proof of this result in Section 6.5. We remark that it was shown in Fine, Gaglione, Rosenberger and Spellman [85] that

$$\langle a_1, \ldots, a_{2n-2}, b_1, b_2; [a_1, a_2] \ldots [a_{2n-3}, a_{2n-2}] = [b_1, b_2] b_1^{2n}\rangle$$

with $n \geq 2$ is an n-free group which is not residually free.

6.2 CSA groups and commutative transitivity

The commutative transitive property plays a large role in the study of fully residually free groups. Closely tied to this concept is the concept of a CSA group. To introduce this concept we need the idea of *malnormality*. Before defining this we first mention the following. B. Baumslag proved in [21] that a residually free group is fully residually free if and only if it is commutative transitive. We will give the proof of this in the next section. Remeslennikov [214] has deduced from Baumslag's result that full residual freeness actually follows from 2-residual freeness. Recall that G is 2-residually free if given any pair x, y of nontrivial elements of G there is a homomorphism ϕ of G onto a free group such that $\phi(x) \neq 1$, $\phi(y) \neq 1$.

Lemma 6.2.1. *Suppose that G is a 2-residually free group. Then G is fully residually free.*

Proof. Suppose that G is 2-residually free. To show that G is fully residually free it suffices to show that G is commutative transitive. Suppose then there are elements

$a, b, c \in G$ with $b \neq 1$ such that $ab = ba$ and $bc = cb$ but $ac \neq ca$. Then $b \neq 1$ and $[a, c] \neq 1$. There is a free group F and an epimorphism $h\colon G \to F, x \to \overline{x}$ such that $\overline{b} \neq 1$ and $[\overline{a}, \overline{c}] \neq 1$. But this is impossible since $\overline{a}\overline{b} = \overline{b}\overline{a}$ and $\overline{b}\overline{c} = \overline{c}\overline{b}$ and $\overline{b} \neq 1$ in the free group F. This contradiction shows that G is commutative transitive, hence fully residually free as claimed. □

Definition 6.2.1. Let G be a group and H a subgroup of G. H is *malnormal* in G or *conjugately separated* in G provided $g^{-1}Hg \cap H = \{1\}$ unless $g \in H$.

Now suppose that G is a fully residually free group with more than one element. Let $u \in G \setminus \{1\}$ and let M be its centralizer which we will denote by $C_G(u)$. Then M is maximal Abelian in G. We claim that M is malnormal in G. If G is Abelian, then $M = G$ and the conclusion follows trivially. Suppose that G is non-Abelian. Suppose that $w = g^{-1}zg \neq 1$ lies in $g^{-1}Mg \cap M$. If $g \notin M$ then $[g, u] \neq 1$. Thus, there is a free group F and an epimorphism $h\colon G \to F, x \to \overline{x}$ such that $\overline{w} \neq 1$ and $[\overline{g}, \overline{u}] \neq 1$. Let $C = C_F(\overline{u})$. Then $\overline{w} \in \overline{g}^{-1}C\overline{g} \cap C$. However the maximal Abelian subgroups in a free group are malnormal. This implies $\overline{g} \in C$, contradicting $[\overline{g}, \overline{u}] \neq 1$. This contradiction shows that $g^{-1}Mg \cap M \neq \{1\}$ implies $g \in M$ and hence the maximal Abelian subgroups in G are malnormal.

Definition 6.2.2. A group G is a *CSA-group* or *conjugately separated Abelian group* provided the maximal Abelian subgroups are malnormal.

From the argument given above we have proved the following.

Lemma 6.2.2. *If G is fully residually free then G is CSA.*

The next lemma shows the relationship between CSA groups and CT groups.

Lemma 6.2.3. *The class of CSA groups is a proper subclass of the class of CT groups.*

Proof. We first show that every CSA-group is commutative transitive. Let G be a group in which maximal Abelian subgroups are malnormal and suppose that M_1 and M_2 are maximal Abelian subgroups in G with $z \neq 1$ lying in $M_1 \cap M_2$. Could we have $M_1 \neq M_2$? Suppose that $w \in M_1 \setminus M_2$. Then $w^{-1}zw = z$ is a non-trivial element of $w^{-1}M_2w \cap M_2$ so that $w \in M_2$. This is impossible and therefore $M_1 \subset M_2$. By maximality we then get $M_1 = M_2$. Hence, G is commutative transitive whenever all maximal Abelian subgroups are malnormal.

We now show that there do exist CT groups that are not CSA. In any non-Abelian CSA-group the only Abelian normal subgroup is the trivial subgroup $\{1\}$. To see this suppose that N is any normal Abelian subgroup of the non-Abelian CSA-group G. Then N is contained in a maximal Abelian subgroup M. Let $g \notin M$. Then

$$N = g^{-1}Ng \cap N \subset g^{-1}Mg \cap M.$$

The fact $N \neq \{1\}$ would imply that $g \in M$ which is a contradiction.

Now let p and q be distinct primes with p a divisor of $q - 1$. Let G be the non-Abelian group of order pq. Then it is not difficult to prove that the centralizer of every non-trivial element of G is cyclic of order either p or q. Thus G is commutative transitive. However, the (necessarily unique) Sylow q-subgroup of G is normal in G. Hence from the argument above G cannot be CSA. □

In the next section we will show that in the presence of residual freeness CSA is actually equivalent to CT.

We mention here an important construction that will play a pivotal role in the structure theory of fully residually free groups. We will return to this construction in more detail in Section 6.4.

Definition 6.2.3 ([196, 197]). Let $G \neq \{1\}$ be a commutative transitive group. Let $u \in G \setminus \{1\}$ and let $M = C_G(u)$. Let $B \neq \{1\}$ be a torsion free Abelian group. Then

$$G(u, B) = \langle G, B;\ \text{rel}\,(G),\ \text{rel}\,(B), [B, M] = 1\rangle$$

is the *B-extension of the centralizer* M of u in G. If $B = \langle t;\quad \rangle$ is infinite cyclic, then

$$G(u, B) = \langle G, t;\ \text{rel}\,(G), t^{-1}zt = z,\ \text{for all } z \in M\rangle$$

is the *free rank one extension of the centralizer* M of u in G.

Observe that if $B \neq \{1\}$ is an arbitrary torsion free Abelian group, then $G(u, B)$ is a direct union of free rank one extensions of centralizers.

One can prove that if $G \neq \{1\}$ is fully residually free, then so is every free rank one extension of a centralizer of G (see Theorem 6.4.1). From this it follows that if $B \neq \{1\}$ is any torsion-free Abelian group and $u \in G \setminus \{1\}$, then $G(u, B)$ is locally fully residually free. In the special case where B is residually $\mathbb{Z}$, in particular if B is free Abelian, then $G(u, B)$ is fully residually free.

Definition 6.2.4. Let $n \in \mathbb{N}$. A group G is *n-free* provided every subgroup generated by n or fewer distinct elements is free.

As was seen earlier the following is immediate;

Lemma 6.2.4. *A two-generator subgroup of a fully residually free group is either free of rank 2 or Abelian. In particular if G is fully residually free and $a, b \in G$ with $[a, b] \neq 1$ then $\langle a, b\rangle$ is free and freely generated by a and b.*

Observe that if G is an n-free group then G is also an m-free group for all $1 \leq m < n$. The 1-free groups are precisely the torsion free groups while 2-free groups are commutative transitive. To see the second assertion, let $G \neq \{1\}$ be 2-free and let u be a nontrivial element in G. Let $M = C_G(u)$. We claim that M is locally cyclic and therefore Abelian. Suppose $a, b \in M$. Since G is 2-free, a, u generate a free group. Since a, u commute this must be cyclic and hence they are both powers of a single element g.

Thus $a = g^\alpha, u = g^\beta$ for some α, β. Similarly, since b and u commute and b, u generate a free group, we have an element h with $b = h^\delta, u = h^\gamma$ for some δ, γ. Now consider $\langle h, g\rangle \subset G$. This is free because G is 2-free however this has the relation $g^\beta = h^\gamma$. Therefore $\langle h, g\rangle$ must be cyclic and h and g are both powers of a single element. Hence a and b are both powers of this element and any 2-generator subgroup of M is cyclic. A straightforward induction then shows that any n-generator subgroup $\langle a_1, \ldots, a_n\rangle \subset M$ is also cyclic. Thus, centralizers of non-trivial elements are locally cyclic and 2-free groups are commutative transitive as claimed.

In addition, 2-free groups are also CSA-groups. Assume as above that $G \neq \{1\}$ is 2-free, $u \in G \setminus \{1\}$ and $M = C_G(u)$. Suppose $g^{-1}Mg \cap M \neq \{1\}$ and suppose $w = g^{-1}zg \in g^{-1}Mg \cap M$ with $w \neq 1$ and with $z \in M$. Certainly $z \in M \setminus \{1\}$. If g commutes with z, then g would commute with u by commutative transitivity and we would have $g \in M$. Suppose that $g \notin M$. Then g cannot commute with z. Consider $F = \langle g, z\rangle$ which is free since G is 2-free. Since g and z do not commute, F is free of rank 2 with basis $\{g, z\}$. Thus, $C = C_F(z) = M \cap F = \langle z\rangle$ is malnormal in F. But $w \in g^{-1}Cg \cap C$ and $w \neq 1$. This implies that g commutes with z contrary to the assumption $g \notin M$. Hence, $g^{-1}Mg \cap M \neq \{1\}$ implies that $g \in M$ and therefore M is malnormal in G. Thus, 2-free groups are CSA-groups as claimed.

For each integer $n \in \mathbb{N}$ the non-orientable surface group

$$\langle a_1, \ldots, a_{n+1}; a_1^2 \ldots a_{n+1}^2 = 1\rangle$$

of genus $n + 1$ is n-free but not $(n + 1)$-free.

6.3 Universally free groups

We have observed that being fully residually free easily implies commutative transitivity. B. Baumslag [21] proved a converse; commutative transitivity together with residually free implies fully residually free. Gaglione and Spellman [111] and independently Remeslennikov [214] then were able to show that these conditions in the presence of residual freeness are equivalent to universal freeness in the non-Abelian case. Thus there is the following rather remarkable theorem tying together the logical condition of universal freeness to the group theoretic conditions of full residual freeness and commutative transitivity.

Theorem 6.3.1 ([21], [111], [214]). *Let G be non-Abelian and residually free. Then the following are equivalent:*
(1) *G is fully residually free;*
(2) *G is commutative transitive;*
(3) *G is CSA;*
(4) *G is universally free.*

Proof. (Theorem 6.3.1) Although the proof can be done as a string of implications we prefer to lay out the proof as a sequence of lemmas doing individual equivalences. Throughout we assume that G is a non-Abelian residually free group.

Lemma 6.3.1. *If G is non-Abelian and residually free then commutative transitivity is equivalent to full residual freeness.*

Proof. (Lemma 6.3.1) This is the theorem of B. Baumslag [21]. We have seen in Lemma 6.1.3 that fully residually free groups are CT. Conversely suppose that G is residually free and CT and we must show that it is fully residually free. We first need the following. □

Lemma 6.3.2. *Let A be an Abelian normal subgroup of a non-Abelian residually free group G. Then A is contained in the center of G. In particular if G is CT then A must be trivial.*

Proof. Recall that a non-Abelian free group has no nontrivial normal Abelian subgroups. Suppose that A is an Abelian normal subgroup of G and suppose that A is not contained in the center of G. Then there exists $a \in A$ and $b \in G$ such that $[a, b] \neq 1$. Then from the residual freeness there exists a normal subgroup N of G with G/N free and $[a, b] \neq 1$ modulo N. However then AN/N is a nontrivial normal Abelian subgroup of the non-Abelian free group G/N a contradiction.

If G is also CT and non-Abelian then it is centerless so from the above it follows that any normal Abelian subgroup must be trivial. □

We now complete the proof of Lemma 6.3.1. Let $g_1, \ldots, g_n$ be a set of nontrivial elements of G. We want to show that there is a free group F and an epimorphism $h: G \to F$ such that $h(g_i) \neq 1$ for all $i = 1, \ldots, n$. We do this by induction on the size n of the set of elements.

Since G is residually free this is true for $n = 1$. Assume that for $n \geq 1$ given nontrivial $g_1, \ldots, g_n$ in G there exists a nontrivial $g \in G$ such that for all normal subgroups N of G if $g \notin N$ then $g_i \notin N$ for $i = 1, \ldots, n$. By the residual freeness this is true for $n = 1$. We show that given $g_1, \ldots, g_n, g_{n+1}$ we can find a $g' \neq 1$ such that if $g' \notin N$ for any normal subgroup N of G then $g_i \notin N$ for $i = 1, \ldots, n, n+1$. Let g be the assumed element for $g_1, \ldots, g_n$ and for each $x \in G$ let

$$c(x) = [g, g_{n+1}^x]$$

where g_{n+1}^x is the conjugate $xg_{n+1}x^{-1}$. If $c(x) = 1$ for all x then each conjugate of g_{n+1} commutes with $g \neq 1$. Then by commutative transitivity the normal closure $N_{g_{n+1}}$ is Abelian and hence here trivial. But g_{n+1} is in it and nontrivial. Therefore $c(x) \neq 1$ for some $x \in G$. Choose $c(x)$ as g'. Then if $g' \notin N$ for a normal subgroup N of G it follows that $g_1, \ldots, g_{n+1} \notin N$. It follows from this induction that for each finite set $g_1, \ldots, g_n \in G$ there is a $g \in G$ such that if $g \notin N$ for some normal subgroup N of G then $g_1, \ldots, g_n$ is also not in N. Since G is residually free it follows that there is such an N and therefore G is fully residually free. □

Lemma 6.3.1 shows that (1) is equivalent to (2) in Theorem 6.3.1. We now show the equivalence to CSA.

Lemma 6.3.3. *If G is a non-Abelian residually free group then CT is equivalent to CSA.*

Proof. We have seen that CSA always implies CT. Suppose that G is residually free and CT. Then from Lemma 6.3.1 G is fully residually free. Hence from Lemma 6.2.3 G is CSA. □

Therefore (2) is equivalent to (3) in Theorem 6.3.1.

Recall (see Chapter 5) that all countable non-Abelian free groups have the same universal theory. A *universally free group* is a group that has the same universal theory as the class of countable non-Abelian free groups.

Lemma 6.3.4. *Let G be a non-Abelian residually free group. Then G is fully residually free if and only if it is universally free.*

Proof. Suppose first that G is universally free. Then it satisfies the same universal sentences as the class of non-Abelian free groups. In particular it satisfies the universal sentence

$$\forall x, y, z((y \neq 1) \wedge (xy = yx) \wedge (yz = zy) \rightarrow (xz = zx)) .$$

Therefore G is CT. Since it is residually free it follows from Lemma 6.3.1 that it is fully residually free.

Conversely we must show that a fully residually free group is universally free. First let a, b be nontrivial noncommuting elements of G. Then as shown, the subgroup $H = \langle a, b\rangle$ is a rank 2 free subgroup of G. Since all countable non-Abelian free groups have the same universal theory, H is clearly universally free. We claim that H and G are universally equivalent, that is they have the same universal theory.

We need the following general results. For groups A and B we say that B is fully residually A if given any finite set $b_1, \dots, b_n$ of nontrivial elements of B there is an epimorphism $\phi\colon B \to A$ such that $\phi(b_i) \neq 1$ for all b_i.

Lemma 6.3.5. *Suppose that A is a subgroup of B. Then if B is fully residually A it follows that A and B are universally equivalent.*

Proof. Let $Th_\forall(A), Th_\forall(B)$ be respectively the universal theories of A and B. Since $A \subset B$ we have

$$Th_\forall(B) \subset Th_\forall(A) .$$

Now two groups have the same universal theory if and only if they have the same existential theory (see Chapter 4), by using negations. Hence it suffices to show that every existential sentence true in the overgroup B is also true in A. From this it is clear that the groups A and B have the same universal theory if and only if the following is true: every finite system

$$p_i(x_1, \dots, x_n) = 1, \quad 1 \leq i \leq I$$
$$q_j(x_1, \dots, x_n) \neq 1, \quad 1 \leq j \leq J$$

of equations and inequations has a solution in A if and only if it has a solution in B (see, e.g., [25]).

Suppose that $(b_1, \ldots, b_n)$ is a solution to a given system in B. Since B is fully residually A there is an epimorphism $h\colon B \to A$ with $h(b_i) \neq 1$ and which preserves the equations and inequations. Therefore the image will give a solution in A. It follows that $Th_\forall(A) \subset Th_\forall(B)$ and hence $A \equiv_\forall B$, that is A and B are universally equivalent. □

We next need what is called the *big powers argument*. It was developed by G. Baumslag in his proof that surface groups are residually free ([7]). Let

$$^{s}\delta^{t}(a_1)a_2 \ldots a_{k_1}(a_k)$$

denote the four expressions obtained by deleting or not deleting a_1 and a_k independently.

Lemma 6.3.6. *Let F be a non-Abelian free group and suppose that $f_1, \ldots f_k$ are elements of F. Let $u \in F$ and suppose that u commutes with none of $f_1, \ldots, f_k$. Then if the absolute values of the integers $n_1, \ldots, n_k$ are sufficiently large, then*

$$^{s}\delta^{t}(f_1)u^{n_1} f_2 \ldots f_k(u^{n_k}) \neq 1$$

Lemma 6.3.7. *Any non-Abelian free group is fully residually a rank 2 free group.*

Proof. Let F be a free group of rank n with $n > 2$. We show that F is fully residually free of rank $n-1$ and hence fully residually free of rank 2.

Suppose that

$$F = A * B$$

where A is freely generated by $a_1, \ldots, a_{n-1}$, and B is infinite cyclic and generated by a_n. Take a nontrivial element $f \in F$. Then f can be expressed as

$$f = ^{s}\delta^{t}(f_1)b_1 \ldots f_k(b_k)$$

where each $f_i \in A$ and each $b_j \in B$ and none is the identity. Corresponding to the set of f_i that appears there exists an element $u \in A$ which generates its own centralizer and does not commute with the f_i, that is an element u cyclically reduced which has length greater than any of the f_i and which generates its own centralizer. Now choose an integer k by Lemma 6.3.6 such that the expression for f does not go to the identity under the homomorphism induced by mapping a_i to a_i for $i = 1, \ldots, a_{n-1}$ and a_n to u^p for $p > k$. Given a finite set S of nontrivial elements of F, each of them determines an integer k, and by choosing the maximum, one can ensure that if p is greater than this maximum then none of the elements is mapped to the identity. Therefore F is fully residually free of rank $n-1$. By iterating the argument F is then fully residually free of rank 2. □

Now we show that G is fully residually H completing the proof of Lemma 6.3.4 and hence of Theorem 6.3.1. Now $H = \langle a, b\rangle$ and let $c = [a, b] \neq 1$. Let $g_1, \ldots, g_n$ be any nontrivial elements of G and let $S = \{g_1, \ldots, g_n, c\}$. Since G is fully residually free there exists a homomorphism $\phi\colon G \to F_S$ with F_S a free group and $\phi(g_i) \neq 1$ for $i = 1, \ldots, n$ and $\phi(c) \neq 1$. Since $\phi(c) \neq 1$ it follows that $\phi(a) \neq 1$ and $\phi(b) \neq 1$ and hence F_S is non-Abelian. Therefore F_S is fully residually a rank 2 free group and hence there is a homomorphism

$$\rho\colon F_S \to \langle x, y\rangle$$

with $\{x, y\}$ a free basis for a rank 2 free group and such that

$$\rho(\phi(g_i)) \neq 1 \quad \text{for } i = 1, \ldots, n\,.$$

Now let α be the isomorphism from $\langle x, y\rangle$ to H given by

$$x \to a, \quad y \to b\,.$$

Combining all these maps we have an epimorphism

$$\alpha(\rho(\phi))\colon G \to H$$

such that

$$\alpha(\rho(\phi))(g_i) \neq 1 \quad \text{for } i = 1, \ldots, n\,.$$

Therefore G is fully residually H. □

This completes the proof of Theorem 6.3.1 which in essence becomes the cornerstone step in both the structure theory of fully residually free groups and the solution of the Tarski problems.

Using the CT property Remeslennikov [214] gives the following corollary.

Corollary 6.3.1. *A countable group G is fully residually free if and only if it is 2-residually free.*

There have been various extensions of this theorem. Gaglione and Spellman [110] proved if G is finitely presented then universally free implies residually free, while Remeslennikov [214] showed finitely generated and universally free implies residually free.

Theorem 6.3.2 ([214]). *Let G be a finitely generated universally free group. Then G is fully residually free.*

Remeslennikov's proof uses an embedding of the fully residually free group G into an ultrapower of a rank 2 free group and then uses the corresponding ultrapower ${}^*\mathbb{Z}$ of the integers $\mathbb{Z}$ to embed G into $SL_2({}^*\mathbb{Z})$. The proof in the finitely presented case is much simpler and direct. We present only the proof in this case where G is finitely presented. Suppose that G is a finitely presented universally free group. Let

$$G = \langle x_1, \ldots, x_n; R_1 = \ldots = R_m = 1\rangle$$

be a finite presentation for G where $R_i = R_i(x_1, \ldots, x_n)$ are words in $x_1, \ldots, x_n$. Since a universally free group is CT it suffices to show that G is residually free. Suppose W is a nontrivial element of G. Then W is given by $W = W(x_1, \ldots, x_n)$ a word in the given generators. Consider now the existential sentence

$$\exists x_1, \ldots, x_n \left(\left(\bigwedge_{i=1}^{m} R_i(x_1, \ldots, x_n) = 1 \right) \wedge (W(x_1, \ldots, x_n) \neq 1) \right).$$

This existential sentence is clearly true in G. Since G is universally free it is also existentially free so this existential sentence must be true in all non-Abelian free groups. Therefore in any non-Abelian free group F there exists elements $a_1, \ldots, a_n$ such that $R_i(a_1, \ldots, a_n) = 1$ for $i = 1, \ldots, n$; and $W(a_1, \ldots, a_n) \neq 1$. The map then from G into F given by $x_i \to a_i$ for $i = 1, \ldots, n$ defines a homomorphism where the image of W is nontrivial. Then from Theorem 6.3.1 G is fully residually free.

Thus together with Theorem 6.3.1 we have the following:

Corollary 6.3.2. *A finitely generated non-Abelian group G is universally free if and only if it is fully residually free.*

Chiswell (see [47, 48]) as a corollary then gives this characterization of universally free groups.

Theorem 6.3.3 ([47]). *A group G is universally free if and only if it is non-Abelian and locally fully residually free.*

Proof. The characterization of arbitary universally free groups is a consequence of the characterization of finitely generated universally free groups as follows.

Suppose first that G is universally free. Then the sentence $\exists x, y(xy \neq yx)$ holds in G and hence G is non-Abelian. Choose $a, b \in G$ with $ab \neq ba$. Since for each feely reduced word $w(x, y)$ the sentnce

$$\forall x, y((xy \neq yx) \to (w(x, y) \neq 1))$$

holds in G it follows that $\langle a, b\rangle$ is free on $\{a, b\}$.

Now let G_0 be any finitley generated non-Abelian subgroup of G. Since the above argument holds for any pair a, b of noncommuting elements of G, it follows that G_0 must contain a copy of F_2 the rank 2 free group. Thus $F_2 \subset G_0 \subset G$.

Now every universal sentence of L_0 true in G must be true in G_0. On the other hand every universal sentence of L_0, true in G_0, must be true in F_2. It follows that G_0 is universally free. Since G_0 is finitely generated if must be fully residually free.

Therefore G must be non-Abelian and locally fully residually free since any finitely generated Abelian subgroup of G is torsion-free and hence free Abelian and therefore fully residually free.

Now suppose that G is non-Abelian and locally fully residually free. Then G is the direct union of it's finitely generated non-Abelian subgroups. Such a subgroup is a finitely generated non-Abelian fully residually free group and therefore universally free and hence satisfies those universal sentences and those existential sentences of L_0 true in the non-Abelian free groups. But universal sentences and existential sentences are preserved in direct unions. Therefore G is universally free completing the proof. □

Remeslennikov further uses the embedding of the fully residually free group G into an ultrapower of a rank 2 free group. In particular an ultrapower of a free group is called a *nonstandard free group* and Remeslennikov proved the following.

Theorem 6.3.4. *A non-Abelian group G is universally free if and only if it can be embedded as a subgroup of a nonstandard free group.*

6.4 Constructions of residually free groups

Magnus asked the question of whether an orientable surface group is residually free. This was answered by G. Baumslag [7] using what is called the *big powers argument* which we touched on in the last section and will explain more fully below. This led to the following construction which is essential to both build examples of fully residually free groups and classify their structure.

Definition 6.4.1. Let G be a CT group, let $a \in G \setminus \{1\}$ and let $M = C_G(a)$ where $C_G(a)$ is the centralizer of a in G. Suppose A is an Abelian group. Then the group

$$H = \langle G, A;\ \text{rel}\,(G),\ \text{rel}\,A, [A, z] = 1, \forall z \in M\rangle$$

is a *centralizer extension* of G by A. If $A = \langle t;\quad\rangle$ is infinite cyclic then $H = G(a, t)$ is the HNN extension

$$G(a, t) = \langle G, t;\ \text{rel}\,(G), t^{-1}zt = z,\ \text{for all } z \in M\rangle$$

and is called the *free rank one extension of the centralizer M* of a in G.

What becomes crucial is the fact that if the base group G is fully residually free and A is a fully residually free Abelian group, then a centralizer extension of G by A is again fully residually free. To prove this result we need to prove it for rank one free extensions of centralizers and to do this we need to apply the following *big powers lemma* of G. Baumslag [7].

Lemma 6.4.1 (Big Powers Lemma). *Let $a_1, a_2, \ldots, a_k$ be elements of a free group. If $a_1 b^{n_1} a_2 b^{n_2} \ldots a_k b^{n_k} = 1$ for infinitely many values of n_1, infinitely many values of $n_2, \ldots,$ and infinitely many values of n_k then there exists i, $1 \le i \le k$, such that $a_i b = b a_i$.*

Theorem 6.4.1. *Let G be a fully residually free group. Let $a \in G$ with $a \ne 1$ and let $M = C_G(a)$. Then the free rank one extension of the centralizer M of a in G,*

$$G(a, t) = \langle G, t;\ rel\,(G), t^{-1}zt = z,\ for\ all\ z \in M\rangle$$

is also fully residually free.

Proof. View the extension of centralizer $G(a,t)$ as the free product with amalgamation

$$G(a,t) = G *_M (M \times \langle t \rangle).$$

Now let $g_1, g_2, \ldots, g_k$ be finitely many nontrivial elements of $G(a,t)$. We may write for $j = 1, 2, \ldots, k$

$$g_j = a_{0,j} t^{m_{1,j}} a_{1,j} t^{m_{2,j}} \ldots a_{N(j)-1,j} t^{m_{N(j),j}} z_j$$

where $N(j) \geq 0$, $a_{i,j} \in G \setminus M$, $m_{i,j} \in \mathbb{Z} \setminus \{0\}$ and $z_j \in M$. Note that $a_{i,j} \in G \setminus M$ is equivalent to $[a_{i,j}, a] \neq 1$. Now since G is fully residually free, there is a free group F and an epimorphism $\phi\colon G \to F$ such that

$$\phi([a_{i,j}, a]) = [\phi(a_{i,j}), \phi(a)] \neq 1 \quad \text{for all } i, j.$$

Then each $\phi(a_{i,j}) \neq 1$ and $\phi(a) \neq 1$.

Let $C_F(\phi(a)) = \langle f \rangle$ be the centralizer of $\phi(a)$ in the free group F. Suppose that $\phi(z_j) = f^{e_j}, 1 \leq j \leq k$. For each positive integer $n \in \mathbb{N}$ we may define an extension of ϕ, $\psi_n\colon G(a,t) \to F$ by $\psi_n|_G = \phi$, $\psi_n(t) = f^n$. Now fix a j, $1 \leq j \leq k$. Could we have $\psi_n(g_j) = 1$ for infinitely many $n \in \mathbb{N}$? If so we would have

$$\phi(a_{0,j}) f^{m_{1,j}n} \ldots \phi(a_{N(j)-1,j}) f^{m_{N(j),j}n} f^{e_j} = 1$$

for infinitely many values of n.

Applying the big powers lemma we conclude that $\phi(a_{i,j}) f = f \phi(a_{i,j})$ for some $i, 0 \leq i \leq N(j) - 1$. For that $\phi(a_{i,j})$ we have

$$\phi(a_{i,j}) \in C_F(f) = \langle f \rangle = C_F(\phi(a))$$

and so $[\phi(a_{i,j}), \phi(a)] = 1$ which contradicts our choice of ϕ. The contradiction shows that the set

$$S_j = \{n \in \mathbb{N}\colon \psi_n(g_j) \neq 1\}$$

is a cofinite subset of the natural numbers $\mathbb{N}$, that is its complement $S_j' = \mathbb{N} \setminus S$ is finite.

Since this is true for all j, $1 \leq j \leq k$ we must have that the finite intersetion

$$S_1 \cap S_2 \cap \cdots \cap S_k \neq \emptyset.$$

To see this, notice that if $S_1 \cap S_2 \cap \cdots \cap S_k = \emptyset$ then

$$(S_1 \cap S_2 \cap \cdots \cap S_k)' = S_1' \cup \cdots \cup S_k' = \mathbb{N}$$

which is impossible since $S_1' \cup \cdots \cup S_k'$ is a finite union of finite sets.

Now choose $n \in S_1 \cap S_2 \cap \cdots \cap S_k$. Then $\psi_n(g_j) \neq 1$ for all j, $1 \leq j \leq k$. Therefore $G(a,t)$ is fully residually free. □

Now we state the general result which is fundamental in all further considerations of fully residually free groups. The proof of the general result depends on the fact that it can be reduced to the case of free rank one extensions of centralizers.

Theorem 6.4.2 ([18]). *Let G be a fully residually free group and A is an Abelian fully residually free group. Then a centralizer extension of G by A is again fully residually free.*

G.Baumslag [7] used this type of argument to show that the orientable surface groups S_g with $g \geq 2$ are all residually free. Recall that for $g \geq 2$ the group S_g which is the fundamental group of an orientable surface of genus g has the presentation

$$S_g = \langle a_1, b_1, \ldots, a_g, b_g; [a_1, b_1] \ldots [a_g, b_g] = 1 \rangle .$$

Baumslag observed that each S_g with $g \geq 2$ embeds in S_2 and residual freeness is inherited by subgroups so it suffices to show that S_2 is residually free. He actually showed more. If F is a non-Abelian free group and $u \in F$ is a nontrivial element which is neither primitive nor a proper power then the group K given by

$$K = \langle F * \overline{F}; u = \overline{u} \rangle ,$$

where $\overline{F}$ is an identical copy of F and $\overline{u}$ is the corresponding element to u in $\overline{F}$, is residually free. A one-relator group of this form is called a *Baumslag double*. In our terminology, he proceeded by embedding K in the free rank one extension of centralizers

$$H = \langle F, t; t^{-1}ut = u \rangle$$

by

$$K = \langle F, t^{-1}Ft \rangle .$$

The group H is then residually free and hence K is residually free. Therefore every Baumslag double is residually free. The group

$$S_2 = \langle a_1, b_1, a_2, b_2; [a_1, b_1] = [a_2, b_2] \rangle$$

is a Baumslag double answering the original question.

Later in this chapter we will give an alternative proof of Baumslag's result.

R. Lyndon in [173] found that the solution of any one-variable equation over a free group F can be obtained by specializing finitely many *parametric words* into the integers. A *parametric word* over F with parameters in a commutative ring A is a formal expression that can be obtained from a basis of F by finitely many concatenations and *exponentiations* from A. We will explain exponentiation from A in Section 6.4.1. In particular Lyndon used the polynomial ring $\mathbb{Z}[t_1, \ldots, t_n]$ over the integers $\mathbb{Z}$ and specialized the parameters into $\mathbb{Z}$. Appel [5] extended Lyndon's results to prove that the solution of a one variable equation over a free group F can be parametrized by finitely many words of the form fg^th where $f, g, h \in F$ and t is a parameter from the polynomial ring $\mathbb{Z}[t]$. This led Lyndon to introduce and study the free exponential group $F^{\mathbb{Z}[t]}$ over the polynomial ring $\mathbb{Z}[t]$. He then proved that the group $F^{\mathbb{Z}[t]}$ is *discriminated* by F, that is given finitely many nontrivial elements $f_1, \ldots, f_n$ in $F^{\mathbb{Z}[t]}$ there is a homomorphism $\phi\colon F^{\mathbb{Z}[t]} \to F$ which is the identity on F and for which $\phi(f_1), \ldots, \phi(f_n)$ are all nontrivial. (The general notion of discrimination and what are called discriminating groups will be examined in Chapter 11).

In particular since F is a free group if follows that $F^{\mathbb{Z}[t]}$ is fully residually free and hence each finitely generated subgroup is fully residually free. Myasnikov and Remeslennikov [196, 197] axiomatized the notion of exponential groups and proved that $F^{\mathbb{Z}[t]}$ can be obtained starting from F by an infinite chain of free extensions of centralizers. The basic idea is that to construct $F^{\mathbb{Z}[t]}$ one must extend each centralizer sufficiently many times so that each centralizer is a free Abelian group of infinite rank and hence isomorphic to the additive group of $\mathbb{Z}[t]$. This then implies that any finitely generated subgroup of $F^{\mathbb{Z}[t]}$ is a subgroup of a group that can be obtained from F by finitely many extensions of centralizers. Since an extension of centralizers is an HNN group, Bass–Serre Theory can be employed to describe the structure of these groups. In particular, each finitely generated subgroup of $F^{\mathbb{Z}[t]}$ has a finite graph of groups decomposition with Abelian edge groups. We will examine this decomposition in detail in Chapter 7. Further we will look at a certain decomposition with Abelian edge groups, called a JSJ-decomposition, which is in a sense canonical. Subsequently Kharlampovich and Myasnikov [153] proved that a finitely generated group G is discriminated by a free group F if and only if it is embeddable in $F^{\mathbb{Z}[t]}$. This implies that any finitely generated fully residually free group can be embedded as a subgroup of $F^{\mathbb{Z}[t]}$ where $F = F_\omega$ is a free group of countably infinite rank. It follows that each finitely generated fully residually free group can be obtained as a subgroup of a group that can be obtained from a free group by finitely many extensions of centralizers.

To make this more precise, let $\mathcal{F}$ be the smallest class of groups containing the infinite cyclic groups and closed under the following four operators:
(1) Isomorphism;
(2) Finitely Generated Subgroups;
(3) Free Products of Finitely Many Factors;
(4) Free Rank One Extensions of Centralizers.

From the structure theory of graphs of groups, any finitely generated subgroup of a group formed from a free group by taking finitely many free rank one extensions of centralizers would be in $\mathcal{F}$.

The remainder of Section 6.4 will be devoted to proving the following fundamental structure theorem for finitely generated fully residually free groups.

Theorem 6.4.3. *Let G be a finitely generated group and $F = F_\omega$ be a free group of countably infinite rank. Then the following are equivalent:*
(1) *G is fully residually free;*
(2) *G can be embedded as a subgroup of $F^{\mathbb{Z}[t]}$;*
(3) *G is in the class $\mathcal{F}$.*

Hence the finitely generated fully residually free groups can be identified with the finitely generated subgroups of the free exponential group $F^{\mathbb{Z}[t]}$ and alternatively can be identified as the finitely generated subgroups in the class $\mathcal{F}$. In subsequent sections we will present several alternative characterizations of the class of finitely generated fully residually free groups.

To begin, we describe the theory of exponential groups.

6.4.1 Exponential and free exponential groups

Let A be an arbitrary commutative ring with unity and G a group. We define an exponentiation from A in G. If such an exponentiation exists we call G an A-group or A-exponential group.

Definition 6.4.2. Let G be a group and A a commutative ring with unity. Then G is an A-group or A-exponential group if G admits operators from A, $G \times A \to G$, $(g, \alpha) \to g^\alpha$ subject to the conditions
(i) $g^\alpha g^\beta = g^{\alpha+\beta}$
(ii) $(g^\alpha)^\beta = g^{\alpha\beta}$
(iii) $g^1 = g$
(iv) $(h^{-1}gh)^\alpha = h^{-1}g^\alpha h$
for all $g, h \in G, \alpha, \beta \in A$.

The rank r, free A-group, that we denote $(F_r)^A$, exists and is unique up to an A-isomorphism. Lyndon showed that this group is formed from a basis for F_r, a free group of rank r, by allowing exponentiation from A. One can show that if the additive group A^+ of A is torsion free, then $(F_r)^A$ is commutative transitive and the centralizer of any non-trivial element $u \neq 1$ in $(F_r)^A$ is a right A-module; in fact a free rank one right A-module. However, Myasnikov and Remeslennikov [196] have given an example (necessarily non-free) of an A-group G containing a maximal Abelian subgroup which is not a right A-module. They study the subclass $\mathcal{M}_A$ of $\mathcal{L}_A$ consisting of those A-groups G satisfying the additional condition

$$gh = hg \text{ implies } (gh)^\alpha = g^\alpha h^\alpha$$

for all $g, h \in G, \alpha \in A$. We remark that $(F_r)^A$ can fail to be commutative transitive if A

is not torsion free. Indeed, if $A = \mathbb{Z}_3$, then the free Burnside group of rank 2 considered as $(F_2)^A$ is a non-Abelian, nilpotent group and no such group can be commutative transitive.

The properties of the free exponential group $(F_r)^A$ depend in part on the properties of the ring A. Residual properties can be applied to rings as well as groups and there is a parallelism between the theories of fully residually free groups and fully residually $\mathbb{Z}$ rings (or ω-residually $\mathbb{Z}$ rings). Below we present a result (Theorem 6.4.1.1) on fully residually $\mathbb{Z}$ rings analogous to the Baumslag result on fully residually free groups. We then exploit this parallelism to shed further light on fully residually free groups by embedding such groups in a hybrid object, a group admitting exponents from a fully residually $\mathbb{Z}$ ring. The prototypical example is the free exponential group $F^{\mathbb{Z}[t]}$, the group admitting exponents from the polynomial ring $\mathbb{Z}[t]$. Myasnikov and Remeslennikov define a *ring of Lyndon type* as a commutative ring A whose additive group A^+ is torsion-free and contains the additive group $\mathbb{Z}$ of its minimum subring as a pure subgroup. In particular any fully residually $\mathbb{Z}$ ring is of Lyndon type. We show that, if A is of Lyndon type, then any finitely generated group embeddable in a free A-group is already embeddable in a free $\mathbb{Z}[t]$-group. We now make all these comments precise.

Definition 6.4.3. Let A be a ring with identity $1 \neq 0$. For a natural number n, A is *n-residually $\mathbb{Z}$* provided to every ordered n-tuple $(\alpha_1, \ldots, \alpha_n) \in (A\backslash\{0\})^n$ there is a ring homomorphism $\rho\colon A \to \mathbb{Z}$ such that $\rho(\alpha_i) \neq 0$ for $i = 1, \ldots, n$. A is *fully residually* $\mathbb{Z}$ or ω-residually $\mathbb{Z}$ provided it is n-residually $\mathbb{Z}$ for every natural number N. A is *residually* $\mathbb{Z}$ if it is 1-residually $\mathbb{Z}$.

Lemma 6.4.2. *Let A be a ring with identity $1 \neq 0$. Suppose that A is residually $\mathbb{Z}$. Then A is commutative and has characteristic zero.*

Proof. Let $\alpha, \beta \in A$. If $\alpha\beta \neq \beta\alpha$ then $\alpha\beta - \beta\alpha \neq 0$ and there is an epimorphism $\rho\colon A \to \mathbb{Z}$ such that $\rho(\alpha)\rho(\beta) - \rho(\beta)\rho(\alpha) = \rho(\alpha\beta - \beta\alpha) \neq 0$. Then $\rho(\alpha)\rho\beta) \neq \rho(\beta)\rho(\alpha)$, which is impossible in $\mathbb{Z}$. Therefore A is commutative.

Suppose $n \cdot 1 = 0$ in A. Since A is residually $\mathbb{Z}$ and $1 \neq 0$ in A there is a retraction $\rho\colon A \to \mathbb{Z}$ such that $\rho(1) = 1 \neq 0$. Then $\rho(n \cdot 1) = n \cdot \rho(1) = n \cdot 1 = 0$ in $\mathbb{Z}$. Hence $n = 0$ and therefore A has characteristic zero. □

If a ring A has characteristic zero we shall identify its minimum subring with $\mathbb{Z}$ and consider an epimorphism $\rho\colon A \to \mathbb{Z}$ to be a retraction. The next result is analogous to the Baumslag result on residually free groups.

Theorem 6.4.4. *Let A be a ring with identity $1 \neq 0$. Suppose that A is residually $\mathbb{Z}$. Then the following three statements are pairwise equivalent.*

(i) *A is 2-residually $\mathbb{Z}$;*

(ii) *A is an integral domain;*

(iii) *A is ω -residually $\mathbb{Z}$.*

Proof. (i) $\Longrightarrow$ (ii) Suppose $\alpha, \beta \neq 0$ and $\alpha\beta = 0$ in A. Then there is a retraction $\rho\colon A \to \mathbb{Z}$ such that $\rho(\alpha) \neq 0, \rho(\beta) \neq 0, \rho(\alpha\beta) = \rho(\alpha)\rho(\beta) = 0$. Since this is impossible in $\mathbb{Z}$ the ring A must be an integral domain.

(ii) $\Longrightarrow$ (iii) Suppose $\alpha_i \neq 0$ for $i = 1,\dots,n$. Let $\alpha = \alpha_1 \dots \alpha_n$. Then $\alpha \neq 0$ and hence there exists a retraction $\rho\colon A \to \mathbb{Z}$ such that $\rho(\alpha) = \rho(\alpha_1 \dots \alpha_n) = \rho(\alpha_1)\dots\rho(\alpha_n) \neq 0$. It follows that $\rho(\alpha_i) \neq 0$ for all $i = 1,\dots,n$.

(iii) $\Longrightarrow$ (i) is obvious. □

Examples of ω-residually $\mathbb{Z}$ rings are the integral polynomial rings $\mathbb{Z}[v_1,\dots,v_n]$. Another example is the following. Let $d > 1$ be an integer which is not a perfect square. Then the residue class ring $A = \mathbb{Z}[v_1, v_2]/(v_1^2 - dv_2^2 - 1)$ is ω-residually $\mathbb{Z}$. Note that if A is any locally ω-residually $\mathbb{Z}$ ring, then the additive group A^+ of A is torsion free and $\mathbb{Z}$ is a pure subgroup of A^+. A subgroup H of an Abelian group G is *pure* if whenever an element of H has an nth root in G then it has an n-th root in H.

Definition 6.4.4. A is a *ring of Lyndon type* if its additive group A^+ is torsion free and $\mathbb{Z}$ is a pure subgroup of A^+.

The following lemma is clear.

Lemma 6.4.3. *The polynomial ring $\mathbb{Z}[t_1,\dots,t_n]$ is of Lyndon type.*

A ring R elementarily equivalent to $\mathbb{Z}$ is called a *Peano ring*. If I is an infinite index set and D is a non-principal ultrafilter on I then from Los's Theorem (see Chapter 4) the ultrapower ${}^*\mathbb{Z} = \mathbb{Z}^I/D$ is an example of a Peano ring.

Lemma 6.4.4. *Every Peano ring is locally ω-residually $\mathbb{Z}$.*

Proof. Let R be any Peano ring. Then R is easily seen to be commutative. Hence, every subring is commutative. Let A be a finitely generated subring of R. A is a homomorphic image of an integral polynomial ring $\mathbb{Z}[v_1,\dots,v_n]$. Let $\theta\colon \mathbb{Z}[v_1,\dots,v_n] \to A$ be an epimorphism. By the Hilbert Basis Theorem, $\mathbb{Z}[v_1,\dots,v_n]$ is Noetherian. (See e.g. Lang [163] or [41]). Thus, if K is the kernel of θ then K is finitely generated as an ideal. Let $f_1,\dots,f_m$ generate K. Suppose $\alpha_1,\dots,\alpha_k$ are finitely many non-zero elements of A. Choose preimages $g_j \in \mathbb{Z}[v_1,\dots,v_n]$, such that $\theta(g_j) = \alpha_j$ for $j = 1,\dots,k$. Let $\theta(\overline{v})$ be the ordered n-tuple $(\theta(v_1),\dots,\theta(v_n))$ of elements of A. Then

$$(\wedge_{i=1}^m (f_i(\theta(\overline{v})) = 0)) \wedge (\wedge_{j=1}^k (g_j(\theta(\overline{v})) \neq 0) \qquad (*)$$

is true in R. Therefore the existential sentence

$$\exists\overline{x}(\bigwedge_{i=1}^m f_i(\overline{x}) = 0) \wedge (\bigwedge_{j=1}^k g_j(\overline{x}) \neq 0) \qquad (**)$$

is true in R where $\overline{x} = (x_1,\dots,x_n)$ is a tuple of distinct variables. We must have that (**) is true in $\mathbb{Z}$ also. Hence there is an ordered n-tuple $\overline{u} = (u_1,\dots,u_n) \in \mathbb{Z}^n$ such that

$$(\wedge_{i=1}^m f_i(\overline{u}) = 0) \wedge (\wedge_{j=1}^k g_j(\overline{u}) \neq 0) \qquad (***)$$

is true in $\mathbb{Z}$. We may then define a retraction $h: \mathbb{Z}[v_1, \ldots, v_n] \to \mathbb{Z}$ determined by $v_i \to u_i$ for $i = 1, \ldots, n$. Since K is contained in the kernel of h there is a retraction ρ such that the triangle pictured in Figure 6.1 commutes.

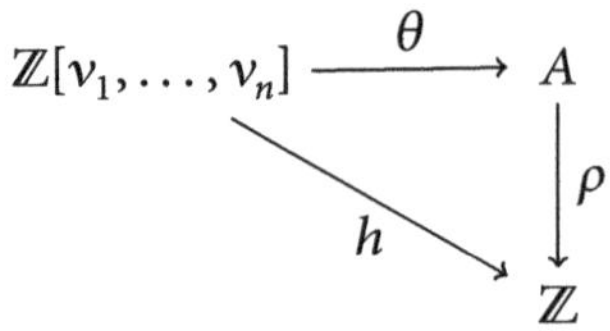

Fig. 6.1. Commutative Triangle.

But then $\rho(\alpha_j) = \rho\theta(g_j) = h(g_j) = g_j(\overline{u}) \neq 0$ for $j = 1, \ldots, k$. Thus A is ω-residually $\mathbb{Z}$. Hence R is locally ω-residually $\mathbb{Z}$ completing the proof. □

In the preceding lemma, the only property of Peano rings that was utilized is that every Peano ring has the same universal theory as $\mathbb{Z}$. Hence every ring having the same universal theory as $\mathbb{Z}$ is locally ω-residually $\mathbb{Z}$. The converse is also true. That is, a ring A with unity $1 \neq 0$ has the same universal theory as $\mathbb{Z}$ if and only if it is locally ω-residually $\mathbb{Z}$. As for groups the universal theory of a ring R consists of all the universal sentences true in R in a basic language L appropriate for ring theory.

Although Lyndon proved results about the free exponential group $F^{\mathbb{Z}[t]}$ most of these results extend immediately to F^A where A is a ring of Lyndon type. We will see this in the next section.

6.4.2 Fully residually free groups embedded in $F^{\mathbb{Z}[t]}$

We now prove Theorem 6.4.2 which we restate.

Theorem 6.4.5. *Let G be a finitely generated group and $F = F_\omega$ be a free group of countably infinite rank. Then the following are equivalent:*
(1) *G is fully residually free;*
(2) *G can be embedded as a subgroup of $F^{\mathbb{Z}[t]}$;*
(3) *G is in the class $\mathcal{F}$.*

The proof will be in several steps. We first prove that $F^{\mathbb{Z}[t]}$ is fully residually free which in turn implies that every finitely generated subgroup of $F^{\mathbb{Z}[t]}$ is fully residually free. Recall that the class $\mathcal{F}$ is the smallest class of groups containing the infinite cyclic groups and closed under the following four operators:
(1) Isomorphism;
(2) Finitely Generated Subgroups;
(3) Free Products of Finitely Many Factors;
(4) Free Rank One Extensions of Centralizers.

We next show that the class of finitely generated subgroups in $\mathcal{F}$ coincides with the finitely generated subgroups of $F^{\mathbb{Z}[t]}$. Finally we show that every finitely generated fully residually free group is embeddable in $F^{\mathbb{Z}[t]}$.

Lemma 6.4.5. *The group $F^{\mathbb{Z}[t]}$ is fully residually free. In particular every finitely generated subgroup of $F^{\mathbb{Z}[t]}$ is fully residually free.*

Proof. Let A be the integral polynomial ring $\mathbb{Z}[t_1, \ldots, t_n]$. Let $r \geq 1$ be a cardinal and let F be a free group of rank r. Let X be a basis for F so that $F = \langle X; \ \rangle$. From the universal property of the A-tensor completion F^A of F (see [166]) it follows that every ring retraction $\rho\colon A \to \mathbb{Z}$ induces a group retraction $\overline{\rho}\colon (F)^A \to F$ such that $\overline{\rho}(g^\alpha) = \overline{\rho}(g)^{\rho(\alpha)}$ for all $g \in (F)^A, \alpha \in A$. Furthermore, Lyndon showed that to every element $g \in F^A$ there corresponds an element $\alpha \in A$ satisfying the following property.

(∗) If $\rho\colon A \to \mathbb{Z}$ is any ring retraction, then $\overline{\rho}(g) = 1$ if and only if $\rho(\alpha) = 0$.

Moreover, he showed that if $g \neq 1$, then any element $\alpha \in A$ corresponding to g in the sense of (∗) is non-zero. Thus, let $g_1, \ldots, g_n$ be finitely many non-trivial elements of F^A. Let $\alpha_i \in A$ correspond to g_i in the sense of (∗) for $i = 1, \ldots, n$. Then since A is ω-residually $\mathbb{Z}$ there is a retraction $\rho\colon A \to \mathbb{Z}$ such that $\rho(\alpha_i) \neq 0$ for $i = 1, \ldots, n$. It follows that $\overline{\rho}(g_i) \neq 1$ for $i = 1, \ldots, n$. Thus, F^A is fully residually free. Since full residual freeness is inherited by subgroups, every finitely generated subgroup of F^A is also fully residually free. □

A careful analysis of Lyndon's paper reveals that his arguments go through if A is any ω-residually $\mathbb{Z}$ ring. The full residual freeness of F^A also follows from residual freeness since F^A is commutative transitive – the centralizers of non-trivial elements being free rank one right A-modules. This result can be extended to further rings of Lyndon type. Specifically:

Lemma 6.4.6. *If A is any ω-residually $\mathbb{Z}$ ring, or more generally a ring of Lyndon type, where A^+, the additive subgroup of A, is fully residually $\mathbb{Z}^+$, where $\mathbb{Z}$ is the additive subgroup of $\mathbb{Z}$. The minimum subring of A is isomorphic to $\mathbb{Z}$. Then every free A-group is fully residually free.*

Now consider the additively written Abelian group $\mathbb{Z}^{\mathbb{N}}$ which is the unrestricted direct power of countably infinite copies of the infinite cyclic group $\mathbb{Z}$. Every countable subgroup of $\mathbb{Z}^{\mathbb{N}}$ is free Abelian. $\mathbb{Z}^{\mathbb{N}}$ is residually $\mathbb{Z}$ as is easily seen by taking projections and hence it is residually free. Clearly it is commutative transitive and therefore it is fully residually free. $\mathbb{Z}^{\mathbb{N}}$ is uncountable so it cannot be embedded in the countable group $(F_\omega)^{Z[t]}$.

We now show that the finitely generated subgroups of $F^{\mathbb{Z}[t]}$ coincide with the finitely generated groups in the class $\mathcal{F}$.

Lemma 6.4.7. *Every element of $\mathcal{F}$ is embeddable in $(F_\omega)^{\mathbb{Z}[t]}$.*

Proof. Certainly every infinite cyclic group is embeddable in $F = F_\omega$ which in turn is embeddable in $F^{\mathbb{Z}[t]}$. Any group isomorphic to one embeddable in $F^{\mathbb{Z}[t]}$ is clearly also embeddable in $F^{\mathbb{Z}[t]}$. Now suppose the finitely generated groups G and H are each embeddable in $F^{\mathbb{Z}[t]}$. Each of the images of G and H in $F^{\mathbb{Z}[t]}$ involve parametric words on a finite set of generators of a free basis X for F_ω. By applying the isomorphism operator, if necessary, we may assume that the images of G and H in $F^{\mathbb{Z}[t]}$ are such that the subsets Y and Z of X containing the generators actually appearing in the parametric words in the images of G and H respectively are disjoint. Then $G * H$ is embeddable in $(F(Y))^{\mathbb{Z}[t]} * (F(Z))^{\mathbb{Z}[t]}$ which in turn is a subgroup of $(F(X))^{\mathbb{Z}[t]} = F^{\mathbb{Z}[t]}$.

Now suppose $G \neq 1$ is embeddable in $F^{\mathbb{Z}[t]}$ and $u \in G \setminus \{1\}$. Let $B = \langle t; \ \rangle$ be infinite cyclic and let $K = G(u, B)$ be the free rank one extension of $Z_G(u)$. Now if F is any free group and A is any ring of Lyndon type then $F^A = \cup_{n<\omega} F(n)$ where $F(0) = F$ and $F(n+1) = \langle g^\alpha : g \in F(n), \alpha \in A\rangle$. Define for $g \in F^A$ the *level* of g by level(g) =min$\{n \in \omega : g \in F(n)\}$. Now every level $n > 0$ element g has a unique canonical form

$$g_0 g_1 v_1^{c_1} \cdots g_k v_k^{c_k} g_{k+1}$$

where $g_0, \ldots, g_{k+1} \in F(n-1)$, level$(v_i) = n-1$, and $c_i \in A \setminus \mathbb{Z}$ for $i = 1, \ldots, k$. Suppose $a_1, \ldots, a_r$ generate G, which we now identify with its image in $F^{\mathbb{Z}[t]}$. Then at each level in the construction of each a_j only finitely many coset representatives $c(t) \in \mathbb{Z}[t] \setminus \mathbb{Z}$ are involved so there is a finite bound $N \in \mathbb{N}$ such that $deg(c(t)) < N$ for all such $c(t)$. Suppose that the centralizer of u in $F^{\mathbb{Z}[t]}$ is generated as a $\mathbb{Z}[t]$-module by u_0. Then $\langle G, u_0^{t^N} \rangle \cong K$. This completes the proof. □

Lemma 6.4.8. *Every finitely generated subgroup of F^A lies in $\mathfrak{F}$ whenever A is of Lyndon type.*

Proof. Suppose $G = \langle a_1, \ldots, a_r \rangle \subset F^A$ where r is finite. We use induction on n =max $\{$level(a_1),...,level$(a_r)\}$. If $n = 0$ then $G \subset F(0) = F_\omega$ so G is a finitely generated free group and certainly lies in $\mathfrak{F}$. Now suppose the result is true for all $0 \leq m < n$. Then at least one generator, $a_i = a$, must have level n. Suppose $a = g_0 g_1 v_1^{c_1} \cdots g_k v_k^{c_k} g_{k+1}$ has level n. Now $\mathbb{Z}$ is a pure subgroup of A^+ so if $c_1, \ldots, c_m$ are all the coset representatives (perhaps with repetitions) involved in all the generators of level n, then by Theorem 13.4.1 of M. Hall [127], $\mathbb{Z} = \langle 1 \rangle$ is a direct summand in the subgroup of A^+ generated by $1, c_1, \ldots, c_m$. Let $1, b_1, \ldots, b_q$ be a basis for that subgroup. Then $c_i = n_{i,0} + n_{i,1} b_1 + \cdots + n_{i,q} b_q, i = 1, \ldots, m$. The element a may be reached from $\langle g_0, g_1, v_1, \ldots, g_k, v_k, g_{k+1} \rangle$ by repeated centralizer extensions. For example $v_1^{c_1}$ lies in the extension of the centralizer of v_1 adjoining one step at a time commuting elements $t_{1,1}, \ldots, t_{1,q}$ to commute with v_1. Then $v_1^{c_1} = v_1^{n_{1,0}} t_{1,1}^{n_{1,1}} \cdots t_{1,q}^{n_{1,q}}$. The same may be done for each distinct v_i in a and for each generator of level n taking care to adjoin new stable letters only for those v_i which have not already appeared. Thus G may be obtained as a finitely generated subgroup of repeated centralizer extensions of a group generated by finitely many elements of levels less than n. This completes the induction and proves the result. □

Since the polynomial ring $\mathbb{Z}[t]$ is of Lyndon type we obtain the following.

Corollary 6.4.1. *Every finitely generated subgroup of $F^{\mathbb{Z}[t]}$ lies in $\mathcal{F}$.*

Combining the last two lemmas and the corollary we have:

Lemma 6.4.9. *The finitely generated subgroups of $F^{\mathbb{Z}[t]}$ coincide with the finitely generated groups in the class $\mathcal{F}$.*

We now get to the final step in the proof of Theorem 6.4.1 which is extremely important in it own right.

Theorem 6.4.6. *Let G be a finitely generated fully residually free group. Then G is discriminated by a non-Abelian free group F_r if and only if G is embeddable in the free exponential group $F_r^{\mathbb{Z}[t]}$.*

Proof. From Lyndon's result if G is embeddable in $F_r^{\mathbb{Z}[t]}$ then G is discriminated by the free group F_r. The converse, that the freely discriminated groups are embeddable in Lyndon's exponential group is the essence of the paper "Irreducible Affine Varieties over a Free group II" by Kharlampovich and Myasnikov [155]. The proof is quite long and involved and we refer the reader to that paper. □

Since any countable free group is embeddable in $F = F_\omega$ we must have that G is fully residually free if and only if G is embeddable in $F^{\mathbb{Z}[t]}$.

Corollary 6.4.2. *A finitely generated group G is fully residually free if and only if G is embeddable in $F^{\mathbb{Z}[t]}$.*

This completes the proof of Theorem 6.4.1 and establishes the fact that finitely generated fully residually free groups can be identified with the finitely generated subgroups of the free exponential group $F^{\mathbb{Z}[t]}$ and alternatively can be identified as the finitely generated subgroups in the class $\mathcal{F}$. Since they can be identified with the finitely generated subgroups of $\mathcal{F}$ it follows as explained earlier that each finitely generated fully residually free group is a subgroup of a group formed from a free group by finitely many free rank one extensions of centralizers. Since each of these is an HNN group with Abelian associated subgroups it follows from Bass–Serre theory (see Chapter 2) that each finitely generated fully residually free group has a finite graph of groups decomposition and further that the edge groups are Abelian. Later in this chapter we will describe a canonical graph of groups decomposition of this form called a JSJ decomposition.

Lemma 6.4.10. *Each finitely generated fully residually free group has a finite graph of groups decomposition with Abelian edge groups.*

As an application of this construction we obtain that finitely generated fully residually free groups must be finitely presented.

Theorem 6.4.7. *A finitely generated fully residually free group G is finitely presented.*

Proof. From the proof of Theorem 6.4.1 a finitely generated fully residually free group can be considered as a subgroup of a group K formed from a free group by finitely many free rank one extensions of centralizers. Clearly K is finitely presented. Since G is finitely generated this implies, as used above, that G has a finite graph of groups decomposition with Abelian edge groups. Since K is finitely presented and G is finitely generated each vertex group is finitely presented so G is finitely presented. □

As a further application we give an alternative proof of Baumslag's result on surface groups. Recall that, if F is a non-Abelian free group and $u \in F$ is a nontrivial element which is neither primitive nor a proper power then the group K given by

$$K = \langle F * \overline{F}; u = \overline{u} \rangle$$

where $\overline{F}$ is an identical copy of F and $\overline{u}$ is the corresponding element to u in $\overline{F}$, is called a *Baumslag double*. A surface group with even genus $g \geq 2$ in the orientable case is a Baumslag double.

Theorem 6.4.8. *A Baumslag double is fully residually free.*

Proof. Suppose that G is a Baumslag double. G is finitely generated and can be expressed as

$$G = \langle F * \overline{F}; u = \overline{u} \rangle$$

where F is a free group and $\overline{F}$ is an identical copy of F and $\overline{u}$ is the corresponding element to u in $\overline{F}$. F can clearly be embedded in $F^{\mathbb{Z}[t]}$. Let $u \in F$. Define $u_1 = u^t \in F^{\mathbb{Z}[t]}$. Let $\overline{G}$ be the subgroup of $F^{\mathbb{Z}[t]}$ generated by F and $F_1 = u^t F (u^t)^{-1}$. Since u commutes with u^t we have $u^t u (u^t)^{-1} = u$ and hence the subgroup generated by F and F_1 is the free product with amalgamation with u equal to its conjugate by u^t. This is then isomorphic to G. Therefore G is embeddable as a subgroup of $F^{\mathbb{Z}[t]}$ and is therefore fully residually free. □

Corollary 6.4.3. *For $g \geq 2$ the group S_g, the fundamental group of an orientable surface of genus g, is fully residually free.*

Proof. Each S_g with $g \geq 2$ is a subgroup of S_2 and S_2 is a Baumslag double. □

6.4.3 A characterization in terms of ultrapowers

A *nonstandard free group* is an ultrapower (see Chapter 4) $^*F$ of a free group F with respect to some ultrafilter. From Los's Theorem, a group G, and any of its ultrapowers, have the same elementary theory, and hence the same universal theory. It follows that any nonstandard free group ${}^*F$ is universally free and so are its finitely generated subgroups. Therefore from the Gaglione, Spellman, Remeslennikov result we have finitely generated subgroups of nonstandard free groups must be fully residually free. This was first pointed out by Remeslennikov. The converse is also true. That is any finitely generated fully residually free group must be embedded as a subgroup of a nonstandard free group. Since for any countable free group F we have $F \subset F_2$, the free group of rank 2, we also have ${}^*F \subset^* F_2$. Therefore we may work with the free group of rank 2.

Theorem 6.4.9. *Let G be a finitely generated group. Then G is fully residually free if and only if G can be embedded as a subgroup of a nonstandard free group, that is a subgroup of some ultrapower ${}^*F_2$ of the free group F_2.*

Proof. Let ${}^*F_2$ be a nonstandard free group so that ${}^*F_2$ is a ultrapower of F_2 with respect to some ultrafilter D. From Los' Theorem ${}^*F_2$ has the same elementary theory and hence the same universal theory as F_2. It follows that finitely generated non-Abelian subgroups of ${}^*F_2$ are universally free. Therefore they are fully residually free and we have that any finitely generated subgroup of a nonstandard free group must be fully residually free since Abelian subgroups are torsion-free.

Conversely suppose that G is a non-Abelian, finitely generated, fully residually free group. Let a and b be a fixed pair of non-commuting elements of G and let $(g_1, \ldots, g_n) \in (G \setminus \{1\})^n$ be an arbitrary finite tuple of non-trivial elements of G. Then there is a free group F and an epimorphism $\Psi\colon G \to F$ such that $[\Psi(a), \Psi(b)] = \Psi([a, b]) \neq 1$ and $\Psi(g_i) \neq 1$ for $i = 1, \ldots, n$. However, every non-Abelian free group is fully residually – F_2, that is fully residually free of rank 2. The free group F is non-Abelian since $[\Psi(a), \Psi(b)] \neq 1$ and therefore, there is an epimorphism $\phi\colon F \to F_2$ such that $\phi(\Psi(g_i)) \neq 1$ for $i = 1, \ldots, n$. It follows that every non-Abelian, fully residually free group is actually fully residually F_2. Let I be the set of all epimorphisms $h\colon G \to F_2$. G may be embedded in F_2^I via $\phi\colon G \to F_2^I, \phi(g)(h) = h(g)$ for all $h \in I$. Now define for each non-trivial element $g \in G \setminus \{1\}$ its *support* by $\mathrm{Supp}(g) = \{h \in I\colon h(g) \neq 1\}$. The fact that G is fully residually F_2 is equivalent to the assertion that the family $\{\mathrm{Supp}(g)\colon g \in G \setminus \{1\}\}$ satisfies the finite intersection property. Hence this family extends to an ultrafilter D on I. The group G may be embedded in the ultrapower ${}^*F_2 = F_2^I/D$ via ${}^*\phi\colon G \to^* F_2$ by ${}^*\phi(g) = \phi(g)/D$. □

6.5 Structure of fully residually free groups

In this section we present a complete classification of fully residually free groups of rank 3 or less. In principle, the methods can be extended to any finite rank. To prove this classification we first prove the rather surprising result that a 2-free residually free group must actually be 3-free. We mention that there are fully residually free groups which are not 2-free and there are 3-free, fully residually free groups which are not 4-free and in general n-free fully residually free groups that are not $(n + 1)$-free.

Theorem 6.5.1. *Every 2-free, residually free group is 3-free.*

The proof depends on the following three results, Lemma 6.5.1, Lemma 6.5.2, and Lemma 6.5.3, which are of interest in their own right. After putting together these results to give a proof of Theorem 6.5.1, we will then return to the proofs of the lemmas, which are quite technical.

Definition 6.5.1. A group G satisfies property $\mathcal{P}$ if every 2-free subgroup of G must be 3-free.

Lemma 6.5.1. *Let G be a 3-free, residually free group, let u be a nontrivial element in G and let $M = Z_G(u)$. Then the HNN extension*

$$K = \langle t, G; \mathrm{rel}(G), t^{-1}zt = z, z \in M\rangle$$

satisfies property $\mathcal{P}$.

Lemma 6.5.2. *Let G be a finitely generated 2-free residually free group and let M be a maximal 2-generator subgroup of G. Then M is malnormal in G.*

Using Lemma 6.5.2 we can remove the restriction that the base be 3-free in preserving property $\mathcal{P}$ under rank one extensions of centralizers.

Lemma 6.5.3. *Let G be a residually free CSA group satisfying property $\mathcal{P}$. Further let $u \in G, u \neq 1$ and $M = Z_G(u)$. Then the HNN extension*

$$K = \langle t, G; \mathrm{rel}(G), t^{-1}zt = z, z \in M\rangle$$

also satisfies property $\mathcal{P}$.

We now use these three results to prove Theorem 6.5.1.

Proof. (Theorem 6.5.1) Since being 2-free implies commutative transitivity, it follows from the theorem of B. Baumslag that a 2-free residually free group is actually fully residually free.

Recall that the class $\mathcal{F}$ is the smallest class of groups containing the infinite cyclic groups and closed under the following four operators:
(1) Isomorphism;
(2) Finitely Generated Subgroups;
(3) Free Products of Finitely Many Factors;
(4) Free Rank One Extensions of Centralizers.

We first claim that every member of $\mathcal{F}$ satisfies property $\mathcal{P}$. This is clear since property $\mathcal{P}$ is easily seen to be preserved by subgroups and free products. Further from Lemma 6.5.3 it is preserved under rank one extensions of centralizers.

Now suppose G is a 2-free, residually free group. From the remarks above G is fully residually free. If H is a 3-generator subgroup of G then H is also a 2-free fully residually free group. But every 3-generator fully residually free group lies in $\mathcal{F}$. Therefore

H lies in $\mathcal{F}$. But since every member of $\mathcal{F}$ satisfies property $\mathcal{P}$ every 2-free subgroup of a member of $\mathcal{F}$ must be 3-free. Since H is a 2-free subgroup of itself and lies in $\mathcal{F}$, it follows that H is 3-free. Since H is 3-generator H is then free. Therefore, every 3-generator subgroup of G is free and consequently every 2-free, residually free group is 3-free. □

We now prove the three required lemmas.

Proof of Lemma 6.5.1. Since G is a CSA group the centralizer M is malnormal in G. K is then an HNN group with malnormal associated subgroups and therefore the Nielsen reduction arguments found in [103] and based on the work of Peczynski and Reiwer [210] (see Chapter 2) can be adapted analogously to this situation.

It suffices to prove that a 3-generator, 2-free subgroup H of K is actually free. Suppose then that H is a 2-free, 3-generator subgroup of K. We may assume that H actually has rank 3, that is H cannot be generated by two elements. Using the Nielsen reduction method described in [103] we may assume that we have a minimal generating system $\{x_1, x_2, x_3\}$ for H and we wish to show that $\{x_1, x_2, x_3\}$ is actually a free generating system for H. This is certainly the case if $H \cap M = \{1\}$ by the theorem of H.Neumann on subgroups in HNN groups ([202]). Hence let $H \cap M \neq \{1\}$ and let $H_1 = G \cap H$. Since $H_1 \subset H$ it is 2-free.

Analogously as in the proof of Theorem 1 of [103], $\{x_1, x_2, x_3\}$ will freely generate a free group (either they are in the base G and so generate a free group by property $\mathcal{P}$ or some t symbols remain uncancelled in forming a reduced word in $\{x_1, x_2, x_3\}$) except, after a suitable conjugation and renumbering, possibly in the following cases:

(1) $x_1, x_2 \in G, x_3 = ht^r$ or $x_3 = t^r h$ with
$r \in \mathbb{Z} \setminus \{0\}, h \in G \setminus M, h \notin \langle x_1, x_2\rangle, \langle x_1, x_2\rangle \cap M = \langle y\rangle, y \neq 1$.

(2) $x_1, x_2 \in G, x_3 = t^r k t^r$ or $x_3 = t^r k t^{-r}$ with
$r \in \mathbb{Z} \setminus \{0\}, k \in G \setminus M, \langle x_1, x_2\rangle \cap M = \langle y\rangle, y \neq 1$.

Consider first case (1) with $x_3 = ht^r$. The argument is identical with $x_3 = t^r h$. Notice that $hyh^{-1} = x_3 t^{-r} y t^r x_3^{-1} = x_3 y x_3^{-1} \in H_1 = H \cap G$. Recall that $x_3, y \in H, x_3 y x_3^{-1} \in H, h \in G, hyh^{-1} \in G$. Since H is 2-free we have $h \notin M$ for otherwise $y = x_3 y x_3^{-1} \in \langle x_1, x_2\rangle$ gives a non-trivial relation. Therefore $\langle y, hyh^{-1}\rangle$ is free of rank two and y, hyh^{-1} constitute a free basis because M is malnormal in G.

Since x_1, x_2 are in the base G of the HNN group K and generate a free group and $x_3 \notin G$ the only cancellations can occur via conjugates of the associated subgroup relations. Therefore a non-trivial relation between x_1, x_2 and x_3 implies a non-trivial relation between x_1, x_2 and hyh^{-1}.

Now we regard the subgroups $N = \langle x_1, x_2, h\rangle \subset G$ and

$$Q = \langle x_1, x_2, hyh^{-1}\rangle \subset G\,.$$

The subgroups N and Q are both free groups because G is 2-free and satisfies property $\mathcal{P}$ and Q is a subgroup of $N \cap H_1$. We have $h \notin Q$ because H is 2-free. If h were in Q

then $h \in H$ and hence $t^r \in H$ and then $t^r yt^{-r} y$ would define a non-trivial relation in the two generator subgroup $\langle y, t^r \rangle$ of H. If x_1, x_2, hyh^{-1} freely generate Q then there is no non-trivial relation between x_1, x_2 and hyh^{-1} and therefore in this case H is freely generated by x_1, x_2, x_3. Now let Q be free of rank two (Q cannot be cyclic since $x_1, x_2 \in Q$.) Let $\{z_1, z_2\}$ be a free generating system of Q. Since Q is a subgroup of the free group N we may now apply Lemma 2.1 of [219]. Let $y = y_1^\alpha, \alpha \geq 1$ where $y_1 \in N$ is not a proper power in N. Since $h \notin Q$ but $y \in Q$ and $hyh^{-1} \in Q$ there is a Nielsen transformation from $\{z_1, z_2\}$ to a system in which an element is conjugate (in N) to a power of y_1. Hence without loss of generality let $y = y_1^\alpha$ be a primitive element in Q. Since $y \in \langle x_1, x_2 \rangle \subset Q$, a conjugate of y is also a primitive element in $\langle x_1, x_2 \rangle$. Without loss of generality we may then assume that $x_1 = z_1 = y$. We cannot have $h = h_1 y_1^\gamma$ for some $h_1 \in Q = \langle z_1, z_2 \rangle \subset H$ and $\gamma \in \mathbb{Z} \setminus \{0\}$ for if so then $y_1^\gamma t^r \in H$ and then $y_1^\gamma t^r yt^{-r} y_1^{-\gamma} = y$ defines a non-trivial relation in the two-generator subgroup $\langle y_1^\gamma t^r, y \rangle$ of H. Therefore we get a Nielsen transformation from $\{z_1, z_2\} = \{y, z_2\}$ to a system $\{y, gy_1^\beta g^{-1}\}$ for some $g \in N, g \notin \langle y_1 \rangle$ and some $\beta \geq 1$. Since $y \in Q, h \notin Q$ but $hyh^{-1} \in Q$ we then also get a Nieslen transformation from $\{z_1, z_2\}$ to $\{y, hy_1^\beta h^{-1}\}$. Also we have necessarily $\alpha = \beta\delta$ for some $\delta \geq 1$, that is $hyh^{-1} = (hy_1^\beta h^{-1})^\delta$.

Now we have

$$H = \langle x_1, x_2, x_3 \rangle = \langle z_1, z_2, x_3 \rangle = \langle y, hy_1^\beta h^{-1}, ht^r \rangle = \langle hy_1^\beta h^{-1}, ht^r \rangle$$

because $y = t^{-r} h^{-1} hyh^{-1} ht^r$. This implies that H is a 2-generator group contradicting the fact that H has rank three. Therefore Q must be free of rank 43 and hence H is freely generated by x_1, x_2 and x_3.

In case (2) suppose first that $x_3 = t^r kt^{-r}$. Then a non-trivial relation between x_1, x_2 and x_3 would imply such a relation between y and k. Let $H_2 = t^r Ht^{-r}$. This is also 2-free being a conjugate of H and contains y and k. Therefore there is no such relation and x_1, x_2, x_3 generate a free group.

In the final situation where $x_3 = t^r kt^r$ let $H_3 = t^{-r} Ht^r, H_4 = t^r Ht^{-r}$. Both H_3 and H_4 are 2-free. A non-trivial relation between x_1, x_2 and x_3 implies a non-trivial relation between y and $k^{-1} yk$ or between y and kyk^{-1}. But these pairs lie in either H_3 or H_4 so no such relation is possible completing the proof. □

Proof of Lemma 6.5.2. 2-free residually free groups are commutative transitive and torsion-free. It follows that the centralizers of non-trivial elements are isolated. This means that for all $n \in \mathbb{N}, b^n \in Z_G(a)$ implies that $b \in Z_G(a)$. Further any such root must necessarily be unique since 2-free residually free groups are universally free and universally free groups have unique roots.

Now suppose that G is a finitely generated 2-free residually free group and let M be a maximal 2-generator subgroup of G. If rank$(G) \leq 2$ then $M = G$ and clearly then M is malnormal. Thus we assume that the rank of G is ≥ 3. Suppose that M is not malnormal in G. Then there exist non-trivial $u, w \in M$ and $g \in G \setminus M$ with $w = g^{-1}ug$. It follows from the work of Gaglione and Spellman [111] that maximal

2-generator subgroups in finitely generated 2-free residually free groups are isolated. Since M is isolated we may assume without loss of generality that neither u nor w is a proper power in G.

First of all u and w cannot be conjugate within M for if so there would exist an $h \in M$ such that $g^{-1}ug = h^{-1}uh$ and so $gh^{-1} \in Z_G(u) = \langle u \rangle$. But then $g = hu^n$ for some $n \in \mathbb{Z}$ and so $g \in M$.

Next M is non-Abelian. If it were Abelian, it would have to be cyclic since G is 2-free. Since G has rank at least 3, G is non-Abelian and therefore a non-Abelian commutative transitive group. It follows that G must have a trivial center. Suppose $M = \langle m \rangle$. Since $Z(G) = \{1\}$ there exists a $g \in G$ with $gm \neq mg$. Then M is properly contained in the rank two free subgroup $\langle g, m \rangle$ contradicting the maximality of M. Since M is non-Abelian and G is 2-free it follows that $M = \langle a, b \rangle$ is free of rank two.

Since G is fully residually free there exists an epimorphism $\phi \colon G \to F$ with F free such that $\phi(a) \neq 1$, $\phi(b) \neq 1$, $\phi([a, b]) \neq 1$. It follows that the image of M is also free of rank two and $\phi_{|M}$ is an isomorphsim of M onto $F_0 = \langle \alpha, \beta \rangle$ with $\alpha = \phi(a)$, $\beta = \phi(b)$. It follows that $u(\alpha, \beta)$, $w(\alpha, \beta)$ are non-trivial conjugate elements in F and hence $w(\alpha, \beta)$ is not conjugate to $u(\alpha, \beta)^{-1}$ in F and therefore also not conjugate in F_0. Moreover since $\phi_{|M}$ is an isomorphism we have that $u(\alpha, \beta)$, $w(\alpha, \beta)$ are not conjugate in F_0 and neither is a proper power in F_0. However since they are conjugate in F we say that they are *virtually conjugate* in F_0.

Next we have that the element $x_3^{-1}u(x_1, x_2)x_3w(x_1, x_2)^{-1}$ cannot be primitive in the rank three free group on x_1, x_2, x_3. If it were consider $\langle M, g \rangle = \langle a, b, g \rangle$. The element $g^{-1}u(a, b)gw(a, b)^{-1}$ is a relator in $\langle M, g \rangle$ so we may rewrite a presentation for $\langle M, g \rangle$ with generators v_1, v_2, v_3 where

$$v_3 = g^{-1}u(a, b)gw(a, b)^{-1}$$

and delete v_3 via a Tietze transformation. Then $M \subset \langle M, g \rangle = \langle v_1, v_2 \rangle$. Then by the maximality of M we have $M = \langle M, g \rangle$ and hence $g \in M$ contradicting $g \notin M$.

Now since G is finitely generated so is $\phi(G) = F$. Consider the group

$$K = \langle F, s; \operatorname{rel}(F), s^{-1}u(\alpha, \beta)s = u(\alpha, \beta) \rangle .$$

Since F is a finitely generated non-Abelian free group it is certainly 3-free, so by Lemma 6.5.1, K satisfies property $\mathcal{P}$. In K we have that

$$w(\alpha, \beta) = \gamma^{-1}u(\alpha, \beta)\gamma = \gamma^{-1}(s^{-1}u(\alpha, \beta)s)\gamma = (s\gamma)^{-1}u(\alpha, \beta)(s\gamma)$$

where $\gamma = \phi(g)$. Letting $t = s\gamma$ we have

$$H = \langle \alpha, \beta, t; t^{-1}u(\alpha, \beta)t = w(\alpha, \beta) \rangle .$$

Now $u(\alpha, \beta)$, $w(\alpha, \beta)$ are non-trivial elements of the free group on α, β and further $w(\alpha, \beta)$ is not conjugate within $\langle \alpha, \beta \rangle$ to $u(\alpha, \beta)^{\pm 1}$. It follows from the work of Fine,

Roehl and Rosenberger [102] that H is 2-free. Since $H \subset K$ and K satisfies property $\mathcal{P}$, H is is also 3-free and hence a free group. However the only way that a group with the presentation

$$\langle \alpha, \beta, t; t^{-1}u(\alpha, \beta)t = w(\alpha, \beta)\rangle$$

could be free is if $x_3^{-1}u(x_1, x_2)x_3w(x_1, x_2)^{-1}$ is primitive in the rank three free group $\langle x_1, x_2, x_3; \ \rangle$ which we have shown to be impossible. This contradiction shows that it is impossible for M not to be malnormal in G. □

Proof of Lemma 6.5.3. The beginning of this proof exactly mirrors that of the proof of Lemma 6.5.1. Since G is a CSA group satisfying the maximal condition on Abelian subgroups, the centralizer M is malnormal in G. K is then an HNN group with malnormal associated subgroups and therefore the Nielsen reduction arguments found in [103] and based on the work of Peczynski and Reiwer [210] can be adapted analogously to this situation.

As in Lemma 6.5.1 it suffices to prove that a 3-generator, 2-free subgroup H of K is actually free. Suppose then that H is a 2-free, 3-generator subgroup of K. We may assume that H actually has rank 3, that is H cannot be generated by two elements. Using the Nielsen reduction method described in [103] we may assume that we have a minimal generating system $\{x_1, x_2, x_3\}$ for H and we wish to show that $\{x_1, x_2, x_3\}$ is actually a free generating system for H. This is certainly the case if $H \cap M = \{1\}$ by the theorem of H.Neumann on subgroups in HNN groups. Hence let $H \cap M \neq \{1\}$ and let $H_1 = G \cap H$. Since $H_1 \subset H$ it is 2-free and therefore 3-free since $H_1 \subset G$ and G satisfies property $\mathcal{P}$.

Analogously as in the proof of Proposition 2 and Theorem 1 of [103], $\{x_1, x_2, x_3\}$ will generate a free group. Either they are in the base G and so generate a free group by property $\mathcal{P}$ or some t symbols remain uncancelled in forming a word in $\{x_1, x_2, x_3\}$ except possibly in the following cases (after a suitable conjugation and renumbering):

(1) $x_1, x_2 \in G, x_3 = ht^r$ or $x_3 = t^rh$ with $r \in \mathbb{Z} \setminus \{0\}$, $h \in G \setminus M$, $h \notin \langle x_1, x_2\rangle$, $\langle x_1, x_2\rangle \cap M = \langle y\rangle$, $y \neq 1$. Note that by the conditions on G, $M \cap \langle x_1, x_2\rangle$ is either cyclic or trivial. If it were trivial there could be no cancellation between x_1, x_2 and x_3 and since x_1, x_2 generate a free group so do $\{x_1, x_2, x_3\}$.

(2) $x_1, x_2 \in G, x_3 = t^rkt^r$ or $x_3 = t^rkt^{-r}$ with $r \in \mathbb{Z} \setminus \{0\}$, $k \in G \setminus M$, $\langle x_1, x_2\rangle \cap M = \langle y\rangle$, $y \neq 1$.

The proof of case (2) in Lemma 6.5.1 did not use that the base was 3-free only the residual freeness and 2-free property of the subgroup H. Therefore case (2) follows exactly as in Lemma 6.5.1.

Now consider case (1) with $x_3 = ht^r$ and we suppose as before that H has rank three. As in the proof of Lemma 6.5.1, H will be free unless there is a non-trivial relation between x_1, x_2 and $hyh^{-1} = x_3yx_3^{-1}$. Consider the subgroup $H_1 = H \cap G$.

Since G is 2-free with property $\mathcal{P}$ and H is 2-free it follows that H_1 is 3-free. Now $x_1, x_2, hyh^{-1} \in H_1$ and therefore $H_2 = \langle x_1, x_2, hyh^{-1}\rangle$ is a non-Abelian free group. If the rank of H_2 is three we are done since then no non-trivial relation is possible. Rank one is impossible since H_1 is non-Abelian so we suppose that H_2 has rank two, that is $\langle x_1, x_2, hyh^{-1}\rangle = \langle z_1, z_2\rangle \subset H$. From work of Gaglione and Spellman [111], 2-free residually free groups the property that any two generator subgroup is contained in a maximal two-generator subgroup. Therefore since H is a 2-free residually free group we have $\langle z_1, z_2\rangle \subset \langle X_1, X_2\rangle$ where $\langle X_1, X_2\rangle$ is maximal in H. Now $\langle x_1, x_2\rangle \subset \langle X_1, X_2\rangle$ and hence $y \in \langle X_1, X_2\rangle$. Further $hyh^{-1} \in \langle X_1, X_2\rangle$ so therefore $x_3 y x_3^{-1} \in \langle X_1, X_2\rangle$. The subgroup $\langle X_1, X_2\rangle$ is a maximal 2-generator subgroup of the 2-free residually free group H and therefore from Lemma 6.5.2 it is malnormal in H. It follows that $x_3 \in \langle X_1, X_2\rangle$. This implies that $\langle x_1, x_2, x_3\rangle = H \subset \langle X_1, X_2\rangle$ and hence $H = \langle X_1, X_2\rangle$ contradicting the fact that H has rank three. Therefore rank two for H_2 is impossible and hence H also is free of rank three. □

We note that there is a converse result to Lemma 6.5.2 which gives another characterization of 2-free fully residually free groups.

Lemma 6.5.4. *Let G be a non-Abelian finitely generated fully residually free group. If every maximal two-generator subgroup of G is malnormal in G then G is 2-free.*

Proof. Suppose G were a non-Abelian finitely generated fully residually free group with the property that every maximal two generator subgroup is malnormal. Since G satisfies the maximal condition for Abelian subgroups it would follow that if G were not 2-free there would exist a non-trivial $a \in G$ such that $M = Z_G(a)$ is free Abelian of finite rank $k > 1$. Suppose $b_1, \ldots, b_k$ is a basis for M. Since G is a non-Abelian commutative transitive group it has a trivial center. Thus there exists a non-trivial $g \in G$ with $gb_1 \neq b_1 g$. Consider $\langle g, b_1\rangle$ and extend to a maximal 2-generator subgroup $H = \langle c_1, c_2\rangle \supset \langle g, b_1\rangle$. Since $gb_1 \neq b_1 g$, H is non-Abelian and therefore $c_1c_2 \neq c_2c_1$. In a fully residually free group every pair of non-commuting elements freely generates a subgroup. Therefore H is free with basis c_1, c_2. But $1 \neq b_1 = b_2^{-1} b_1 b_2 \in H$ and hence it follows that $b_2 \in H$ since H is malnormal. But then the free Abelian group of rank two, $\langle b_1, b_2\rangle$, is contained in a non-Abelian free group which is impossible. This contradiction shows that G must be 2-free. □

We now present the complete classification. The proof is similar to the proof of Theorem 6.5.1.

Theorem 6.5.2. *Let G be a fully residually free group. Then*
(1) *if rank$(G) = 1$ then G is infinite cyclic;*
(2) *if rank$(G) = 2$ then either G is free of rank 2 or free Abelian of rank 2;*
(3) *if rank$(G) = 3$ then either G is free of rank 3, free Abelian of rank 3 or a free rank one extension of centralizers of a free group of rank 2. That is G has a one-relator*

presentation

$$G = \langle x_1, x_2, x_3; x_3^{-1} v x_3 = v \rangle$$

where $v = v(x_1, x_2)$ *is a non-trivial element of the free group on* x_1, x_2 *which is not a proper power.*

Proof. If $\text{rank}(G) = 1$ then clearly G must be infinite cyclic since residually free groups are torsion-free.

If $\text{rank}(G) = 2$ then suppose $G = \langle a, b \rangle$. If G is non-Abelian then $[a, b] \neq 1$ and since G is fully residually free there exists a free group F and an epimorphism $\phi\colon G \to F$ with $\phi(a) \neq 1, \phi(b) \neq 1, \phi[a, b] \neq 1$. Let $\alpha = \phi(a)$, $\beta = \phi(b)$. Since $\phi[a, b] = [\alpha, \beta] \neq 1$ it follows that $\langle \alpha, \beta \rangle$ freely generate a free subgroup of rank 2. Hence ϕ is a monomorphism and G is isomorphic to $\langle \alpha, \beta \rangle$ and hence is also free of rank 2. □

This handles ranks 1 and 2. We now concentrate on the case where $\text{rank}(G) = 3$. We first extend property $\mathcal{P}$.

Definition 6.5.2. A group G has *property-*$\mathcal{Q}$ if every non-Abelian 3-generator subgroup of G is either free or has a one-relator presentation $\langle x, y, t; t^{-1} v t = v \rangle$ where $1 \neq v = v(x, y)$ is not a proper power in $\langle x, y; \ \rangle$.

If a finitely generated group G is fully residually free then it falls in the class $\mathcal{F}$. Clearly infinite cyclic groups satisfy property-$\mathcal{Q}$ as do finitely generated subgroups of groups with property-$\mathcal{Q}$. Further property-$\mathcal{Q}$ is preserved under taking free products. The proof of Theorem 6.5.2 then follows directly from the following crucial lemma which says that property-$\mathcal{Q}$ is preserved under free rank one extensions of centralizers of fully residually free groups. This is analogous to our proof of Theorem 6.5.1.

Lemma 6.5.5. *Let* G *be a fully residually free group of rank at least 2 which satisfies property-*$\mathcal{Q}$*. Let* $1 \neq u \in G$ *and* $M = Z_G(u)$*. Let*

$$K = \langle t, G; \text{rel}(G), t^{-1} z t = z, \text{ for all } z \in M \rangle.$$

Then K *satisfies property-*$\mathcal{Q}$*.*

Proof of Lemma 6.5.5. Since G is fully residually free it follows as before that M is malnormal in G and that K is also fully residually free. Let $H = \langle x_1, x_2, x_3 \rangle$ be a non-Abelian 3-generator subgroup of K. Since K is fully residually free we may without loss of generality assume that the rank of H is 3 since if the rank is less than three the result follows as above. Since K is an HNN group with both associated subgroups malnormal and equal to M we can get analagous results as in [103] and hence use the Nielsen reduction method for HNN groups with malnormal associated subgroups as described in [103] and based on the work of Peczynski and Reiwer [210] (See Chapter 2). In particular each $x \in K$ may be written in the reduced form $x = h \in G$ or $x = ht^r h_1 t^{r_1} \dots h_n t^{r_n} k$ with $n \geq 0, h, k \in G, r \in \mathbb{Z} \setminus \{0\}$ and $h_i \in G \setminus M, r_i \in \mathbb{Z} \setminus \{0\}, i = 1, \dots, n$ if $n \geq 1$.

If H is 2-free then since H is fully residually free it follows from Theorem 6.5.1 that H is free and hence free of rank 3. Therefore suppose that H is not 2-free. Then there is a 2-generator subgroup N of H which is free Abelian of rank 2. By the Nielsen reduction method we may assume – possibly after a Nielsen transformation and a suitable conjugation – that one of the following cases holds:

(A) $x_1, x_2 \in G$ and $x_1x_2 = x_2x_1$;

(B) $x_1, x_2 \in G \setminus M$, $\langle x_1, x_2\rangle$ free of rank 2, $\langle x_1, x_2\rangle \cap M = \langle y\rangle$ and $x_3 = t^s g$ with $s \in \mathbb{N}, g \in M$;

(C) $x_1 \in M$ and $x_2 = t^s g$ with $s \in \mathbb{N}, g \in M$.

We first consider Case (A). If $x_3 \in G$ then the result follows since G has property-$\mathcal{Q}$. Therefore we suppose that $x_3 \notin G$.

If $x_1, x_2 \notin M$ then there can be no non-trivial relations between $\langle x_1, x_2\rangle$ and x_3. Therefore

$$H = \langle x_1, x_2, x_3; x_1^{-1}x_2x_1 = x_2\rangle = \mathbb{Z} * (\mathbb{Z} \times \mathbb{Z})$$

which is a free rank one extension of centralizers of the free group $\langle x_2, x_3; \;\rangle$ If either x_1 or x_2 is in M they both are, since M is malnormal in G. Now suppose $x_1, x_2 \in M$. Let $x_3 = ht^r h_1 t^{r_1} \dots h_n t^{r_n} k$ with $n \geq 0, h, k \in G$ and r a nontrivial integer, and $h_i \in G \setminus M, r_i \in \mathbb{Z} \setminus \{0\}, i = 1, \dots, n$ if $n \geq 1$. If $n \geq 2$ or $n = 1$ and $h \notin M$ or $n = 1$ and $k \notin M$ then again we have

$$H = \langle x_1, x_2, x_3; x_1^{-1}x_2x_1 = x_2\rangle = \mathbb{Z} * (\mathbb{Z} \times \mathbb{Z}).$$

So we may assume that these don't hold.

Suppose first that $n = 1$ and $h, k \in M$. Then we may assume that $h = k = 1$ that is $x_3 = t^r h_1 t^{r_1}$. In forming a cyclically reduced word in x_1, x_2, x_3 which contains x_3 some t-symbols will remain uncancelled and hence H will form a free group except possibly if (1) $r = -r_1$ or (2) $r = r_1$.

If $r = -r_1$ then there is a non-trivial relation between x_1, x_2, x_3 which is not a consequence of $[x_1, x_2] = 1$ if and only if there is a non-trivial relation of the form

$$x_1^{\alpha_1}x_2^{\beta_1}h_1^{\gamma_1}\dots x_1^{\alpha_m}x_2^{\beta_m}h_1^{\gamma_m} = 1,$$

$m \geq 1$, all $\alpha_i, \beta_i, \gamma_i \in \mathbb{Z}$ which is not a consequence of the relation $[x_1, x_2] = 1$. However this is impossible since $h_1 \notin M$ and G has property-$\mathcal{Q}$. Hence if $r = -r_1$ again we have that

$$H = \langle x_1, x_2, x_3; x_1^{-1}x_2x_1 = x_2\rangle = \mathbb{Z} * (\mathbb{Z} \times \mathbb{Z}).$$

If $r = r_1$ then there is a non-trivial relation between x_1, x_2, x_3 which is not a consequence of $[x_1, x_2] = 1$ if and only if there is a non-trivial relation of the form

$$x_1^{\alpha_1}x_2^{\beta_1}h_1^{\epsilon}\dots x_1^{\gamma_1}x_2^{\delta_1}\dots x_1^{\alpha_m}x_2^{\beta_m}h_1^{\epsilon}x_1^{\gamma_m}x_2^{\delta_m}h_1^{-\epsilon} = 1,$$

$m \geq 1, \epsilon = 1$ or $\epsilon = -1$, all $\alpha_i, \beta_i, \gamma_i \in \mathbb{Z}$ which is not a consequence of the relation $[x_1, x_2] = 1$. But this is also impossible since $h_1 \notin M$ and M is malnormal in G. Therefore in this case again we have that

$$H = \langle x_1, x_2, x_3; x_1^{-1} x_2 x_1 = x_2 \rangle = \mathbb{Z} * (\mathbb{Z} \times \mathbb{Z}).$$

Now let $n = 0$ so that $x_3 = ht^r k$ with $r \neq 0, h, k \in G$. If $h, k \in M$ then H is Abelian. If both $h, k \notin M$ then as before we have that

$$H = \langle x_1, x_2, x_3; x_1^{-1} x_2 x_1 = x_2 \rangle = \mathbb{Z} * (\mathbb{Z} \times \mathbb{Z}).$$

Hence without loss of generality we may assume that $k \in M, h \notin M$. We may assume that $k = 1$ so that $x_3 = ht^r$. Then there is a non-trivial relation between x_1, x_2, x_3 which is not a consequence of $[x_1, x_2] = 1$ if and only if there is a non-trivial relation of the form

$$x_1^{\alpha_1} x_2^{\beta_1} h x_1^{\gamma_1} x_2^{\delta_1} h^{-1} \ldots x_1^{\alpha_m} x_2^{\beta_m} h x_1^{\gamma_m} x_2^{\delta_m} h^{-1} = 1,$$

$m \geq 1$, all $\alpha_i, \beta_i, \gamma_i \in \mathbb{Z}$ which is not a consequence of the relation $[x_1, x_2] = 1$ But this is also impossible since $h \notin M$ and M is malnormal in G. Therefore in this case again we have that

$$H = \langle x_1, x_2, x_3; x_1^{-1} x_2 x_1 = x_2 \rangle = \mathbb{Z} * (\mathbb{Z} \times \mathbb{Z})$$

which completes Case (A).

We now consider case (B). Here we have immediately that

$$H = \langle x_1, x_2, x_3; x_3^{-1} y x_3 = y \rangle.$$

Further, M must be isolated in G (see Section 6) since G is fully residually free so that y is not a proper power and hence H is a free rank one extension of centralizers of the desired form.

Finally we consider case (C). We may assume that $g = 1$ that is $x_2 = t^s$. The case $g \neq 1, g \in M$ is analogous. If $x_3 \in M$ then H would be Abelian. If $x_3 \in G \setminus M$ then both $\langle x_1, x_3 \rangle$ and $\langle x_2, x_3 \rangle$ are necessarily free of rank 2 and therefore

$$H = \langle x_1, x_2, x_3; x_1^{-1} x_2 x_1 = x_2 \rangle.$$

Therefore we assume that $x_3 \notin G$ and let

$$x_3 = ht^r h_1 t^{r_1} \ldots h_n t^{r_n} k$$

in reduced form, with $n \geq 0, h, k \in G, r \in \mathbb{Z} \setminus \{0\}$ and $h_i \in G \setminus M, r_i \in \mathbb{Z} \setminus \{0\}, i = 1, \ldots, n$ if $n \geq 1$.

If $h, k \notin M$ then as in Case (A) and as above we have

$$H = \langle x_1, x_2, x_3; x_1^{-1} x_2 x_1 = x_2 \rangle.$$

If $n \geq 1$ and $h, k \in M$ then we may assume that $h = k = 1$ and using the facts that $t^{-r} x_1 t^r = x_1, t^{-r} x_2 t^r = x_2$ we may replace x_3 by

$$x_3' = t^r x_3 t^r = h_1 t^{r_1} \dots h_n t^{r_n + r}.$$

If in this case $x_3' \in G$ then we are done since it reverts to the previous argument. If $x_3' \notin G$ then we continue with $H' = t^r H t^r = \langle x_1, x_2, x_3' \rangle$. Since each $h_i \in G \setminus M$ and $n \geq 1$ there can be no non-trivial relation between x_1, x_2, x_3' which is not a consequence of $[x_1, x_2] = 1$. Hence

$$H' = \langle x_1, x_2, x_3'; x_1^{-1} x_2 x_1 = x_2 \rangle$$

is free rank one extension of centralizers and therefore so is its isomorphic conjugate H.

Hence suppose $h \notin M$ and $k \in M$ or $h \in M$ and $k \notin M$ if $n \geq 1$. Without loss of generality we may assume that $h \notin M$ and $k \in M$ and thus $k = 1$ if $n \geq 1$. If $s | r_n$, that is $r_n = \alpha s$ for some $\alpha \in \mathbb{Z}$, then we may replace x_3 by $x_3 x_2^{-\alpha}$ to get a shorter system and we continue with this new one. Hence let $n \geq 1$ and $s \nmid r_n$. Then in forming a cyclically reduced word in x_1, x_2, x_3 which contains x_3 some t--symbols remain uncancelled and again we get that

$$H = \langle x_1, x_2, x_3; x_1^{-1} x_2 x_1 = x_2 \rangle.$$

Therefore we now suppose that $n = 0$ so that $x_3 = h t^r k$. Since the case $h, k \notin M$ has already been done, we may assume as above that either $h \notin M$ and $k = 1$, so that $x_3 = h t^r$ with $h \notin M$, or $h, k \in M$. If $h, k \in M$ then H would be Abelian. Hence let $x_3 = h t^r$ with $h \notin M$. As before if $s \mid r_n$ we may pass to a shorter system and so we may assume that $s \nmid r_n$. Then there is a non-trivial relation between x_1, x_2, x_3 which is not a consequence of the relation $[x_1, t^s] = 1$ if and only if there is a non-trivial relation of the form

$$x_1^{\alpha_1} t^{s\beta_1} h x_1^{\gamma_1} t^{s\delta_1} h^{-1} \dots x_1^{\alpha_m} t^{s\beta_m} h x_1^{\gamma_m} t^{s\delta_m} h^{-1} = 1,$$

$m \geq 1$, all $\alpha_i, \beta_i, \gamma_i, \delta_i \in \mathbb{Z}$ which is not a consequence of the relation $[x_1, t^s] = 1$. From cancellation arguments this is possible only if $\beta_i = \delta_i = 0$ for all $i = 1, \dots, m$, that is we would have a non-trivial relation of the form

$$x_1^{\alpha_1} h x_1^{\gamma_1} h^{-1} \dots x_1^{\alpha_m} h x_1^{\gamma_m} h^{-1} = 1$$

which is impossible since necessarily $\langle x_1, h \rangle$ is free of rank 2 since $h \notin M$. Therefore again we get that H must have the presentation

$$H = \langle x_1, x_2, x_3; x_1^{-1} x_2 x_1 = x_2 \rangle$$

completing case (C) and the proof of the lemma. □

The verification of this lemma then completes the proof of Theorem 6.5.2.

The cancellation arguments used in the proof of the above lemma together with Lemma 2.1 of [219] also gives the following result which is a corollary of the lemma.

Corollary 6.5.1. *Let $K = \langle a, b, t; t^{-1}vt = v \rangle$ where $1 \neq v = v(a, b)$ is not a proper power in the free group $\langle a, b; \ \rangle$. Let $\{x_1, x_2, x_3\}$ be a generating system of K. Then $\{x_1, x_2, x_3\}$ is Nielsen equivalent to $\{a, b, t\}$.*

We now give some applications to determining when two elements of a free group are virtually conjugate.

Definition 6.5.3 (Remeslennikov). Let F_0 be a free group and let u, w be two elements of F_0. Then u and w are *virtually conjugate* in F_0 if there exists a free overgroup $F \supset F_0$ in which u and w are conjugate.

We have the following two observations.

(1) If F_0 is a countable free group and u and w are virtually conjugate elements in F_0 then there is a free group F of rank two, $F \supset F_0$ in which u and w are conjugate.
(2) If F_0 is a free group and $u, w \in F \setminus \{1\}$ are virtually conjugate in F_0, then w cannot be conjugate to u^{-1} in F_0.

Now let $F_2 = \langle b_1, b_2; \ \rangle$ be free of rank two and suppose that $u(b_1, b_2), w(b_1, b_2) \in F_2 \setminus \{1\}$ and suppose that neither u nor w is a proper power in F_2. Then the following theorem is a consequence of the proof of Lemma 6.5.2.

Theorem 6.5.3. *Under the above assumptions, if u and w are virtually conjugate in F_2, then either u and w are already conjugate in F_2 or*

$$b_3^{-1}u(b_1, b_2)b_3w(b_1, b_2)^{-1}$$

is primitive in

$$F_3 = \langle b_1, b_2, b_3; \ \rangle.$$

Now for each $u(b_1, b_2), w(b_1, b_2) \in F_2 \setminus \{1\}$ such that neither is a proper power in F_2 and w is not conjugate to u in F_2 and

$$b_3^{-1}u(b_1, b_2)b_3w(b_1, b_2)^{-1}$$

is not primitive in $F_3 = \langle b_1, b_2, b_3; \ \rangle$ set $G(u, w)$ to be the HNN group

$$G(u, w) = \langle b_1, b_2, b_3; w(b_1, b_2) = b_3^{-1}u(b_1, b_2)b_3 \rangle .$$

Corollary 6.5.2. *Under the above assumptions the images of b_1 and b_2 commute in any free homomorphic image of $G(u, w)$.*

Let $e_i(u)$ respectively $e_i(w)$ be the exponent sum in u (respectively in w) of b_i, $i = 1, 2$. Let $\Delta(u, w)$ be the determinant $\Delta(u, w) = e_1(u)e_2(w) - e_1(w)e_2(u)$.

Corollary 6.5.3. *If, in addition to the hypotheses of the previous corollary we assume that $\Delta(u, w) \neq 0$, then $G(u, w)$ has only Abelian free homomorphic images.*

Proof. Suppose $h\colon G(u, w) \to F$ is an epimorphism from $g(u, w)$ onto the free group F. Then f is generated by $h(b_i), i = 1, 2, 3$. Moreover $h(b_1), h(b_2)$ commute.

If $h(b_1) = h(b_2) = 1$, then $F = \langle h(b_3)\rangle$ is Abelian. Hence assume that $\langle h(b_1), h(b_2)\rangle = \langle c\rangle$ with $c \neq 1$. Let $h(b_i) = c^{\beta_i}, i = 1, 2$. Applying h to $w(b_1, b_2) = b_3^{-1}u(b_1, b_2)b_3$ we get

$$c^{e_1(w)\beta_1+e_2(w)\beta_2} = h(b_3)^{-1}c^{e_1(u)\beta_1+e_2(u)\beta_2}h(b_3).$$

If $h(b_3)$ commutes with c then F is Abelian. Suppose not. Since the centralizer $Z_F(c)$ is malnormal in F we infer the following system of two linear equations in the two unknowns β_1, β_2.

$$e_1(u)\beta_1 + e_2(u)\beta_2 = 0$$
$$e_1(w)\beta_1 + e_2(w)\beta_2 = 0$$

Since $\Delta(u, w) \neq 0$ this implies that $\beta_1 = \beta_2 = 0$ giving that $h(b_1) = h(b_2) = 1$ and contradicting that $c \neq 1$. From this contradiction we have that $h(b_3)$ commutes with c and hence F is Abelian. □

7 Fully residually free groups II

7.1 Fully residually free groups: limit groups

As we will see in the next chapter, fully residually free groups appear as the coordinate groups of solutions of certain systems of equations over free groups. Hence they will play a large role in the solution of the Tarski problems. In Sela's solution to the Tarski problems [233, 234, 235, 236, 237] finitely generated fully residually free groups are termed *limit groups* because they arise is his treatment via limits of sequences of homomorphisms into free groups. In this section we describe this approach and then prove that the two approaches are equivalent, that is a finitely generated group is a limit group if and only if it is fully residually free. After this we will briefly describe an equivalent geometric-topological definition of a limit group.

In Sela's approach the class of fully residually free groups arises in the study of $\mathrm{Hom}(G, F)$ where G is a finitely generated group and F is a free group. This arises in the study of equations over free groups in the following manner. Let

$$w_i(x_1, \ldots, x_n) = 1, \quad i = 1, \ldots$$

be a system of equations where the w_i are words in a free group $F = F(x_1, \ldots, x_n)$. Then the solutions are in one-to-one correspondence with $\mathrm{Hom}(G, F)$ where G is the group with the presentation

$$G = \langle x_1, \ldots, x_n; w_1 = 1, w_2 = 1, \ldots \rangle .$$

Definition 7.1.1. Let G be a finitely generated group and $F = F_\omega$ a free group of countable rank. A sequence $\{f_i\}$ in $\mathrm{Hom}(G, F)$ is *stable* if for all $g \in G$ the sequence $\{f_i(g)\}$ is eventually always $\{1\}$ or eventually never $\{1\}$. The *stable kernel* of $\{f_i\}$, which is denoted by $\underrightarrow{\mathrm{Ker}}\, f_i$, is

$$\{g \in G\colon\ f_i(g) = 1 \ \text{ for almost all } \ i\} .$$

A finitely generated group H is a *limit group* if there is a finitely generated group G and a stable sequence $\{f_i\}$ in $\mathrm{Hom}(G, F)$ such that

$$H = G/\underrightarrow{\mathrm{Ker}}\, f_i .$$

Limit groups can be viewed geometrically. For a stable sequence $\{f_i\}$ we can consider each f_i as acting on the Cayley graph of G. On a subsequence we can pass to an action on the limiting $\mathbb{R}$-tree (see Chapter 3). If the limiting tree is not a line, then $\underrightarrow{\mathrm{Ker}}\, f_i$ is the kernel of this action and so H acts faithfully on the resulting $\mathbb{R}$-tree. Later we will describe a more precise geometric description of limit groups.

The next theorem shows that the limit groups correspond exactly to the finitely generated fully residually free groups.

Theorem 7.1.1. *Let G be a finitely generated group. Then G is a limit group if and only if G is fully residually free.*

Proof. Suppose that G is a finitely generated fully residually free group. Let

$$S_1 \subset S_2 \subset \cdots \subset G$$

be a covering of G by an increasing sequence of finite sets of elements of G. Then since G is fully residually free for each i there is a homomorphism $f_i\colon G \to F$ which is injective on S_i. Since the S_i cover G this sequence is stable with trivial stable kernel so that $G = G/\underrightarrow{\mathrm{Ker}}\, f_i$. Therefore G is a limit group.

Conversely suppose that G is a limit group. Let H be a finitely generated group with stable sequence $\{f_i\}$ such that $G = H/\underrightarrow{\mathrm{Ker}}\, f_i$. Consider a sequence of epimorphisms

$$H = H_0 \to H_1 \to \cdots \to G$$

obtained by adding one relation at a time. Let H' be a group such that

$$\mathrm{Hom}(H', F) = \mathrm{Hom}(G, F).$$

Then all but finitely many of the homomorphisms f_i factor through H' since each added relation is trivialized by almost all f_i. Therefore all but finitely many f_i factor through G and each nontrivial element of G is trivialized by only finitely many f_i. This implies that if $g_1, \ldots, g_n$ are finitely many nontrivial elements in G then there must exist an f_i in the stable sequence such that $f_i(g_j) \neq 1$ for $i = 1, \ldots, n$. Therefore G is fully residually free. □

We have seen that any finitely generated fully residually free group lies in the class $\mathcal{F}$. In Sela's terminology this arises somewhat differently and they are called CLG's or *constructible limit groups*. The precise definition of a constructible limit group depends upon the generalized Abelian decomposition of a fully residually free group which in turn was a consequence of the construction of a fully residually free group as an iterated extension of centralizers. We will discuss this in more detail in Section 7.2. However here we point out that as a consequence of Theorems 6.4.1 and 6.6.1 we obtain the equivalence of limits groups and constructible limit groups.

Theorem 7.1.2. *Let G be a finitely generated group. Then G is a limit group if and only if G is a CLG.*

We now describe an equivalent geometric-topological definition of a limit group.

7.1.1 Geometric limit groups

Residual properties have their roots in pure group theory and purely algebraic considerations. The theorem of Benjamin Baumslag and especially the extensions by

Remeslennikov and Gaglione and Spellman are very surprising in that they tie together the algebraic residual properties and the logical property of universal freeness. Further though, from Sela's approach, limit groups can be considered geometrically. As we mentioned, for a stable sequence $\{f_i\}$, we can consider each f_i as acting on the Cayley graph of G. On a subsequence we can pass to an action on the limiting $\mathbb{R}$-tree (see Chapter 3). If the limiting $\mathbb{R}$-tree is not a line, then $\underrightarrow{\mathrm{Ker}}\, f_i$ is the kernel of this action and so H acts faithfully on the resulting $\mathbb{R}$-tree. Here we describe in more detail, geometric limit groups.

Definition 7.1.2. A *real tree* T is a geodesic metric space satisfying the condition

$$d(w,x) + d(y,z) \geq \min(d(w,y) + d(x,z), d(w,z) + d(x,y))$$

where d is the metric on T. It is a *real G-tree* if G is a group of isometries on T. If G has a fixed point on T then it is a trivial G-tree and T is a minimal G-tree if T contains no proper G-invariant subtree.

Lemma 7.1.1. *If T is a nontrivial real G-tree then it contains a unique minimal G-subtree.*

Now consider the set $\mathcal{A}(G)$ of all nontrivial minimal real G-trees. On $\mathcal{A}(G)$ we define a topology in the following manner. A sequence T_n (with corresponding metrics d_n) of real G-trees converges to a real G-free T with metric d if and only if for all $\epsilon > 0$ and finite subsets $K \subset T$, $P \subset G$ there exists $K_n \subset T_n$ and bijections $\mu_n \colon K_n \to K$ such that

$$|d_n(g\mu_n(x_n), \mu_n(y_n) - d(gx_n, y_n)| < \epsilon$$

whenever $x_n, y_n \in K_n$ and $g \in P$. This is called the *equivariant Gromov–Hausdorf topology* and can be interpreted in the following manner. As subtrees T_i of T get larger they coincide with T.

On $\mathcal{A}(G)$ identify the G-tree T with metric d with the G-tree T with metric λd for $\lambda > 0$. Let $P\mathcal{A}(G)$ be the set of equivalence classes of G-trees under the identification above.

Now let F be a free group. Fix a set of generators for F so that the Cayley graph Γ is a tree. If we take d to be the word metric then (Γ, d) is a real tree. If G is a group then a homomorphism $f \in \mathrm{Hom}(G, F)$ defines an action of G on (Γ, d). Hence the Cayley graph becomes a real G-tree. Call it T_f. Since $G/\ker(f)$ acts freely it follows that if f is nontrivial then T_f is nontrivial. For T_f, constructed in this manner, consider the equivalence class in $P\mathcal{A}(G)$ of the minimal G-invariant subtree for T_f. Denote this also by T_f. Now consider the space

$$\mathcal{T}(\mathcal{G}) = \text{closure of } \{T_f \colon 1 \neq f \in \mathrm{Hom}(G, F)\}$$

with respect to the equivariant Gromov–Hausdorf topology as defined above. It was proved by Paulin that this space is actually compact.

Now we can formally define a geometric limit group.

Definition 7.1.3. Let T be a real G-tree in $\mathcal{T}(G)$. Let $\ker(T)$ be the kernel of the action of G on T. Then a *geometric limit group* is any group of the form

$$G/\ker(T)$$

for such a T.

The following theorem describes some of the structure properties of geometric limit groups.

Theorem 7.1.3. *Let T_n be a sequence of nontrivial minimal G-trees arising from homomorphism $f_n\colon G \to F$ that converge to a nontrivial minimal G-tree T. Then:*
(a) *If T is not a line then* $\underrightarrow{\ker}(f_n) = \ker(T)$,
(b) *Stabilizers in* $G/\ker(T)$ *are free Abelian,*
(c) *T is a line if and only if f_n has a nontrivial Abelian image for all sufficiently large n.*

It follows from this theorem that the concepts of limit group and geometric limit group are actually equivalent.

Theorem 7.1.4. *Let G be a finitely generated group. Then G is a limit group if and only if G is a geometric limit group.*

Proof. If G is a finitely generated free Abelian group then G is both a limit group and a geometric limit group.

Suppose that G_1 is non-Abelian. Then from parts (a) and (c) of the theorem it follows that T is not a line. Then if G_1 is a geometric limit group $G_1 = G/\ker(T)$. From part (a) then $G_1 = G/\underrightarrow{\ker}(f_n)$ and so G_1 is a limit group. The converse follows in an analogous manner. □

In Sela's approach to the solution of the Tarski problem, the fact that the two notions of limit groups coincide, allowed for the use of the Rips machinery on $\mathbb{R}$-trees. In this approach we need the following ideas which we mention here.

Definition 7.1.4. A real G-tree is *stable* if for every descending sequence of nondegenerate subtrees

$$T \supset T_1 \supset T_2 \supset \dots$$

the corresponding sequence of point stabilizers

$$\operatorname{Stab}_G(T_1) \subset \dots$$

eventually stabilizes.

A real G-tree is *very small* if it is nontrivial, minimal, stable and has nondegenerate Abelian arc stabilizers.

Theorem 7.1.5. *For a geometric limit group $G_1 = G/\ker(T)$ the tree T is a very small G_1-tree.*

7.2 JSJ-decompositions and automorphisms

Each finitely generated fully residually free group G has a finite graph of groups decomposition with Abelian edge groups. We say that G has an *Abelian splitting*. This result is an immediate consequence of the construction of a finitely generated fully residually free group as a subgroup of iterated extensions of centralizers. Further the edge groups can be taken to be cyclic.

In this section we show that there is a cyclic splitting of a finitely generated fully residually free group that is canonical among cyclic splittings and can be used to study automorphisms of such groups and solve the isomorphism problem. Rips and Sela [217] introduced a special type of cyclic splitting of a group, called a JSJ decomposition, which contains in an appropriate interpretation all other cyclic splittings of the group. He showed that JSJ decompositions exist for torsion-free hyperbolic groups and used this to solve the isomorphism problem for such groups. Rips and Sela [217] using group actions on $\mathbb{R}$-trees determined JSJ decompositions for arbitrary finitely presented groups. Other authors (see [32, 217]) have looked at JSJ decompositions for different classes of groups. JSJ decompositions of groups are analogs of the characteristic submanifold constructions for irreducible 3-manifolds described by Jaco and Shalen [136, 137, 138] and Johannson [139]. This is the genesis of the name JSJ.

In this section we will describe the JSJ decomposition of a finitely generated fully residually free group due to Kharlampovich and Myasnikov [155]. They also give an effective algorithm to construct the JSJ decomposition of a freely indecomposable fully residually free group.

The existence of an effective JSJ decomposition allows one to both solve the isomorphism problem for finitely generated fully residually free groups and to describe the automorphism group. We will also prove that for a finitely generated fully residually free group G the automorphism group $\mathrm{Aut}(G)$ is tame (see Section 7.2.2).

Before stating the main result we must give several relevant definitions, some of which have already been touched upon.

A group G is *freely indecomposable* if it does not split as a nontrivial free product. A *splitting* for G is a nontrivial graph of groups decomposition for G. It is an *Abelian splitting* if each edge group is Abelian and a *cyclic splitting* if each edge group is cyclic. A *one-edge splitting* is a splitting with a single edge. Hence a one edge splitting of G is either a free product with amalgamation $G = A *_C B$ or an HNN group with a single associated subgroup. If A is the base and C is the associated subgroup we denote this latter construction by $G = A*_C$. Since fully residually free groups are built up from extensions of centralizers it follows that for a fully residually free group one-edge Abelian splittings are particularly simple.

Theorem 7.2.1. *Let G be a finitely generated fully residually free group, Consider a one edge Abelian splitting of G. Suppose that M is a noncyclic maximal Abelian subgroup of G. Then*

(1) *If $G = A *_C B$ then M is conjugate to a subgroup of either A or B.*
(2) *If $G = A*_C$ and M is not conjugate into A then there is a conjugate M^g such that*

$$G = A *_C M^g.$$

Proof. Since G is fully residually free it is CSA and hence commutative transitive. Suppose first that $G = A *_C B$ with C Abelian. Since G is a free product with amalgamation G acts on a Bass–Serre tree $\mathcal{T}$ (see Chapter 3) with vertex stabilizers A, B. Let M be a maximal Abelian subgroup of G and suppose that M is not conjugate into either A or B. Since G is CSA the subgroup M is malnormal. Since M is Abelian it follows that M either fixes a line $\mathcal{L}$ in $\mathcal{T}$ or a point on the boundary (see [29]). If M fixed a point on the boundary there would be an increasing sequence of edge stabilizers

$$C_1 \subset C_2 \subset \cdots \subset M.$$

However M is malnormal and each C_i is conjugate to C hence a contradiction. Therefore M fixes a line $\mathcal{L}$ in $\mathcal{T}$. We call $\mathcal{L}$ the *axis* of M. Conjugating if necessary the amalgamated subgroup C must then be the stabilizer of an edge in the axis. But then M acts as $M = M' \oplus \mathbb{Z}$ where M' fixes the axis. Since M' fixes the axis it follows that $M' \subset C$ and hence by commutative transitivity and the maximality of M we have $C \subset M$. Therefore C fixes the whole axis. However there is only one orbit of edges so there exists an $a \in A \setminus C$ with $a^{-1}Ca = C$. Since M is malnormal we have $a \in M$ a contradiction. Therefore in the free product with amalgamation case M must be conjugate to a subgroup of either A or B.

In the HNN case suppose that $G = A*_C$ an HNN group with base A. As in the first case G acts on a Bass–Serre tree. Let M be a maximal Abelian subgroup of G which is not conjugate into A. As in the first case M must preserve a line in the tree. The associated subgroup C could then be assumed to lie in M and fix this whole line and hence $A \cap M = C$. If t is the stable letter in the HNN extension then we can assume without loss of generality that $t^{-1}Ct = C$. Then as before since M is malnormal in G it would follow that $t \in M$ a contradiction. Then if M is not conjugate into A there is a conjugate M^g with $G = A *_C M^g$. □

The following types of subgroups play a crucial role in the JSJ decompositions of fully residually free groups. First in a graph of groups a vertex is *flexible* if there is a refinement of the graph replacing this vertex by another non-trivial graph of groups. Otherwise the vertex is *rigid*.

Definition 7.2.1 (QH-vertex). Let P be a group which admits one of the following presentations
(1) $\langle p_1, \dots, p_m, a_1, b_1, \dots, a_g, b_g; \prod_{k=1}^m p_k \prod_{i=1}^g [a_i, b_i] = 1 \rangle$,
(2) $\langle p_1, \dots, p_m, v_1, \dots, v_g; \prod_{k=1}^m p_k \prod_{i=1}^g v_i^2 = 1 \rangle$.
Suppose $\Gamma(V, E)$ is a splitting of a group G as a graph of groups and suppose $P = G_v$ where $v \in V$. Then G is a *QH-vertex* if $e_1, \dots e_n$ are the all edges with initial vertex v then $\alpha(e_i) = p_i$ where α is the boundary monomorphism taking G_{e_i} into G_v (see Chapter 3).

QH stands for *quadratically hanging*. Notice that a QH-vertex is a free group and geometrically is the fundamental group of a punctured surface.

Definition 7.2.2 (QH-subgroup). A subgroup P of G is a *QH-subgroup* if there is a splitting of G as a graph of groups and a QH-vertex G_v such that P can be conjugated into a stabilizer of G_v. It is a *maximal QH-subgroup* or *MQH-subgroup* if for each elementary cyclic splitting $G = G_1 *_C G_2$ either P can be conjugated into G_1 or G_2 or C can be conjugated into P in such a way that there is an elementary splitting of P over a cyclic subgroup C_1 so that this splitting extends to an elementary splitting of the whole group G and C is hyperbolic with respect to the splitting of G over C_1.

We need a few more concepts and then we can define a JSJ decomposition. Let G have a one-edge splitting Γ. Then an element $g \in G$ is *elliptic* if g is conjugate into a vertex group in Γ. We say that g acts elliptically on the Bass–Serre tree. If Δ is a generalized Abelian decomposition of G then an element $g \in G$ is Δ-elliptic if
(1) g is conjugate into a vertex of Δ,
(2) if the vertex is a surface vertex then g is conjugate to a boundary component,
(3) if the vertex is Abelian then g is conjugate to a periperal subgroup.

Now we define a JSJ decomposition.

Definition 7.2.3. Let G be a finitely generated group. Let $\mathcal{A}$ be the class of all 1-edge splittings of G satisfying
(JS 1)the edge group is Abelian,
(JS 2)the edge group E is *isolated* in G. That is for each integer $n > 1$ and for each $g \in G$ if $g^n \in E$ then $g \in E$,
(JS 3)every noncyclic Abelian subgroup $A \subset G$ is elliptic.
Then a *JSJ Decomposition* for G is a generalized Abelian decomposition whose set of elliptic elements coincides with the intersection of all elliptics in $\mathcal{A}$.

The relevant fact for fully residually free groups is the following result due to Kharlampovich and Myasnikov and Sela (see [39] and the references there, [233, 234, 235, 236, 237]).

Theorem 7.2.2.
(a) *A finitely generated freely indecomposable fully residually free group which is indecomposable relative to JSJ-decompositions is either the fundamental group of a closed surface, a free group or a free Abelian group.*
(b) *A finitely generated freely indecomposable fully residually free group admits a nontrivial cyclic JSJ-decomposition if it is not Abelian or a surface group.*

We need a more technical version of the above result.

The following is a modified version of Theorem 3.13 in [39] which summarizes the most relevant facts about JSJ splittings for fully residually free groups.

Theorem 7.2.3 ([39]). *Let G be a finitely generated fully residually free group not Abelian or a surface group. Then G has a cyclic JSJ-decomposition Γ that satisfies the following properties*

(1) *The decomposition Γ is universal in that given a minimal G-tree T one can find refinements at flexible vertices and obtain a decomposition Γ_r of G so that there exists a G-equivariant simplicial map from the Bass–Serre tree of Γ_r onto T.*
(2) *The Bass–Serre tree corresponding to Γ is 2-acylindrical (see [39]).*
(3) *A maximal Abelian vertex is called an elementary vertex.*
(4) *The graph (V, E) of Γ is a bipartite graph: two elementary vertices and two non-elementary vertices are never joined by an edge $e \in E$.*
(5) *Each flexible vertex is a maximal QH-vertex. Let $G = G_1 *_C G_2$ or $G = G_1 *_C$ be a cyclic splitting of G. Then C can be conjugated into the stabilizer of a flexible vertex of Γ if and only if the splitting in question is hyperbolic with respect to another splitting of G.*

Moreover there is an algorithm to determine the JSJ decomposition Γ.

7.2.1 Automorphisms of fully residually free groups

The existence of the JSJ decomposition tightly controls the structure of automorphisms and the structure of the automorphism group of a finitely generated fully residually free group.

From Theorem 7.2.2, Bumagin, Kharlampovich and Myasnikov deduce the following result which rigidly controls the structure of automorphisms.

Theorem 7.2.4 ([39]). *Let G be a finitely generated freely indecomposable fully residually free group not Abelian or a surface group and let Γ be the cyclic JSJ-decomposition satisfying the properties of Theorem 7.2.2. Suppose that Δ is another cyclic JSJ-decomposition of G satisfying the same properties. Then Δ can be obtained from Γ by conjugations and modifying boundary homomorphisms.*

From this result the following is clear.

Corollary 7.2.1. *Let G be a finitely generated fully residually free group, not Abelian or a surface group, and let Γ be the cyclic JSJ-decomposition satisfying the properties of Theorem 1.7.3. Then an automorphism α of G must take vertices to isomorphic vertices of the same type, that is flexible QH-vertices to isomorphic QH-vertices and rigid (free or free Abelian) vertices to ridid vertices.*

A fundamental idea is the existence of *canonical automorphisms*. Since free groups, free Abelian groups and surface groups all have tame automorphism groups, that is their automorphisms all come from free group automorphisms (see next section) we will assume throughout that our fully residually free group G is not Abelian or a surface group. Let G then be a finitely generated fully residually free group and

$\Gamma = \Gamma(V, E)$ a cyclic JSJ-decomposiiton for G. By an *e-automorphsim of a vertex group* G_v we mean an outer automorphism of G_v that maps each edge subgroup of G_v into a conjugate of itself.

Definition 7.2.4. Let $\Gamma(V, E)$ be an Abelian splitting of a group G and let $e \in E$ be an edge with endpoints $i(e) = v, \tau(e) = u$ and edge stabilizer G_e. A *generalized Dehn twist along the edge e* is an automorphism $\beta_a : G \to G$ with $a \in C_{G_v}(G_e)$ satisfying

(1) If e is a separating edge, let Δ_v (or Δ_u) denote the connected component of $(V, E) \setminus \{e\}$ that contains v or u. Then $\beta_a(g) = g$ for $g \in G_w$ with $w \in \Delta - v$ and $\beta_a(g) = aga^{-1}$ for $g \in G_w$ with $w \in \Delta_u$.
(2) If e is a non-separating edge, then one can choose a maximal tree T in (V, E) so that $e \notin T$. Let t be the stable letter that corresponds to e. Then $\beta_a(t) = at$ and $\beta(g) = g$ for all $g \neq t$.

Now we define the canonical automorphisms of a finitely generated fully residually free group G. Throughout we assume that G is not Abelian or a surface group.

Definition 7.2.5. Let G be a finitely generated fully residually free group not Abelian or a surface group and let Γ be the cyclic JSJ-decomposition satisfying the properties of Theorem 7.2.1. Then a *canonical automorphism* of G is one of the following

(1) a generalized Dehn twist along an edge in Γ,
(2) an e-automorphism of an elementary vertex group,
(3) an e-automorphism of a flexible QH-vertex group.

The following ideas are crucial in studying both the structure of Aut(G) for a fully residually free group G and for considering the equivalence of fully residually free gorups and constructible limit groups.

Definition 7.2.6. Let G be a fully residually free group and Δ a generalized Abelian decomposition for G. Then the associated modular group Mod(Δ) is the subgroup of Aut(G) generated by

(1) inner automorphisms of G,
(2) Dehn twists of edges of Δ,
(3) unimodular automorphisms of Abelian vertices A that are the identity on the periperal subgroup,
(4) automorphisms of QH-vertices that arise from automorphisms of the underlying surface that fix boundary components.

The modular group of G denoted Mod(G) is the group of G generated by Mod(Δ) for all generalized Abelian decompositions of G.

Definition 7.2.7. Let G be a fully residually free group and S a fixed generating system for G. Let $f : G \mapsto F$ with F a free group. The length of f is

$$|f| = \max_{g \in S} |f(g)|$$

where $|\cdot|$ is the word metric length in F. A homomorphism $f\colon G \mapsto F$ is *short* if its length is minimal in its equivalence class.

The following was proved by Rips and Sela [217]:

Theorem 7.2.5. *Suppose that G is freely indecomposable and that f_i is a convergent sequence of homomorphisms from G to a free group F. Then* $\underrightarrow{\mathrm{Ker}} f_i \neq \{1\}$.

In analyzing the structure of Aut(G), Bumagin, Kharlampovich and Myasnikov (see [39]), proved the following.

Theorem 7.2.6 ([39]). *Let G be a finitely generated fully residually free group not Abelian or a surface group and let Γ be a cyclic JSJ-decomposition for G. Let* $\mathrm{Out}_\Gamma(G)$ *be the outer automorphism group of G generated by the canonical automorphisms. Then* $\mathrm{Out}_\Gamma(G)$ *has finite index in* $\mathrm{Out}(G)$.

These structure results can be used to prove that the automorphism group of a finitley generated fully residualy free group is tame (see the next section).

7.2.2 Tame automorphism groups

A minimal finite presentation of a finitely presented group G is a presentation that is minimal with respect to the number of generators. Hence a presentation $G = \langle x_1, \ldots, x_n; r_1 = \cdots = r_m = 1\rangle$ is a minimal finite presentation for G if $n = \mathrm{rank}(G)$ the minimal number of generators necessary to present G. Now suppose that $G = \langle x_1, \ldots, x_n; r_1 = \cdots = r_m = 1\rangle$ with $1 \le n, m < \infty$ is minimal finite presentation of G. Let $F = \langle x_1, \ldots, x_n; \ \rangle$ be the free group of rank n on $\{x_1, \ldots, x_n\}$. An automorphsim $\alpha\colon G \to G$ is *tame* if it is induced by or lifts to an automorphism on F (considered as the generators of G). If each automorphism of G is tame we say that the automorphism group Aut(G) is *tame*. In [125] Gupta and Shpilrain give a survey of some of the known general results on tame automorphisms and tame automorphism groups. If G is a surface group, a result of Zieschang [258], and improved upon by Rosenberger [219, 220], shows that G has only one Nielsen class of minimal generating systems. An easy consequence of this is that Aut(G) is tame. Rosenberger (see [220]) uses the term *almost quasifree* for a finitely presented group which has a tame automorphism group. If G is almost quasifree, $G = \langle x_1, \ldots, x_n; r_1 = \cdots = r_m = 1\rangle$, $1 \le n, m < \infty$ a minimal finite presentation of G and, in addition, each automorphism of $F = \langle x_1, \ldots, x_n; \ \rangle$ induces an automorphism of G, G is called *quasifree*. Rosenberger observed that a non-cyclic, nonSfree one-relator group is quasifree only if it has a presentation $\langle a, b; [a, b]^n = 1\rangle$ for $n \ge 1$. This is a Fuchsian group if $n \ge 2$ and isomorphic to a free Abelian group of rank 2 if $n = 1$.

As a consequnce of the JSJ decomposition of a fully residually free group we can prove the following.

Theorem 7.2.7. *The automorphism group* $\mathrm{Aut}(G)$ *of a finitely generated freely indecomposable fully residually free group* G *is tame.*

Proof. If G is either free Abelian or a surface group then as mentioned its automorphism group is tame. Therefore we may assume that G is neither Abelian nor a surface group and hence has a non-trivial cyclic JSJ decomposition as in Theorem 7.2.2. Fix a presentation for this JSJ decomposition say on generators $x_1, \ldots, x_n$. We may assume that algebraically this generating system is minimal. This set of generators will then be partitioned into sets of generators for the various vertices.

First we need the following.

Lemma 7.2.1. *Let* G *be a finitely generated fully residually free group not Abelian or a surface group and let* Γ *be a cyclic JSJ-decomposition for* G*. Then a canonical automorphism relative to this JSJ decomposition is tame.*

Proof of Lemma 7.2.1. An e-automorphism of an elementary vertex group is just an automorphism of a free Abelian group which is tame. An e-automorphism of a flexible vertex is an e-automorphism of a QH-vertex. Geometrically the e-automorphisms of QH-vertices are Dehn twists along simple closed curves on the punctured surface Σ with $\pi_1(\Sigma) = G_u$. These can then be pictured as surface group automorphisms which are tame or free group automorphisms which are tame. Finally a generalized Dehn twist along an edge, since the edges are cyclic is just a free group automorphism which is tame. □

Clearly any conjugation and hence any inner automorphism is tame. Let $\mathrm{Aut}_\Gamma(G)$ be the subgroup of $\mathrm{Aut}(G)$ generated by the canonical automorphisms and the inner automorphisms. Since products of tame automorphisms are clearly also tame it follows as a corollary of Lemma 7.2.1.

Corollary 7.2.2. $\mathrm{Aut}_\Gamma(G)$ *is tame.*

We now complete the proof of Theorem 7.2.1. Suppose $\phi\colon G \to G$ is any automorphism. From Corollary 7.2.1, ϕ must map vertices to isomorphic vertices of the same type, that is QH-vertices to isomorphic QH-vertices, elementary vertices to isomorphic elementary vertices and free vertices to isomorphic free vertices. This automorphism induces also an automorphism of a rigid group sending edge groups to conjugates of edge groups. Delzant and Potyagailo have proved that a finitely presented group admits a finite hierarchy obtained by successive splittings along a family of elementary subgroups (see [63]). The automorphism given above is then tame by induction on the number of steps in the Delzant–Potyagailo hierarchy. We have fixed a presentation for G in terms of the JSJ decomposition. Suppose that

$$H = \langle h_1, \ldots, h_n \rangle$$

is a vertex group with

$$\phi(H) = \langle \phi(h_1), \ldots, \phi(h_n) \rangle = \langle \alpha_1, \ldots, \alpha_n \rangle$$

where $\alpha_1, \ldots, \alpha_n$ are the generators for $\phi(H)$ in the fixed canonical presentation for G. Then by conjugations and canonical automorphisms on $\phi(H)$ we can reexpress the image $\phi(H)$ in terms of the canonical generators $\alpha_1, \ldots, \alpha_n$. That is modulo conjugations and canonical automorphisms, ϕ is just a permutation ϕ_π of the original set of generators. Hence

$$\phi = \phi_\pi \phi_c$$

where ϕ_c is in $\mathrm{Aut}_\Gamma(g)$ and ϕ_π is a permutation of the original set of generators. ϕ_π is tame since permutations of generators are Nielsen transformations and hence free group automorphisms. ϕ_c is tame since it is in $\mathrm{Aut}_\Gamma(G)$. Therefore ϕ is tame completing the proof of the theorem. □

Since each elementary free group is universally free and hence fully residually free the proof of the next corollary is immediate.

Corollary 7.2.3. *The automorphism group of a finitely generated freely indecomposable elementary free group G is tame.*

We note that the converse of Theorem 7.2.1 is false. That is there do exist groups (in fact hyperbolic groups) where every automorphism is tame but which are not fully residually free. As an example the groups

$$G = \langle a_1, \ldots, a_n; a_1^{\alpha_1} \ldots a_n^{\alpha_n} = 1 \rangle$$

with $n \geq 3, 2 \leq \alpha_1, \ldots, \alpha_n$, and

$$G = \langle s_1, \ldots, s_n; s_1^2 = \cdots = s_{n-1}^2 = s_n^{2k+1} = s_1 \ldots s_n = 1 \rangle$$

with $n = 2\ell$ even and $k \geq 1$, are all hyperbolic. Further every automorphism is tame (see [219, 220] and [104]). However not all of these groups are fully residually free.

7.2.3 The isomorphism problem for limit groups

As an application of the study of automorphisms of fully residually free groups, Bumagin, Kharlampovich and Myasnikov [39] were able to prove that the isomorphism problem is solvable for finitely generated fully residually free groups. This is actually part of the study of the algorithmic study of this class of groups. In particular Kharlampovich, Myasnikov, Remeslennikov and Serbin [158] were able to translate the method of Stallings foldings to fully residually free groups by considering *infinite words* in $F^{\mathbb{Z}[t]}$. Doing this they were able to algorithmically solve many problems in fully residullay free groups mirroring the algorithmic solutions in absolutely free groups (see [158]). We mention here the isomorphism problem and discuss the algorithmic theory in more detail in Section 7.3.

Theorem 7.2.8. *The isomorphism problem is solvable in the class of finitely generated fully residually free groups. That is given two finite presentations, that are known to define fully residually free groups, there is an effective algorithm to determine if the defined groups are isomorphic.*

7.2.4 Constructible limit groups

In this chapter we outlined the construction of a finitely generated fully residually free group as a subgroup of an iterated extensions of centralizers starting with free groups. This allowed for inductive arguments on the level of a fully residually free group in studying various properties. This was utilized in Section 6.5 to determine the complete structure of fully residually free groups of rank 3 or less. In Sela's approach this hierarchy arises somewhat differently in terms of what are called *constructible limit groups* or *CLGs*.

Definition 7.2.8. The class of *constructible limit groups* is defined inductively as follows:
(1) The level 0 groups consists of the finitely generated free groups.
(2) A group G is of level n if and only if either
  (a) $G = G_1 * G_2$ with G_1 and G_2 groups of level $\leq n-1$ or
  (b) there exists a homomorphism $\rho\colon G \mapsto G'$ with G' of level $\leq n-1$ and further G has a generalized Abelian decomposition such that
    (i) ρ is injective on the peripheral subgroup of each Abelian vertex,
    (ii) ρ is injective on each edge group E and at least one of the images of E in a vertex group of the one-edged splitting induced by E is a maximal Abelian subgroup,
    (iii) the image of each QH-vertex is a non-Abelian subgroup,
    (iv) for each rigid vertex group B the homomorpshism ρ is injective on the envelope $\overline{B}$ of B. This is defined by first replacing each Abelian vertex with the periperal subgroup and then letting $\overline{B}$ be the subgroup of the resulting group generated by G and by the centralizers of incident edge groups.

A CLG is any group that falls into this class.

As a straightforward example we have that an orientable surface group of genus 2 is a CLG. The orientable surface can be viewed as the union of two punctured tori. The retraction to one of the two tori determines the homomorphism ρ in the definition. When we show that CLG's are equivalent to limit groups this provides another proof that surface groups are fully residually free. As an extension of this consider a Baumslag double

$$G = F \underset{W=\overline{W}}{*} \overline{F}$$

where W is maximal cyclic. Then both vertices are rigid and there is a retraction $G \mapsto F$ satisfying the definition. Hence G is a CLG. As in the surface group case this provides another proof that Baumslag doubles are fully residuallly free.

The hierarchy in the definition of a CLG provides an inductive way to study this class. The following results can all be proved by using induction on the level of a CLG.

Lemma 7.2.2. *A CLG G must be finitely presented.*

Lemma 7.2.3. *A finitely generated subgroup H of a CLG G is also a CLG.*

Recall that a group is *coherent* if any finitely generated subgroup must be finitely presented. Both Kharlampovich and Myasnikov and Sela proved that fully residually free groups must be coherent. This follows as a corollary of the above two lemmas.

Corollary 7.2.4. *A CLG G is coherent.*

The main result is that being a CLG is equivalent to being a limit group and hence being fully residually free.

Theorem 7.2.9. *For a finitely generated group G the following are equivalent:*
(a) *G is a CLG,*
(b) *G is a limit group,*
(c) *G is fully residually free.*

Proof. Since CLG's are extensions of centralizers starting with free groups it follows that a CLG G must be fully residually free. We have already seen that being fully residually free implies that G must be a limit group. We show that if G is a limit group then G must be a CLG.

Let G be a limit group with fixed free group F and let S be a generating set for G. Let f_i be a sequence of homomorphisms from G to F with f_i injective on elements of length at most i in the word metric relative to S. Then the stable kernel of the f_i is trivial. Choose $\widehat{f}_i$ to be short maps equivalent to f_i so then the map

$$\rho\colon G \to G' = G/\underset{\to}{\ker}$$

is an epimorphism. Assume for an inductive argument that G' is a CLG. Let Δ be an Abelian decomposition for G. We show that relative to Δ the map ρ satisfies the conditions to make G a CLG.

Let E be an edge group of the JSJ decomposition. Then it is elliptic in every one-edge splitting so all generalized Dehn twists conincide with some inner automorphism on E. It follows that if $g \in E \setminus \{1\}$ the image $\widehat{f}_i(g)$ is conjugate to $f_i(g)$. Since the stable kernel is trivial this must be nontrivial for all sufficiently large i so ρ restricted to E is injective. Further E must be maximal Abelian in a vertex of the associated one-edge splitting from the CSA property of limit groups.

Abelian vertices are free and finitely generated. Let $P(A)$ be the periperal subgroup. As for the edge group $P(A)$ is elliptic in all relevant one-dege splittings and hence Mod(G) acts as inner automorphisms and therefore ρ restricted to $P(A)$ is injective.

Now let S be a QH-vertex and suppose that $\rho(S)$ is Abelian. Then for all sufficiently large i, $\widehat{f_i}(S)$ is Abelian. Every element of Mod(G) maps S to a conjugate of itself so eventually $f_i(S)$ is nontrivial Abelian contradicting the triviality of the stable kernel.

Finally the envelope of a rigid vertex B is elliptic in every splitting of G and so is preserved up to conjugacy by Mod(G), hence ρ restricted to B is injective. Therefore ρ and Δ satisfy the conditions for G to be a CLG. □

7.2.5 Factor sets and MR-diagrams

The following ideas are crucial to the solution of the Tarski problems. The language in this section for the most part follows from Sela's approach. The ideas originate in work of Makanin and Razborov concerning solutions of systems of equations over free groups.

Definition 7.2.9. A *factor set* for a group G is a finite collection of proper quotients of G, $\{\rho_i : G \to G_i\}$ such that if $f \in \hom(G, F)$ for a fixed non-Abelian free group F then there exists an $\alpha \in \mathrm{Mod}(G)$ such that $f\alpha$ factors through some ρ_i.

The main result is that any finitely generated nonfree group has a factor set where each quotient is a limit group.

Theorem 7.2.10. *Let G be a finitely generated nonfree group. Then G has a factor set $\{\rho_i : G \to G_i\}$ with each G_i a limit group. If the group G is not itself a limit group then we can always take α as the identity map.*

A fairly straightforward proof of this theorem using work of Sela can be found in [29] and depends on the following restricted form of the theorem.

Theorem 7.2.11. *Let G be a finitely generated torsion-free, nonfree limit group that is freely indecomposable, non-Abelian and not a surface group. Then G has a factor set.*

The main Theorem 7.2.10 implies that if construction of the factor set is iterated for each limit group G_i what will be obtained is a finite tree of groups terminating in free groups. This tree of groups and epimorphisms is called a *Makanin–Razborov Diagram* or *MR-digaram* for G (with respect to the fixed non-Abelian free group F). If

$$G \xrightarrow{\rho} G_1 \xrightarrow{\rho_1} G_2 \to \dots \xrightarrow{\rho_{m-1}} G_m$$

is a branch of a MR-diagram and if $f \in \mathrm{Hom}(G, F)$ then f *MR factors* through this branch if there exists $\alpha \in \mathrm{Mod}(G)$, that is the identity of G is not a limit group, $\alpha_i \in \mathrm{Mod}(G_i)$ for $1 \le i \le n$ and $f_m \in \mathrm{Hom}(G_m, F)$ such that $f = f_m\rho_{m-1}\cdots\rho_1\alpha_1\rho\alpha$.

The crucial property of an MR-diagram for G is that given $f \in \mathrm{Hom}(G, F)$ there is a branch of the diagram through which f factors (see [29]).

7.3 Faithful representations of limit groups

It is well known that orientable surface groups of genus ≥ 2 have faithful representations in $PSL(2, \mathbb{C})$. Further the theory of Fuchsian and Kleinian groups occurs also within $PSL(2, \mathbb{C})$. In [212] it was proved that surface groups can be embedded in complex semisimple Lie groups. In this section we prove that any limit group has a faithful representation in $PSL(2, \mathbb{C})$. Hence the full theory of fully residually free groups occurs within this one amazing group. We give two proofs of this result. The first is a short elegant proof that depends on the fact that any limit group can be embedded in ${}^*F$ where ${}^*F$ is the ultrapower of F the free group of countable rank. This representation is constructible which in turns depends on the fact that a limit group can be effectively embedded into the Lyndon completion $F^{\mathbb{Z}[t]}$. After proving these we mention a separate theorem due to Wilton [255] showing that any limit group can actually be embedded in $PSL(2, \mathbb{R})$ and in fact can be embedded in any algebraic group over $\mathbb{R}$ in which a countably generated free group can be embedded.

The second proof depends on showing that certain amalgams of groups admitting faithful representations into $PSL(2, \mathbb{C})$ also admit such faithful representations. These results were interesting their own right. Originally this was proved with the restriction to hyperbolic limit groups, however this restiction is unnecessary. The second proof depends upon the JSJ-decomposition of a limit group.

Theorem 7.3.1. *Let G be a limit group. Then G has a faithful representation $\rho: G \to PSL(2, \mathbb{C})$. Further a faithful representation can be effectively constructed.*

We need the following results (see Section 6.4).

Theorem 7.3.2 (see Section 6.4). *Any non-Abelian limit group can be embedded in a nonstandard free group F^*.*

Theorem 7.3.3 (see Section 6.4). *Let G be a finitely generated group and $F = F_\omega$ be a free group of countably infinite rank. Then G is fully residually free and hence a limit group if and only if G can be embedded as a subgroup of $F^{\mathbb{Z}[t]}$. Further the embedding of G into $F^{\mathbb{Z}[t]}$ is effective.*

The final ingredient we need for the main proof is the following.

Lemma 7.3.1. *Any countable free group F can be embedded into both $SL(2, \mathbb{C})$ and $PSL(2, \mathbb{C})$.*

We now give the proof of Theorem 7.3.1 which is very elegant.

Proof. Let G be a limit group. Then as pointed out in Theorem 7.3.2 G can be embedded in ${}^*F$ where ${}^*F$ is a nonstandard free group, that is an ultrapower of F, a fixed free group of countable rank.

A countable free group F can be embedded into $SL(2, \mathbb{Q})$. Hence the ultrapower ${}^*F$ can be embedded into the ultrapower ${}^*SL(2, \mathbb{Q}) = SL(2, {}^*\mathbb{Q})$. Hence G can be embedded into $SL(2,^* \mathbb{Q})$. However since G is finitely generated it follows that G can be embedded in a finite power of $SL(2,\mathbb{Q})$ and hence into $SL(2,\mathbb{C})$. We note that it is also true from the same argument that G can be embedded into $SL(2,\mathcal{F})$ where $\mathcal{F}$ is any algebraicially closed field of characteristic zero but this doesn't concern us here.

Hence G can be embedded into $SL(2,\mathbb{C})$ and we must show that this faithful representation can be pushed down to $PSL(2,\mathbb{C}) = SL(2,\mathbb{C})/Z(SL(2,\mathbb{C})$ where $Z(SL(2,\mathbb{C})$ is its center. However the center of $SL(2,\mathbb{C})$ is precisely $\{I,-I\}$ where I is the 2×2 identity matrix. A limit group is torsion-free so the image of G in $SL(2,\mathbb{C})$ cannot contain $-I$. Therefore G embeds into $PSL(2,\mathbb{C})$ completing the first part of the result.

Now we must show that this embedding is constructive knowing the JSJ decomposition of the limit group G. As described in Section 2, the Lyndon free exponential group $H = F^{\mathbb{Z}[t]}$ is fully residually free and so embeds into $PSL(2,\mathbb{C})$ from the first part. This is a fixed free exponential group so choose one fixed embedding of H into $PSL(2,\mathbb{C})$. Let G be a limit group. Then given knowledge of the JSJ decomposition of G there exists an effective procedure for embedding G into H the free Lyndon exponential group. Combining this embedding with the fixed embedding of H into $PSL(2,\mathbb{C})$ provides the effective faithful representation of G. □

We now give a second proof which uses the JSJ-decomposition of a limit group. We do this first with the restriction that the limit group be hyperbolic and hence have cyclic centralizers. We will then point out that this restriction is really unnecessary. This proof depends on the faithful representation of amalgams of groups having faithful representations into $PSL(2,\mathbb{C})$. These results are interesting in their own right.

Theorem 7.3.4. *Let G be a hyperbolic limit group. Then G has a faithful representation $\rho\colon G \mapsto PSL(2,\mathbb{C})$.*

The proof follows directly from the following sequence of lemmas together with the structure theory of limit groups.

Lemma 7.3.2. *Let H_1, H_2 be groups with cyclic centralizers and u_1, u_2 maximal cyclic elements in H_1, H_2 respectively. Suppose that H_1, H_2 admit faithful representations in $PSL(2,\mathbb{C})$ then the amalgamated product $H = H_1 \underset{u_1=u_2}{*} H_2$ also admits a faithful representation in $PSL(2,\mathbb{C})$.*

This lemma is a direct consequence of a result in [92].

This lemma implies that free products of groups with cyclic centralizers admitting faithful representations in $PSL(2,\mathbb{C})$ also admit a faithful representation in $PSL(2,\mathbb{C})$.

Corollary 7.3.1. *Free products of groups with cyclic centralizers admitting faithful representations in $PSL(2,\mathbb{C})$ also admits a faithful representation in $PSL(2,\mathbb{C})$.*

Next we have an extension of centralizer analog of the above.

Lemma 7.3.3. *Let G be torsion-free, finitely generated and admit a faithful representation into $PSL(2,\mathbb{C})$. Suppose further that G has cyclic centralizers. Let $\langle w\rangle$ be a maximal cyclic subgroup of G. Then the extension of centralizers*

$$\Gamma = \langle G, t; \mathrm{rel}(G), twt^{-1} = w\rangle$$

admits a faithful representation into $PSL(2,\mathbb{C})$.

Proof. Let G be finitely generated and admit a faithful representation into $PSL(2,\mathbb{C})$, let $\langle w\rangle$ be a maximal cyclic subgroup of G and let

$$\Gamma = \langle G, t; \mathrm{rel}(G), twt^{-1} = w\rangle.$$

We claim that if the centralizer of w is $\langle w\rangle$ then $\langle w\rangle$ is malnormal in G. Suppose that $gw^m g^{-1} = w^n$. Using the faithful representation into $PSL(2,\mathbb{C})$ it would follow that g and w have the same fix points as mappings on $\mathbb{C}\cup\{\infty\}$ and hence g, w commute. It follows that $g \in \langle w\rangle$ and hence $\langle w\rangle$ is malnormal in G.

Choose a representation ϕ into $PSL(2,\mathbb{C})$ so that

$$\phi(w) = \begin{pmatrix} \alpha & 0 \\ 0 & \alpha^{-1} \end{pmatrix}$$

or

$$\phi(w) = \begin{pmatrix} 1 & 1 \\ 0 & 1 \end{pmatrix}.$$

Suppose first that $\phi(w) = \left(\begin{smallmatrix} \alpha & 0 \\ 0 & \alpha^{-1} \end{smallmatrix}\right)$. Let $tr(G) = \{tr(g); g \in \phi(G)\}$. Choose a transcendental number τ which is not algebraic over $K = \mathbb{Q}(tr(G))$. Extend ϕ to ζ such that $\phi = \zeta_{|G}$ and

$$\zeta(t) = \begin{pmatrix} \tau & 0 \\ 0 & \tau^{-1} \end{pmatrix}.$$

Then the map $\zeta\colon \Gamma \mapsto PSL(2,\mathbb{C})$ is injective.

Suppose next that $\phi(w) = \left(\begin{smallmatrix} 1 & 1 \\ 0 & 1 \end{smallmatrix}\right)$. Choose a transcendental number τ which is not algebraic over $K = \mathbb{Q}(tr(G))$. Extend ϕ to ζ so that $\phi = \zeta_{|G}$ and

$$\zeta(t) = \begin{pmatrix} 1 & \tau \\ 0 & 1 \end{pmatrix}.$$

Then the map $\zeta\colon \Gamma \mapsto PSL(2,\mathbb{C})$ is injective. □

The main result necessary for the proof of Theorem 7.3.3 is the following. We consider extensions of centralizers of groups B with the property that any subgroup of B which has cyclic centralizers admits a faithful representation into $PSL(2,\mathbb{C})$.

Lemma 7.3.4. *Let B be a torsion-free, finitely generated group with the property that every subgroup with cyclic centralizers admits a faithful representation into $PSL(2, \mathbb{C})$. Let u be an element that generates its own centralizer. Then the extension of centralizers of B,*

$$\Gamma = \langle t, B; \mathrm{rel}(B), t^{-1}ut = u \rangle$$

also has the property that any subgroup with cyclic centralizers admits a faithful representation into $PSL(2, \mathbb{C})$.

Proof. Let B be a finitely generated group with the property that every subgroup with cyclic centralizers admits a faithful representation into $PSL(2, \mathbb{C})$ and let u be an element that generates its own centralizer. Consider the group

$$\Gamma = \langle t, B; \mathrm{rel}(B), t^{-1}ut = u \rangle$$

and let G be a finitely generated subgroup of Γ that has cyclic centralizers.

We analyze the structure of G in terms of the Bass–Serre tree structure it inherits as a subgroup of the HNN group Γ. We use the more detailed Karrass–Solitar subgroup theorems for HNN groups (see Chapter 2 and [90]).

From the Karrass–Solitar results, since G is a subgroup of an HNN group Γ, G is itself an HNN group with base S_G and free part $t_1, \ldots, t_n$. The base S_G is a linear tree product with each vertex a conjugate of the base B of Γ intersected with G. Further the amalgamated subgroups in the base S_G are conjugates of the associated subgroup $\langle u \rangle$ intersected with G and the associated relation from the HNN structure of G are also given by conjugates of the associated subgroup $\langle u \rangle$ intersected with G. The free part $t_1, \ldots, t_n$ and the base S_G are constructed from a special type of double coset representative system for G within Γ called a CRESS. The details can be found in [146, 147] and [149]. For this book we call such an HNN decomposition of a subgroup of an HNN group a *Karrass–Solitar decomposition.* As with many amalgam decompostions this decomposition is not unique and there may be other amalgam decompositions for the subgroup G. Again from Karrass–Solitar, since G is finitely generated we may assume that the base S_G in a Karrass–Solitar decomposition has only finitely many vertices.

First if G intersects each conjugate of the base B trivially then G is a free group and hence has a faithful representation in $PSL(2, \mathbb{C})$.

Next if G intersects each conjugate of the associated subgroup $\langle u \rangle$ trivially then G is a free product of the base S and a free group. The base S_G is a free product of its vertices each of which is a conjugate of the base B intersected with G. Since G has cyclic centralizers each vertex then admits a faithful representation in $PSL(2, \mathbb{C})$. Then from the corollary to Lemma 7.3.2 it follows that the base has a faithful representation in $PSL(2, \mathbb{C})$ and hence so does all of G.

Finally we may assume that G intersects nontrivially with some conjugate of the cyclic subgroup $\langle u \rangle$. Since u generates its own centralizer in B it follows that the centralizer of u in Γ is $\langle u, t \rangle$. Without loss of generality we may assume that $G \cap \langle u \rangle = \langle u^k \rangle$ for some natural number k with k minimal. If $G \cap \langle t \rangle = \{1\}$ then since u generates its own centralizer in B and G has cyclic centralizers it follows that u^k generates its own centralizer in G and hence $\langle u^k \rangle$ is maximal in G.

From the Karrass–Solitar results it follows that the associated relations have the form $t_i^{-1}w^{-1}u^k w t_i = v^{-1}u^k v$ for some $w, v \in \Gamma$. If $t_i \neq 1$ this would imply that $vt_i^{-1}w^{-1}$ is in the centralizer of u. Since the centralizer of u is $\langle u, t \rangle$ it follows that either t_i must be trivial or $t_i = t^m$ for some m. We then have two possible situations. First, the free part in the Karrass–Solitar decomposition for G, is trivial, or second, the free part is t^m for some natural number m.

In the first case, where the free part is trivial, it follows that G is equal to the base S_G in the Karrass–Solitar decomposition and the amalgamated relations in the base have the form $u^k = u^k$. In this case $G \cap \langle t \rangle = \{1\}$ since otherwise the free part would not be trivial. Therefore as explained above u^k must be maximal in G and hence maximal in each vertex in S_G. Therefore in this situation G is a tree product with maximal cyclic amalgamations and each vertex a conjugate of B intersected with G. Each vertex, being a conjugate of a subgroup of B, has a faithful representation in $PSL(2, \mathbb{C})$ and therefore by iterating Lemma 7.3.4, G has a faithful representation in $PSL(2, \mathbb{C})$.

In the second case, where the free part is $t_1 = t^m$, we have from Karrass–Solitar that the base $S_G = G_1 \underset{u^k = t_1 u^k t_1^{-1}}{*} G_1^{t_1}$ where $G_1 = G \cap B$ and $G_1^{t_1}$ is the conjugate of G_1 by t_1. The associated relation is then $t_1 u^k t_1^{-1} = u^k$ and therefore G is the HNN group

$$G = \langle t_1, G_1; \mathrm{rel}(G_1), t_1 u^k t_1^{-1} = u^k \rangle.$$

The subgroup G_1 has a faithful representation in $PSL(2, \mathbb{C})$ and u^k is maximal in G_1 so we can use Lemma 2.2 to get that G has a faithful representation in $PSL(2, \mathbb{C})$ completing the proof. □

Now let $\mathfrak{F}$ be the class of groups formed by taking iterated extensions of centralizers by maximal cyclic subgroups starting with finitely generated free groups. The groups in $\mathfrak{F}$ are all torsion-free. Since finitely generated free groups can be faithfully represented in $PSL(2, \mathbb{C})$, Lemma 7.3.4 implies that any finitely generated group in $\mathfrak{F}$ with cyclic centralizers can also be faithfully represented.

Corollary 7.3.2. *Any finitely generated subgroup of $\mathfrak{F}$ with cyclic centralizers admits a faithful representation in $PSL(2, \mathbb{C})$.*

We can now give the proof of Theorem 7.3.1.

Proof. Let G be a hyperbolic limit group. From the structure theorems for fully residually free groups G can be embedded as a subgroup of a group in the class $\mathfrak{F}$. Since G is hyperbolic it has cyclic centralizers and hence G admits a faithful representation into $PSL(2, \mathbb{C})$ be corollary 2.2. □

If G is a hyperbolic limit group then any Abelian vertex is cyclic. If the JSJ decomposition is effectively known then by repeated use of the technique to prove lemma 7.3.1 the actual faithful representation of G can be constructed.

The method of proof can be extended to nonhyperbolic limit groups by faithfully representing any free Abelian factors of rank n in the JSJ decomposition of a given limit group by up to conjugacy parabolic matrices of the form

$$\begin{pmatrix} 1 & t_i \\ 0 & 1 \end{pmatrix}, \quad i = 1, 2, \ldots, n$$

or matrices of the form

$$\begin{pmatrix} t_i & 0 \\ 0 & t_i^{-1} \end{pmatrix}, \quad i = 1, 2, \ldots, n$$

respectively, where the t_i are algebraically independent transcendentals.

In [254], answering questions posed by Bestvina, Wilton gave the following result.

Theorem 7.3.5. *A limit group can be embedded in any algebraic group over $\mathbb{R}$ which contains non-Abelian free groups. In particular any limit group has a faithful representation in $PSL(2, \mathbb{R})$.*

Wilton's proof uses the algebraic geometry over groups. We discuss this algebraic geometry in the next chapter. Here in the spirit of this chapter we give a separate proof for $PSL(2, \mathbb{R})$ suggested by A. Lubotzky [167].

Let G be a limit group, say generated by $x_1, \ldots, x_n$. If G is Abelian then just take matrices

$$\begin{pmatrix} 1 & t_i \\ 0 & 1 \end{pmatrix}, \quad i = 1, \ldots, n,$$

with n algebraically independent real transcendentals t_i. This will provide a faithful representation of G.

Now let G be non-Abelian. Then two of the generators, say $x = x_1$ and $y = x_2$, do not commute. Since G is a limit group if follows that the subgroup $\langle x, y\rangle$ is free of rank 2, and there is a faithful representation of $\langle x, y\rangle$ into $PSL(2, \mathbb{R})$. Since G is a limit group $PSL(2, \mathbb{R})$ contains free subgroups of each countable rank, this representation can then be extended to a representation of G into $PSL(2, \mathbb{R})$ (see for example [90]).

Hence the character variety for all representations of G into $PSL(2, \mathbb{R})$ is not empty and is of the form $S[z_1, \ldots, z_s]$ with indeterminates z_i, and we have $s > 0$ since G is non-Abelian and $\langle x, y\rangle$ is free of rank 2. Further S is a subring of the reals. Now replace the $z_1, \ldots, z_s$ by algebraically independent real transcendentals. This gives a faithful representation of G into $PSL(2, \mathbb{R})$.

This proof and also Wilton's proof is not constructive as the proof of the faithfulness of the representation into $PSL(2, \mathbb{C})$.

We mention that a constructive proof of the existence of a faithful representation in $PSL(2, \mathbb{R})$ of a cyclically pinched one-relator group along the lines of Theorem 7.3.2 was given by Fine, Kreuzer and Rosenberger [89].

Theorem 7.3.6 ([89]). *Let $G = \langle a_1, \dots, a_n, b_1, \dots, b_m; u = v\rangle$ with $n \geq 2, m \geq 2$ and $u = u(a_1, \dots, a_n)$ a nontrivial, not primitive and not a proper power element in the free group $F_1 = \langle a_1, \dots, a_n\rangle$ and $v = v(b_1, \dots, b_m)$ a nontrivial, not primitive and not a proper power element in the free group $F_2 = \langle b_1, \dots, b_m\rangle$.*

Then there exists a faithful representation $\phi\colon G \to PSL(2, \mathbb{R})$.

7.4 Infinite words and algorithmic theory

We have examined the class of finitely generated fully residually free groups in a variety of ways. First purely algebraically in terms of residual properties and next logically as universally free groups. We then looked at them as limit groups in terms of classes of homomorphisms into free groups. This was placed in a geometric context as geometric limit groups and then in terms of an iterated construction as constructible limit groups. The important idea is that all of these approaches are equivalent and which definition is used in studying a particular property of this class of groups depends on what is most convenient for that particular setting. Myasnikov, Remeslennikov and Serbin (see [198, 199], [158]) introduced an additional way to approach this class. By considering a finitely generated fully residually free group G as a subgroup of the free exponential group $F^{\mathbb{Z}[t]}$ over the commutative ring $\mathbb{Z}[t]$ they show that each element of G can be represented as an *infinite word* over $\mathbb{Z}[t]$. By using this representation the Stallings folding method (see Chapter 3) can be adapted to fully residually free groups and can be employed to study the subgroup structure. In particular many algorithmic properties, such as algorithmically determining if two subgroups are conjugate, can be handled in a manner analogous to the the solution within ordinary free groups.

In this section we will describe this whole approach. Further it can be extended to the more general class of finitely generated $\mathbb{Z}^n$-free groups. This will be defined and explained in the next subsection.

We first define the concept of an infinite word over an ordered Abelian group. In an ordered Abelian group a closed segment $[a, b]$ is defined for $a, b \in A$ by

$$[a, b] = \{x \in A\colon a \leq x \leq b\}\,.$$

Definition 7.4.1. Let A be a discretely ordered Abelian group with a minimal positive element 1_A and $X = \{x_i; i \in I\}$ a set. An A-word or *infinite word* over A is a function

$$w\colon [1_A, \alpha_w] \mapsto X^{\pm 1} \quad \text{where } \alpha_w \in A, \alpha_w \geq 0.$$

The value in A given by $|w| = \alpha_w$ is called the length of w. The infinite word w is *reduced* if and only if there are no subwords xx^{-1} or $x^{-1}x$ with $x \in X$. Equivalently w is reduced if $w(\alpha) \neq w(\alpha + 1)^{-1}$ for any α with $1 \leq \alpha < \alpha_w$. We let $R(A, X)$ denote the set of reduced A-words.

We can define a partial multiplication on the set of reduced infinite words. Let u, v be infinite A-words with respective intervals $[1_A, \alpha]$, $[1_A, \beta]$. The concatentaion uv is the infinite word with interval $[1_A, \alpha + \beta]$. Pictorially this is

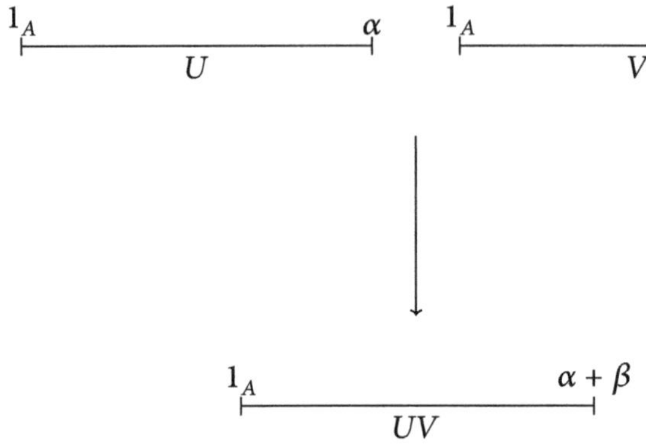

Fig. 7.1. The Concatenation of Infinite Words.

The concatenation may not be reduced, We write $u \circ v$ instead of uv in the case when uv is reduced.

The inverse of an infinite word is formed by taking the X labeling in the reverse direction in terms of inverses. That is if U has the form $x_{1_A} x_{2_A} \ldots x_\alpha$ then u^{-1} also has interval $[1, \alpha]$ but with the labeling $x_\alpha^{-1} \ldots x_{1_A}^{-1}$. Pictorially this is represented by

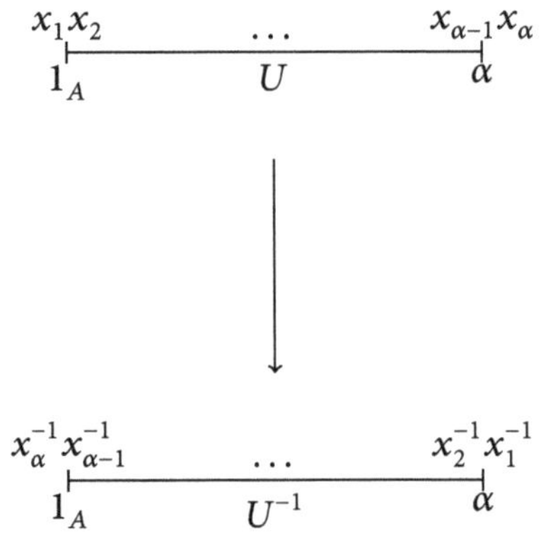

Fig. 7.2. The Inverse of an Infinite Word.

If u has the form $\overline{u}c^{-1}$ and v has the form $c\overline{v}$ then the multiplication of u and v is defined as

$$u * v = \overline{uv}$$

and then reduced. This is only a partial multiplication on $R(A, X)$ and is only defined in the case described above.

An infinite reduced word v is *cyclically reduced* if $v(1_A)^{-1} \neq v(|v|)$. $v \in R(A, X)$ admits a *cyclic decomposition* if $v = c^{-1} \circ u \circ c$, where c is a reduced infinite word and u is cyclically reduced. We denote by $CR(A, X)$ the set of all words from $R(A, X)$ which admit a cyclic decomposition.

As we have seen, the free exponential group $F^{\mathbb{Z}[t]}$ can be constructed as an increasing union of extensions of centralizers starting with the free group F. That is $F^{\mathbb{Z}[t]}$ can be constructed as

$$F = G_0 \subset G_1 \subset \cdots \subset G_n \subset \ldots$$

where G_{i+1} is formed by taking all extensions of centralizers in G_i. By an inductive argument on the G_i it was proved ([198, 199]) that each G_i and hence $F^{\mathbb{Z}[t]}$ can be embedded in the set of reduced words $R(\mathbb{Z}[t], X)$ where X is the set of generators of the free group F. This would then imply that any finitely generated subgroup of $F^{\mathbb{Z}[t]}$ also can be embedded into $R(\mathbb{Z}[t], X)$. Since each finitely generated fully residually free group can be canonically embedded into $F^{\mathbb{Z}[t]}$ we obtain the following result. What is important is that these embeddings are effective allowing for algorithmic solutions to certain decision problems (see [198, 199]).

Theorem 7.4.1. *Given a finite presentation of a finitely generated fully residually free group G then there is an effective embedding $\Psi: G \to R(\mathbb{Z}[t])$ of G into the set of reduced infinite words over $\mathbb{Z}[t]$. This embedding follows from specifying the images of the generators of G.*

In Chapter 3 we saw that the Stallings folding method provided a powerful and algorithmic tool to study subgroups of free groups. Using infinite $\mathbb{Z}[t]$-words and the above embedding this method can be extended to finitely generated fully residually free groups.

A $(\mathbb{Z}[t], X)$-*labeled directed graph* is a combinatorial graph Γ where each edge has a direction and is labeled either by a letter from X or by an infinite word $u^\alpha \in F^{\mathbb{Z}[t]}$ with $u \in U, \alpha \in \mathbb{Z}[t], \alpha > 0$. Recall that $\mathbb{Z}[t]$ is an ordered Abelian group. We denoted the label of an edge e by $\mu(e)$.

As in an ordinary labeled graph for each edge $e \in \Gamma$ we denote by $o(e), t(e)$ the origin and terminus of e repsectively. Further for each edge e we introduce a formal inverse e^{-1} of e as an edge with the label $\mu(e)^{-1}$ and with $o(e^{-1}) = t(e)$ and $t(e^{-1}) = o(e)$. For the new edge we set $(e^{-1})^{-1} = e$ and denote the new graph with the formal inverses adjoined by $\tilde{\Gamma}$.

Again in analogy with ordinary graphs, a *path* in Γ is a sequence of edges $p = e_1 \ldots e_n$ where the terminus of e_i is the origin of e_{i+1}. A path is $p = e_1 \ldots e_k$ is *reduced* if $e_i \neq e_{i+1}^{-1}$ for all $i = 1, \ldots, k-1$. Further it is *label reduced* if the path is reduced and if $e_{k_1} \ldots e_{k_t}$ with $k_1 \le k_t$ is a subpath of p such that $\mu(e_i) = u^{\alpha_i}, \alpha_i \in \mathbb{Z}[t], i \in$

$[k_1, k_t]$ and $\mu(e_{k_1-1}) \neq u^\beta, u_{k_1+1} \neq u^\beta$ for any $\beta \in \mathbb{Z}[t]$ for any $\beta \in \mathbb{Z}[t]$ provided $k_1 - 1, k_2 + 1 \in [1, k]$, then $\alpha = \alpha_{k_1} + \cdots + \alpha_{k_t} \neq 0$ and $\mu(e_{k_1-1}) = u^\beta$ is reduced and $u^\alpha * \mu(e_{k_2+1})$ is reduced.

We are now going to tie reduced paths in a $(\mathbb{Z}[t], X)$-labeled directed graph to subgroups of $\mathbb{Z}[t]$. First we associate a language to $(\mathbb{Z}[t], x)$-graphs.

Definition 7.4.2. Let Γ be a $(\mathbb{Z}[t], X)$-graph and let v be a vertex of Γ. Then the language of Γ with respect to v is

$$L(\Gamma, v) = \{\mu(p) \colon p \text{ is a reduced path in } \Gamma \text{ from } v \text{ to } v\}.$$

What is important is the following lemma proved in [198].

Lemma 7.4.1. *Let Γ be a $(\mathbb{Z}[t], X)$-graph and let v be a vertex of Γ. Then $L(\Gamma, v)$ is a subgroup of $F^{\mathbb{Z}[t]}$.*

In analogy with the Stallings folding method in free groups, foldings and partial foldings can be defined on $(\mathbb{Z}[t], X)$-graph to obtain folded graphs with label reduced edges. As in the Stallings folding situation in free groups subgroup information can then be read off of the folded graph. The details as well as the definition of U-folded can be found in the work of Kharlampovich, Myasnikov, Remeslennikov and Serbin [198, 199], [158] and [238]. What is important is that the folding constructions are effective.

Lemma 7.4.2. *Let Γ be a finite connected $(\mathbb{Z}[t], X)$-graph. Then there exists a U-folded $(\mathbb{Z}[t], X)$-graph Δ that can be found effectively from Γ by a finite sequence of partial and U-foldings.*

Lemma 7.4.3. *Let H be a finitely generated subgroup of $F^{\mathbb{Z}[t]}$. Then there exists a U-folded $(\mathbb{Z}[t], X)$-graph Γ and a vertex v of Γ such that $L(\Gamma, v) = H$.*

Lemma 7.4.4. *There is an algorithm which, given finitely many elements $h_1, \ldots, h_k$ in $F^{\mathbb{Z}[t]}$ in standard form constructs a U-folded $(\mathbb{Z}[t], X)$-graph Γ such that $L(\Gamma, v) = \langle h_1, \ldots, h_k\rangle$.*

From this graph construction several algorithmic solution of problems in free groups can be extended to finitely generated fully residually free groups. We mention several of these and refer to [158] and [159] for further details.

Theorem 7.4.2 ([198, 199]). *Every finitely generated subgroup of $F^{\mathbb{Z}[t]}$ and in particular a finitely generated fully residually free group, has a solvable membership problem. That is there exists an algorithm which givem finitley many standard decomposiitons of elements $g, h_1, \ldots, h_k$ from $F^{\mathbb{Z}[t]}$ decides whether or not $g \in \langle h_1, \ldots, h_k\rangle$.*

Theorem 7.4.3 ([198, 199]). *There exists an algorithm which, given finitely many standard decompositions of elements $h_1, \ldots, h_k, g_1, \ldots, g_m$ from $F^{\mathbb{Z}[t]}$ finds the generators of $H \cap K$ where $H = \langle h_1, \ldots, h_k\rangle, K = \langle g_1, \ldots, g_m\rangle$.*

As a consequence of the last theorem it follows that the Howson property holds in finitely generated fully residually free groups.

Corollary 7.4.1. *The intersection of two finitely generated subgroups of a finitely generated fully residually free group is finitely generated.*

The effectiveness of the construction of the intersection of finitely generated subgroups of a finitely generated fully residually free group leads directly to the solution of the conjugacy problem.

Theorem 7.4.4. *Any finitely generated fully residually free group has a solvable conjugacy problem.*

Further algorithmic solutions can be found in [158] and algorithmic properties involving subgroups of finite index, again analogous to the situation in free groups, can be found in [159].

7.4.1 $\mathbb{Z}^n$-free groups

Much of the material in the previous section can be recast in the wider context of free actions on $\mathbb{Z}^n$-trees. First we must recall from Chapter 3 some basic ideas on groups acting freely on various types of trees.

First a *simplicial tree* is an ordinary graph without circuits (closed paths). A simplical tree can be made into a metric space by giving each edge a length of one. We say that *G acts freely on the graph X* if it acts without inversion and fixes no vertex. The most basic result (and the starting off point for Bass–Serre theory) is the following.

Theorem 7.4.5. *A group G is a free group if and only if G acts freely on a simplicial tree X.*

Chiswell and independently Tits (see [53]) introduced the construction of an $\mathbb{R}$-tree. An $\mathbb{R}$-tree T is a non-empty metric space with metric d such that there is no subspace homeomorphic to a circle and such that if $u, v \in T$ with $r = d(u, v)$ then there exists a unique isometry $\alpha\colon [0, r] \to T$ with $\alpha(0) = u$ and $\alpha(r) = v$.

A *segment* in a $\mathbb{R}$-tree is the image of an isometry $\alpha\colon [0, r] \to T$. $\alpha(0), \alpha(r)$ are the endpoints of the segment. To see that such a structure is "tree-like" it can be proved that the above definition is equivalent to the following, which says that there is a type of branching at every point. A non-empty metric space is an $\mathbb{R}$-tree if

(1) given $u, v \in T$ there is a segment with endpoints u, v,
(2) the intersection of two segments with common endpoint is a segment,
(3) if two segments intersect in a single point which is an endpoint of both then their union is a segment.

Now suppose a group G acts on an $\mathbb{R}$-tree T. We say $g \in G$ is an *inversion* if g leaves a segment invariant but g has no fixed points. As before G acts freely on an $\mathbb{R}$-tree if there are no fixed points. An $\mathbb{R}$-*free group* is a group which acts freely and without inversions on an $\mathbb{R}$-tree. Clearly free groups are $\mathbb{R}$-free. Further free Abelian groups and all orientable surface groups of any genus as well as all non-orientable surface groups of genus ≥ 4 also act freely on $\mathbb{R}$-trees. In fact, in a sense these are the only finitely generated examples. The following result, given by Rips, gives the classification of finitely generated groups acting on $\mathbb{R}$-trees.

Theorem 7.4.6 (Rips Theorem). *A finitely generated group acts freely on an $\mathbb{R}$-tree if and only if it is a free product of surface groups (except for the non-orientable surfaces of genus 1, 2, 3) and free Abelian groups of finite rank.*

Morgan and Shalen [188] further extended the concept of an $\mathbb{R}$-tree to a Λ-tree where Λ is an arbitrary ordered Abelian group. This concept arose in their paper from an example of a valuation on a field where the ordered Abelian group is the valuation group. In Serre's original work on groups acting on trees [238], certain trees were constructed from discrete valuations and Morgan and Shalen's construction can be considered a generalization of this. A complete discussion of Λ-trees can be found in the survey articles by Morgan [187], Shalen [239], [240], the work of Bass [6] and Alperin and Bass [4] or the papers of Chiswell [49], [50], [51] and the book of Chiswell [47].

Let Λ be an ordered Abelian group written additively. If for each $a, b \in \Lambda$ with $a > 0$ there exists a positive integer n such that $b < na$ then Λ is an archimedean ordered Abelian group. In particular all additive subgroups of the reals $\mathbb{R}$ are archimedean. If Λ_1 and Λ_2 are ordered Abelian groups then the direct sum $\Lambda_1 \oplus \Lambda_2$ can also be made into an ordered Abelian group with the lexicographic ordering. However the archimedean property is not necessarily preserved under this construction. For example under the lexicographic ordering $\mathbb{Z} \oplus \mathbb{Z}$ is a non-archimedan ordered Abelian group. In particular if $a = (1, 0)$ and $b = (0, 1)$ then $b < na$ for all positive integers n.

If Λ is an ordered Abelian group and X is a set then an Λ-*metric* on X is a Λ-valued function $d\colon X \times X \to \Lambda$ satisfying the ususal metric space properties; (i) $d(x, y) \geq 0$ and $d(x, y) = 0$ iff $x = y$ (ii) $d(x, y) = d(y, x)$ and (iii) $d(x, y) \leq d(x, z) + d(z, y)$. The pair (X, d) is then an Λ-metric space. In this context an ordinary metric space is just an $\mathbb{R}$-metric space.

The construction of a Λ-tree now mimics the definition of an $\mathbb{R}$-tree. A closed interval in Λ is defined as for $\mathbb{R}$: $[a, b]_\Lambda = \{x \in \Lambda; a \leq x \leq b\}$ for $a, b \in \Lambda$ with $a \leq b$ and $[b, a]_\Lambda = [a, b]_\Lambda$. As for $\mathbb{R}$ a *segment* in a Λ-metric space is the image of an isometry $\alpha\colon [a, b]_\Lambda \to X$ for some $a, b \in \Lambda$ with $a \leq b$. $\alpha(a), \alpha(b)$ are the endpoints of the segment and $a = b$ is allowed. A non-empty Λ-metric space (X, d) is an Λ-tree if

(1) (X, d) is *geodesic*. This means that if $x, y \in X$ then there is a segment in X with endpoints x, y.
(2) The intersection of two segments with common endpoint is a segment.
(3) If two segments intersect in a single point which is an endpoint of both then their union is a segment.

Chiswell proves that actions on Λ-trees are closely tied to Λ-valued Lyndon length functions. If G is a group and Λ an ordered Abelian group then a Λ-valued *Lyndon length function* is a map $L\colon G \to \Lambda$ satisfying
(1) $L(1) = 0$,
(2) for all $g \in G, L(g) = L(g^{-1})$,
(3) for all $g, h, k \in G$, $c(g, h) \geq \min(c(h, k), c(k, g))$ where

$$c(g, h) = 1/2(L(g) + L(h) - L(g^{-1}h)),$$

(4) for all $g, h \in G$, $c(g, h) \in \Lambda$.

Suppose the group G acts by isometries on the Λ-tree (X, d). Let $x \in X$. Define for each $g \in G$, $L_x(g) = d(x, gx)$. This will then define a Λ-valued Lyndon length function of G. In fact Chiswell's theorem [47] actually shows that all Λ-valued Lyndon length functions arise from such actions of the group G on Λ-trees.

Theorem 7.4.7 ([47]). *Let G be a group and $L\colon G \to \Lambda$ a Λ-valued Lyndon length function. Then there is a Λ-tree (X, d), an action of G on X and a point $x \in X$ such that $L = L_x$.*

As with simplicial trees and $\mathbb{R}$-trees the first paramount question is to examine the structure of groups which admit free actions on Λ-trees. The definitions again mimic those for $\mathbb{R}$-trees. We suppose that G acts as a group of isometries on a Λ-tree (X, d). An element $g \in G$ is an *inversion* if g leaves a segment invariant but g has no fixed points. As before G acts freely on a Λ-tree if there are no fixed points. A ***Λ-free group*** is a group which acts freely and without inversions on an Λ-tree. More generally a group is *tree-free* if it is Λ-free for some ordered Abelian group Λ.

It is easy to see that subgroups of Λ are Λ-free since they act on Λ by translations. Harrison [129] using Lyndon length functions proved that free products of Λ-free groups are again Λ-free. More generally the following result of Bass [6] gives a structure theorem for free Λ-groups.

Theorem 7.4.8 ([22]). *A finitely generated $\Lambda \oplus \mathbb{Z}$-free group is the fundamental group of a finite graph of groups with the properties:*
(1) *the vertex groups are Λ-free,*
(2) *the edge groups are maximal Abelian (in the vertex groups),*
(3) *the edge groups embed into Λ.*

Guirardel extends this to $\mathbb{R}^n$-free groups.

Theorem 7.4.9 ([123]). *A finitely generated freely indecomposable $\mathbb{R}^n$-free group is isomorphic to the fundamental group of a finite graph of groups, where each vertex group is a finitely generated $\mathbb{R}^{n-1}$-free group, and each edge group is cyclic.*

Since the structure of $\mathbb{R}$-free groups is known from Rips Theorem this provides an inductive description of $\mathbb{R}^n$-free groups. The converse of this theorem is not true. Notice, that $\mathbb{Z}^n$-free groups are $\mathbb{R}^n$-free.

What ties this to the preceding section is that fully residually free groups are $\mathbb{Z}^n$-free and further each $\mathbb{Z}^n$-free group can be embedded in the set of reduced $\mathbb{Z}^n$ words. Here $\mathbb{Z}^n$ is considered as an ordered Abelian group with the order being lexicographic order.

Theorem 7.4.10 ([158]). *Finitely generated fully residually free groups are $\mathbb{Z}^n$-free for some n depending on the group.*

Martino and Rourke [184] showed that being $\mathbb{Z}^n$-free for some n is preserved under certain amalgamated free products.

Theorem 7.4.11 ([184]). *Let G_1 and G_2 be $\mathbb{Z}^n$-free groups. Then the amalgamated product $G_1 *_C G_2$ is $\mathbb{Z}^m$-free for some natural number m, provided C is cyclic and maximal Abelian in both factors.*

Any $\mathbb{Z}^n$-free groups can be embedded in $CR(\Lambda, X)$ where Λ is $\mathbb{Z}^n$ ordered lexicographically.

Theorem 7.4.12 ([198, 199]). *Let Λ be a discretely ordered Abelian group and X a set. If G is a subgroup of $CR(\Lambda, X)$ then the function $L_G\colon G \mapsto \Lambda$, defined by $L_G(g) = |g|$, is a free Lyndon length function.*

Now a result of Chiswell (see [52]) shows that this implies an embedding into the set of cyclically reduced infinite words.

Theorem 7.4.13 ([47, 48, 49, 50, 51, 52]). *Let Λ be a discretely ordered Abelian group. If $L\colon G \mapsto \Lambda$ is a free Lyndon length function on a group G then there exists an embedding $\phi\colon G \mapsto CR(\Lambda, X)$ such that $|(g)| = L(g)$ for every $g \in G$.*

This is then tied to the structure theorem for $\mathbb{Z}^n$-free groups by the following corollary.

Corollary 7.4.2. *Let Λ be an arbitrary ordered Abelian group. If $L\colon G \to \Lambda$ is a free Lyndon length function on a group G then there exists a length preserving embedding $\phi\colon G \to CR(\Lambda_0, X)$, where $\Lambda_0 = \mathbb{Z} \oplus \Lambda$ with the lexicographic order.*

We close this chapter by mentioning that that every $\mathbb{Z}^n$-free group admits an iterated construction analogous to the extension of centralizer construction of fully residually free groups.

Let $G \subset CR(\Lambda, X)$ be a group of infinite words. Then G is *complete* if it contains the common initial segment $c(g, h)$ for every pair of elements $g, h \in G$. A Lyndon length function $L\colon G \to \Lambda$ is complete if there exists a length preserving embedding $G \to CR(\Lambda, X)$.

Theorem 7.4.14 ([159]). *If G is finitely presented and has a complete free length function in Λ, then G has an index two subgroup that can be represented as a union of a finite series of groups*

$$G_1 \subset G_2 \subset \cdots \subset G_n = G$$

where

(1) G_1 *is a free group,*
(2) G_{i+1} *is obtained from* G_i *by finitely many HNN-extensions in which associated subgroups are maximal Abelian and length-isomorphic.*

This can then be applied to $\mathbb{Z}^n$-free groups via the following result of Kharlampovich, Myasnikov and Serbin [158]

Theorem 7.4.15. *Every finitely generated* $\mathbb{Z}^n$*-free group is a subgroup of a finitely generated complete* $\mathbb{Z}^k$*-free group.*

8 Algebraic geometry over groups

There are three major components in the solution to the Tarski problems: the structure theory of fully residually free groups, algebraic geometry over groups and a reduction theory that allows each system of equations over a free group to be reduced to what is called a triangular system. In this chapter we discuss the basic ideas of *algebraic geometry over groups*. This theory recasts classical algebraic geometry within the guise of infinite groups especially free groups. To begin, in the next section we recall the basic notions of classical algebraic geometry. These concepts will be then be reinterpreted within the context of group theory. Standard treatments of algebraic geometry and commutative algebra can be found in [41].

8.1 Algebraic geometry

Classical algebraic geometry was concerned with the study of *algebraic curves*. We start with an algebraically closed field k and affine n-space k^n over k. At the simplest level an algebraic curve is the locus of points $(g_1, \ldots, g_n) \in k^n$ satisfying a polynomial equation over k. That is the locus of points in k^n such that $p(g_1, \ldots, g_n) = 0$ where $p(x_1, \ldots, x_n) \in k[x_1, \ldots, x_n]$. The standard conic sections, circles, parabolas, ellipses and hyperbolas, over the real number $\mathbb{R}$ are all algebraic curves being the loci of

$$ax^2 + by^2 + cxy + dx + ey + f = 0$$

for various values of $a, b, c, d, e, f \in \mathbb{R}$. Another example are the elliptic curves, the loci of equations

$$y^2 = ax^3 + bx + c\,.$$

Elliptic curves played a role in the final proof of Fermat's last theorem and also are applicable to cryptography (see [160]).

In studying algebraic curves it is often preferable to consider points in projective n-space $P^n(k)$ and the corresponding projective algebraic curves. A compact Riemann surface can be equated with a smooth projective algebraic curve over the complex numbers.

Modern algebraic geometry axiomatizes and unifies many of these concepts, and places the whole study within the context of commutative algebra. As above, the starting point is affine n-space k^n over an algebraically closed field k. An *affine algebraic set* is the set of common zeros in k^n of a set of polynomials in $k[x_1, \ldots, x_n]$. Suppose that $S \subset k[x_1, \ldots, x_n]$. Then we let

$$V(S) = \{P \in k^n : f(P) = 0\}$$

for all $f \in S$. A subset $X \subset k^n$ is an *affine algebraic set* or *algebraic set* if $X = V(S)$ for some $S \subset k[x_1, \ldots, x_n]$. The algebraic sets are closely tied to the ideal structure of the ring $k[x_1, \ldots, x_n]$. The following properties are easy and clear and will be important in forming what is called the Zariski topology

(1) If $S \subset k[X_1, \ldots, X_n]$ and I is the ideal generated by S, then $V(S) = V(I)$. It follows that any algebraic set is equal to $V(I)$ for some ideal I in $k[x_1, \ldots, x_n]$.
(2) If $\{I_\alpha\}$ is any collection of ideals, then $V(\cup_\alpha I_\alpha) = \cap_\alpha(V(I_\alpha))$. It follows that the intersection of any collection of algebraic sets is an algebraic set.
(3) If $I \subset J$, then $V(I) \supset V(J)$.
(4) $V(fg) = V(f) \cup V(g)$ for any polynomials $f, g \in k[x_1, \ldots, x_n]$ and $V(I) \cup V(J) = V(\{fg; f \in I, g \in J\})$. Hence any finite union of algebraic sets is an algebraic set.
(5) $V(0) = k^n$, $V(1) = \emptyset$, $V(X_1 - a_1, \ldots, X_n - a_n) = \{(a_1, \ldots, a_n)\}$ for $a_i \in k$. Hence any finite subset of affine n-space k^n is an algebraic set.

If R is a commutative ring and I is an ideal in R then the *radical* of I, denoted $\sqrt{I}$ or $\mathrm{Rad}(I)$, is the set

$$\{a \in R : a^n \in I \text{ for some some integer } n > 0\} .$$

The radical of an ideal is an ideal.

Lemma 8.1.1. *If R is a commutative ring and $I \subset R$ is an ideal then* $\mathrm{rad}(I)$ *is also an ideal with $I \subset \mathrm{rad}(I)$.*

If X is a subet of affine space then $I(X)$ is ideal of polynomials than vanish on X. It is called the *vanishing ideal*. An ideal I is called a *radical ideal* if $I = \mathrm{rad}(I)$.

Lemma 8.1.2. *If X is a subset of the affine n-space k^n then $I(X)$ is a radical ideal, that is* $\mathrm{rad}(I(X)) = I(X)$.

Recall that a ring R is *Noetherian* (see [41]) if every ideal in R is finitely generated. The Hilbert Basis theorem says that polynomial rings over Noetherian rings are still Noetherian.

Theorem 8.1.1 (Hilbert Basis Theorem). *If R is a Noetherian ring then $R[x_1, \ldots, x_n]$ is also Noetherian. In particular if k is a field then $k[x_1, \ldots, x_n]$ is Noetherian.*

Applying this to algebraic sets this says that any algebraic set is finitely based. Formally:

Corollary 8.1.1. *Let S be a set of polynomials in $k[x_1, \ldots, x_n]$. Then $V(S) = V(S_1)$ for some finite set of polynomials.*

This result is fundamental in algebraic geometry. In extending to algebraic geometry over groups to have this property we will have to introduce what are called equationally noetherian groups (see Section 8.3).

An algebraic set V is *reducible* if $V = V_1 \cup V_2$ for two algebraic sets V_1, V_2 neither equal to V. If V is not reducible it is *irreducible*. We call an irreducible algebraic set an *affine variety* or just a *variety*. Irreducibility of V is tied to the ideal struture of $I(V)$. In particular:

Lemma 8.1.3. *An algebraic set V is irreducible if and only if $I(V)$ is a prime ideal.*

Combining this with the Noetherian property of $k[x_1, \dots, x_n]$ we obtain a unique decomposition of algebraic sets.

Theorem 8.1.2. *Let V be an algebraic set in k^n. Then there are unique irreducible algebraic sets $V_1, \dots, V_m$ such that $V = V_1 \cup \dots \cup V_m$. Further V_i is not contained in V_j if $i \neq j$. The algebraic sets V_i are the irreducible components of V.*

Hilbert's Nullstellensatz describes the full relationship between algebraic sets and their corresponding ideals.

Theorem 8.1.3 (Hilbert's Nullstellensatz). *Let k be an algebraically closed field and let I be an ideal in $k[x_1, \dots, x_n]$. Then $I(V(I)) = \mathrm{rad}(I)$.*

In particular this implies the following. If $f_1, \dots, f_m \in k[x_1, \dots, x_n]$ and g is another polynomial that vanishes whenever $f_1, \dots, f_m$ vanish then there is an integer $r > 0$ and polynomials $h_1, \dots, h_m \in k[x_1, \dots, x_n]$ such that

$$g^r = h_1 f_1 + \dots + h_m f_m .$$

The Nullstellensatz provides a correspondence between algebraic sets and radical ideals and between prime ideals and afffine varieties.

Corollary 8.1.2. *If I is a radical ideal in $k[x_1, \dots, x_n]$ then $I(V(I)) = I$ and hence there is a one-to-one correspondnece between radical ideals and algebraic sets. Further if I is a prime ideal then $V(I)$ is irreducible and hence there is a one-to-one correspondence between prime ideals and affine varieties.*

If $V \subset k^n$ is a nonempty affine variety then $I(V)$ is a prime ideal in $k[x_1, \dots, x_n]$. It follows that (see [41]) the quotient ring $k[x_1, \dots, x_n]/I(V)$ is an integral domain. This integral domain is called the *coordinate ring* of V.

The final idea we mention from classical geometry is the *Zariski topology*. On affine n-space we form a topology by letting the closed sets be precisely the affine varieties. It is easily checked that this does form a topology on k^n. Further this topology is compact but not Hausdorf (see [163]).

8.2 The category of G-groups

Algebraic geometry over groups is a translation of the basic concepts of algebraic geometry to the context of non-Abelian groups. As described in the previous section the starting point for classical algebraic geometry is an algebraically closed field k and the polynomial ring $k[x_1, \dots, x_n]$ over k. What will take the place of this polynomial ring is the concept of a G-group. We rely on the papers [16] and [193] for most of the material in this chapter.

The impetus for the development of algebraic geometry over groups and the connecetion to the solution of the Tarski problems comes mainly from the study of equations over groups.

Definition 8.2.1. Let G be a fixed group. Then a group H is a G-group if it contains a designated copy of G, which we will identify with G. We allow that $G = 1$ and also that $G = H$ so that in particular G is itself a G-group.

G-groups are analogous to algebras over a unitary commutative ring, more specifically a field, with G playing the role of the coefficient ring.

The class of G-groups forms a category in the obvious way. A morphism from a G-group H to a G-group H' is a group homomorphism $\phi\colon H \longrightarrow H'$ which is the identity on G. We call these morphisms *G-homomorphisms*. The kernels of G-homomorphims are termed *ideals*. Ideals are simply the normal subgroups which meet G in the identity. The usual notions of group theory carry over to this category, allowing us to talk about free G-groups, finitely generated and finitely presented G-groups and so on. In particular, it is not hard to identify the finitely generated free G-groups. They take the form

$$G[X] = G[x_1, \dots, x_n] = G * F(X)\,,$$

the free product of G and the free group $F(X) = \langle x_1, \dots, x_n\rangle$ freely generated by $\{x_1, \dots, x_n\}$. We sometimes say that $X = \{x_1, \dots, x_n\}$ freely generates the free G-group $G[X]$. We view $G[X]$ as a non-commutative analogue of a polynomial algebra over a unitary commutative ring in finitely many variables. We think of the elements of $G[X]$ as non-commutative polynomials with coefficients in G. Similarly, if we stay inside the category of G-groups, the free product of two G-groups A and B is their amalgamated product $A *_G B$, with G the amalgamated subgroup.

In dealing with various products, it is sometimes useful to let the *coefficient group* G vary. In particular, if H_i is a G_i-group for each i in some index set I, then the unrestricted direct product $\overline{\prod}_{i\in I} H_i$ can be viewed as a $\overline{\prod}_{i\in I} G_i$-group, in the obvious way. If H_i is a G-group for each i, then we will sometimes think of the unrestricted direct product P of the groups H_i as a G-group by taking the designated copy of G in P to be the diagonal subgroup of the unrestricted direct product of all of the copies of G in the various factors. In the case of, say, the standard wreath product, if U is a G-group and if T is a G'-group, then their (standard) wreath product $A \wr T$ can be viewed as a $G \wr G'$-group in the obvious way again.

8.3 Domains and equationally Noetherian groups

In this section we introduce group-theoretic counterparts to the classical notions of integral domain and Noetherian ring; ideas that are crucial to algebraic geometry.

8.3.1 Zero divisors and G-domains

Definition 8.3.1. Let H be a G-group. Then we term a non-trivial element $x \in H$ *a G-zero divisor* if there exists a non-trivial element $y \in H$ such that

$$[x, g^{-1}yg] = 1 \quad \text{for all } g \in G.$$

Notice that if $G = \{1\}$ then every non-trivial element of the G-group H is a G-zero divisor. We then term a G-group H a *G-domain* if it does not contain any G-zero divisors; in the event that $G = H$ we simply say that H is a domain.

We use the following exponential notation for conjugation; $x^g = g^{-1}xg$.

If X is a subset of a G-group H then $gp_G(X)$ denotes the G-subgroup of H generated by X. If $X = \{x\}$ we just write $gp_G(x)$. Notice then that the equation defining a zero divisor is equivalent to the equation

$$[gp_G(x), gp_G(y)] = 1\,.$$

We will call the subgroup $[gp_G(x), gp_G(y)]$ *the ◇-product* of x and y and denote it by $x \diamond y$. So a non-trivial element x in the G-group H is a G-zero divisor if and only if $x \diamond y = 1$ for some non-trivial $y \in H$. Obviously,

$$x \diamond y = y \diamond x$$

and therefore the element y in the definition of a G-zero divisor is also a G-zero divisor. Notice also that if x is a G-zero divisor, then all the non-trivial elements in $gp_G(x)$ are also G-zero divisors.

Definition 8.3.2. Let H be a G-group. Then we say that an element $x \in H$ is G-nilpotent of degree $\leq k$ if

$$[x^{g_1}, x^{g_2}, \ldots, x^{g_k}] = 1 \quad \text{for all } g_i \in G\,,$$

This means that $gp_G(x)$ is a nilpotent subgroup of class $\leq k$. In the event that $gp_G(x)$ is nilpotent of class exactly k, then we say that x is G-nilpotent of degree k.

Again, if $G = H$ then we omit all mention of G and say that the element x is *nilpotent*. Notice, that if a group H is nilpotent of class c, then every element is nilpotent of degree at most c.

The following lemma shows that G-nilpotent elements are G-zero divisors.

Lemma 8.3.1. *Every non-trivial G-nilpotent element in a G-group H is a G-zero divisor.*

Proof. Let $x \in H$ be a non-trivial G-nilpotent element. Thus $gp_G(x)$ is nilpotent. If z is a non-trivial element in the center of $gp_G(x)$, then $[x, z^g] = 1$ for every choice of $g \in G$, as needed.

In the case of an associative ring, invertible elements are never zero divisors. Similar result holds for groups.

Definition 8.3.3. Let H be a G-group. An element $h \in H$ is termed G-invertible if $gp_G(h) \cap G \neq \{1\}$.

The following lemma then holds.

Lemma 8.3.2. *Let H be a G-group and assume that the subgroup G in H does not contain any G-zero divisors from H. Then any G-invertible element in H is not a G-zero divisor.*

Proof. Let x be a G-invertible element in H. So there exists a non-trivial element $g \in gp_G(x) \cap G$. Suppose that x is a G-zero divisor. Then there exists $y \in H$ such that $[gp_G(x), gp_G(y)] = 1$. It follows that $[gp_G(g), gp_G(y)] = 1$, i.e., g is a G-zero divisor, a contradiction.

G-domains have a somewhat restricted normal subgroup structure, as the following lemma shows.

Lemma 8.3.3. *Let H be a G-domain. Then the following hold:*
(1) *G is a non-Abelian group;*
(2) *Every G-subgroup of H is a G-domain;*
(3) *Every Abelian normal subgroup of H is trivial; in particular, if $H \neq \{1\}$, then H is not solvable and hence not nilpotent;*
(4) *H is directly indecomposable.*

The proof is straightforward.

Our next objective is to show that the class of G-domains is fairly extensive. This class contains all non-Abelian G-groups which are also CSA-groups; hence it contains all non-Abelian, torsion-free hyperbolic groups and all groups acting freely on Λ-trees.

Lemma 8.3.4. *Let G be a non-Abelian group and let H be a CSA G-group. Then H is a G-domain.*

Proof. Let $a, b \in G$, $[a, b] \neq 1$ and suppose x, y are non-trivial elements of H. If

$$[x, y^a] = [x, y^b] = [x, y^{ab}] = 1\,,$$

then by the transitivity of commutation $[y^b, y^{ab}] = 1$ and $[y^a, y^b] = 1$. The first relation implies that $[y, y^a] = 1$ and since a maximal Abelian subgroup M of H containing y is malnormal in H, we have $[y, a] = 1$. Now from $[y^a, y^b] = 1$ it follows that

$[y, y^b] = 1$ and consequently, $[y, b] = 1$. This implies $[a, b] = 1$, a contradiction, which completes the proof of the lemma. □

It follows directly from the argument above that we have also proved the following corollary.

Corollary 8.3.1. *Let H be a G-group and suppose that H is a CSA-group. If a and b are elements of G and if $[a, b] \neq 1$, then for every choice of the non-trivial elements x and y in H, at least one of the following hold:*

$$[x, y^a] \neq 1, \quad \text{or} \quad [x, y^b] \neq 1 \quad \text{or} \quad [x, y^{ab}] \neq 1 .$$

A related result holds for G-domains.

Lemma 8.3.5. *Let H be a G-domain and let $a_1, \ldots, a_n$ be any given non-trivial elements of H. Then there exist elements $g_2, \ldots, g_n$ in G such that*

$$[a_1, a_2^{g_2}, \ldots, a_n^{g_n}] \neq 1.$$

Proof. Since H has no G-zero divisors, there exists an element $g_2 \in G$ such that $[a_1, a_2^{g_2}] \neq 1$. The same argument applies now to $[a_1, a_2^{g_2}]$ and a_3 and so the desired conclusion follows inductively. □

There are many other G-domains besides these CSA-groups, as the following theorem shows.

Theorem 8.3.1. *Let T be a torsion-free domain and let U be a G-domain. Then the (standard) wreath product $W = U \wr T$ of U by T is a $G \wr T$-domain.*

Proof. In order to prove that W is a $G \wr T$-domain, suppose that $x, y \in W$ are a pair of non-trivial elements. Now W is the semi-direct product of B and T, where B, the normal closure of U in W, is the direct product of the conjugates U^t of U by the elements $t \in T$. We have to find an element $z \in G \wr T$ such that $[x, y^z] \neq 1$. If $x, y \in B$, then we can find an element $t \in T$ such that the supports of x and y^t overlap. In view of the fact that U is a G-domain, we can find an element $g \in G$ such that $[x, y^{tg}] \neq 1$ and so we can take $z = tg \in G \wr T$. If x, y are non-trivial modulo B, then simply by going over to the quotient group $W/B(\simeq T)$, the existence of $z \in T$ and hence in $G \wr T$, is immediate. Finally, suppose that $y \in B$ and $x \notin B$. Now $x = tb$, where $t \in T, t \neq 1, b \in B$. We need to express y in the form

$$y = b_1 \ldots b_m$$

where b_i is a non-trivial element in U^{t_i} and $t_1, \ldots, t_m$ are distinct elements of T. We claim that x does not commute with y (and hence that we can choose z=1). Otherwise

$$\{t_1 t, \ldots, t_m t\} = \{t_1, \ldots, t_m\} .$$

This means that right multiplication by t gives rise to a permutation of the finite set $\{t_1, \ldots, t_m\}$. Consequently a big enough power of t induces the identity permutation on $\{t_1, \ldots, t_m\}$ which implies that $t_1 t^k = t_1$ for a large enough choice of k. But T is torsion-free and so this is impossible. This completes the proof of the theorem. □

We now assume the hypothesis and notation of the above theorem. Then we have the following simple result.

Lemma 8.3.6. *Let $W = U \wr T$. Then the following hold:*
(1) *For an arbitrary finite set Z of elements from $G \wr T$ there exists a pair of non-trivial elements $x, y \in B$ such that $[x, y^z] = 1$ for each $z \in Z$;*
(2) *W is not a CSA-group; indeed, it is not even commutative transitive.*

Proof. (1) Let B be the normal closure of U in W. Then B is the direct product of the conjugates U^t of U by the elements $t \in T$. If we fix a finite set Z, then $[x, y^z] = 1$ for each $z \in Z$, provided the supports of all of the elements involved are disjoint. This is easy to arrange since T is torsion-free.

(2) Since U is a G-domain, it is non-Abelian. It follows that B is not commutative transitive and hence neither is W.

It follows that we can use wreath products to construct G-domains which are not CSA-groups. □

We prove next the following theorem which we will see again in Chapter 11.

Theorem 8.3.2. *Let A and B be domains. Suppose that C is a subgroup of both A and B satisfying the following condition:*

$$\text{if } c \in C,\, c \neq 1, \quad \text{either} \quad [c, A] \nsubseteq C \quad \text{or} \quad [c, B] \nsubseteq C. \tag{$\star$}$$

Then the amalgamated free product $H = A *_C B$ is a domain.

Proof. We will make use of the usual notation and terminology for working with elements of amalgamated products, described in Chapter 2. Let us note that if $C = A$ or $C = B$ then H is one of the factors and we have nothing to prove. We can assume therefore that C is a proper subgroup of both factors. The condition (*) implies that if $C \neq \{1\}$ then C is not simultaneously normal in both factors; in particular, its index is greater than 2 in at least one of the factors. Notice also that since both A and B are domains, neither of them is Abelian, unless they are trivial.

Suppose that x is a zero divisor in H. Then there exists a non-trivial element $y \in H$ such that $[y^h, x^{h'}] = 1$, for every choice of elements h, h' in H. So, in particular, $[x, y] = 1$. It follows then from a theorem of A. Karrass and D. Solitar (see [178], Theorem 4.5) that this is impossible unless one of the following conditions holds:
(1) Either x or y belongs to some conjugate of the amalgamated subgroup C;
(2) Neither x nor y is in a conjugate of C, but x is in a conjugate of a factor (A or B) and y is in the same conjugate of that factor;

(3) Neither x nor y is in a conjugate of a factor and $x = g^{-1}cgz^n$, $y = g^{-1}c^*gz^m$, where $c, c^* \in C$, and $g^{-1}cg, g^{-1}c^*g$ and z commute pairwise.

We will consider these three cases in turn.

If $x \in C^g$ then we can assume that $x \in C$. If $y \in A \cup B$, then $[y, x^w] \neq 1$ for a suitable choice of $w \in A \cup B$, because A and B are domains. So $y \notin A \cup B$. Choose $h \in H$ so that $y' = y^h$ is cyclically reduced. By the remark above we can assume that y' is of length at least two. So, replacing y' by its inverse if necessary, we find that

$$y' = a_1b_1 \ldots a_nb_n (n \geq 1) ,$$

where $a_i \in A - C, b_i \in B - C$, for each choice of i. As we mentioned above C is not normal in both A and B. Therefore there exists an element $v \in A \cup B$ such that $x^v \notin C$. For definiteness, suppose that $x^v \in A$. Then

$$x^v y' \neq y' x^v$$

because $x^v y'$ is of length at most n, whereas $y' x^v$ is of length $n + 1$. Thus the first case cannot arise.

We consider next, the second case. We can assume here that both x and y belong to one and the same factor. Since each factor is a domain, there exists an element t in the appropriate factor such that $[y, x^t] \neq 1$, which means that x is not an H-zero divisor. Hence this case cannot arise.

We are left with the third possibility. We can assume here that $x = cz^n$, $y = c_1z^m$, $c, c_1 \in C$ and c, c_1, z commute pairwise, and that the length of z is at least 2. We claim that there exists an element $f \in H$ such that if we write the elements z and f in reduced form then the products $cz^nf^{-1}cz^mf$ and $f^{-1}cz^mfcz^n$ are also in reduced form and not equal to one another. Indeed, if we choose the first syllable of f appropriately, then we can make sure that the product $f^{-1}cz^mf$ is reduced. Similarly, if we choose the last syllable of f appropriately, we can arrange that the products cz^nf^{-1} and fcz^n are reduced and moreover, that the last syllables in f and z either lie in different factors or lie in the same factor but have different right coset representatives. It follows then that $[y, x^{f^{-1}}] \neq 1$. This completes the proof of the theorem. □

We mention here two consequences of this theorem

Corollary 8.3.2. *Let A and B be domains. Then $A * B$ is a domain.*

Corollary 8.3.3. *Let A and B be domains and let C be a subgroup of both of them. If C is malnormal either in A or in B, then $A *_C B$ is a domain.*

It is worth pointing out that the condition $(*)$, above, is essential.

Example 8.3.1. Let F be free on x and y and $\bar{F}$ free on $\bar{x}$ and $\bar{y}$. Let, furthermore, $C = gp_F(x)$ and $\overline{C} = gp_{\bar{F}}(\bar{x})$. Then

$$H = F \underset{C=\overline{C}}{*} \bar{F}$$

is not an H-domain.

Proof. The elements y and $\bar{y}$ have exactly the same action (by conjugation) on the normal subgroup C in the group H. Hence the element $y^{-1}\bar{y}$ centralizes the subgroup C. It follows that $[y^{-1}\bar{y}, gp_H(x)] = 1$, i.e., x is an H-zero divisor.

Finally we have the following

Theorem 8.3.3. *The free product of two G-domains A and B, in the category of G-groups, is a G-domain whenever G is malnormal in both A and B.*

The proof can be carried out along the same lines as the proof of the theorem above. We note only that the first two cases are completely analogous, while the last case is even easier. It suffices here to note only that a non-trivial element of G cannot commute with an element of either A or B which is not contained in G.

It follows immediately then from this theorem that if A and B are free and G is a maximal cyclic subgroup of both of them, then the free product of A and B in the category of G-groups is a G-domain.

We will look again at domains in Chapter 11 when we consider co-discriminating groups.

8.3.2 Equationally Noetherian groups

We now look at the group theoretic counterpart of the Noetherian condition. The Noetherian condition in rings is defined in terms of the ascending chain condition (see [41]) and implies that every ideal is finitely generated. What is important about this condition in algebraic geometry is the Hilbert Basis theorem that asserts that every algebraic set is finitely based. That is if S be a set of polynomials in $k[x_1, \ldots, x_n]$ then $V(S) = V(S_1)$ for some finite set of polynomials. This is what is recast in terms of group theory.

Definition 8.3.4. A G-group H is said to be *G-equationally Noetherian* if for every $n > 0$ and every subset S of $G[x_1, \ldots, x_n]$ there exists a finite subset S_0 of S such that

$$V(S) = V(S_0).$$

In the case $H = G$ we omit all mention of G and simply say that H is an *equationally Noetherian group.*

If $S, S_0 \subseteq G[x_1, \ldots, x_n]$ and $V_H(S) = V_H(S_0)$ we will say that *the systems of equations $S = 1$ and $S_0 = 1$ are equivalent over H*. If $G = 1$, then a subset S of $G[X]$ is termed *coefficient-free* and we refer to the system $S = 1$ of equations as a *coefficient-free system.*

We respectively denote by $\mathcal{E}$ and $\mathcal{E}_G$ the class of all equationally Noetherian groups and the class of all G-equationally Noetherian groups. In particular, $\mathcal{E}_1$ is the class of all G-equationally Noetherian groups with $G = 1$ (i.e., the class of all groups that satisfy the Noetherian condition with respect to coefficient-free systems of equations). If $G' \subseteq G$ then every G-group H can be viewed also as a G'-group; clearly $\mathcal{E}_G \subseteq \mathcal{E}_{G'}$. It follows that an equationally Noetherian group H is G-equationally Noetherian for every choice of the subgroup G of H. The converse is also true.

Lemma 8.3.7. *Let H be a G-group. If G is finitely generated and $H \in \mathcal{E}_1$, then $H \in \mathcal{E}_G$.*

Proof. Let S be a subset of $G[x_1, \ldots, x_n]$. Each element $w \in S$ can be expressed functionally in the form

$$w = w(\bar{x}, \bar{a}) = w(x_1, \ldots, x_n, a_1, \ldots, a_k),$$

where $a_1, \ldots, a_k$ is a finite generating set for G. Let us replace the generators a_i by new variables, say y_i. The new set $S(\bar{x}, \bar{y}) \subseteq G[\bar{x}, \bar{y}]$ is coefficient-free; hence $V_H(S(\bar{x}, \bar{y})) = V_H(S_0(\bar{x}, \bar{y}))$ for some finite subset $S_0(\bar{x}, \bar{y}) \subseteq S(\bar{x}, \bar{y})$. Now consider the set of solutions of the system $S_0(\bar{x}, \bar{y}) = 1$ for which $y_i = a_i$. This is exactly the algebraic set $V_H(S_0(\bar{x}, \bar{a}))$. But these solutions also satisfy the whole system $S(\bar{x}, \bar{y}) = 1$; therefore $V_H(S_0(\bar{x}, \bar{a})) = V_H(S(\bar{x}, \bar{a})) = V_H(S)$. □

It is unknown if H is a G-group with the property that every one-variable system S contained in $G[x]$ is equivalent over H to a finite subset of itself, does this imply that H is G-equationally Noetherian?

Before providing examples of equationally noetherian groups we show that the class of equationally Noetherian groups is closed under subgroups, finite direct products and ultrapowers.

Theorem 8.3.4. *Let $\mathcal{E}_G$ be the class of all G-equationally Noetherian G-groups. Then the following hold:*

(1) *$\mathcal{E}_G$ is closed under G-subgroups, finite direct products and ultrapowers;*
(2) *$\mathcal{E}_G$ is closed under G-universal (G-existential) equivalence, i.e., if $H \in \mathcal{E}_G$ and H' is G-universally (G-existentially) equivalent to H, then $H' \in \mathcal{E}_G$;*
(3) *$\mathcal{E}_G$ is closed under separation, i.e., if $H \in \mathcal{E}_G$ and H' is G-separated by H, then $H' \in \mathcal{E}_G$.*

Proof. (1) Suppose, first, that a G-group H is G-equationally Noetherian and that H' is a G-subgroup of H. Let S be a subset of $G[x_1, \ldots, x_n]$. Since H is G-equationally Noetherian, there is a finite subset S_0 of S such that $V_H(S) = V_H(S_0)$. We claim that $V_{H'}(S) = V_{H'}(S_0)$. Indeed, if $v \in V_{H'}(S_0)$, then $v \in V_H(S_0) = V_H(S)$. Hence $v \in V_{H'}(S)$. It follows that $V_{H'}(S) = V_{H'}(S_0)$, as desired.

Suppose next that $H_1, \ldots, H_k$ are G-equationally Noetherian G-groups and that D is their direct product. According to the remarks made in the introduction, we can turn D into a G-group by choosing the copy of G in D to be the diagonal subgroup $\{(g, \ldots, g) : g \in G\}$ of D. Now let S be a subset of $G[x_1, \ldots, x_n]$. Then for each i, $V_{H_i}(S) = V_{H_i}(S_i)$, where S_i is a finite subset of S. Put $S_0 = S_1 \cup \cdots \cup S_k$. Then $V_D(S) =$

$V_D(S_0)$. This completes the proof of the most of (1). The statement about ultrapowers follows from (2) as indicated below.

(2) Suppose that $S \subseteq G_n$. Now $V_H(S) = V_H(S_0)$, where $S_0 = \{f_1, \ldots, f_k\}$ is a finite subset of S. For each $f \in S$, let

$$\phi_f = \forall x_1 \forall x_2 \ldots \forall x_n \, (\bigwedge_{i=1}^{k} f_i(x_1, \ldots, x_n) = 1 \rightarrow f(x_1, \ldots, x_n) = 1) .$$

Let Φ be the set of all such sentences. Then all of the sentences in Φ are satisfied in H and hence also in H'. This translates into $V_{H'}(S) = V_{H'}(S_0)$, as desired.

Notice that an ultrapower U of a G-group H is G-elementarily equivalent to H (see Chapter 4 or [40]). So, in particular, U and H are G-universally equivalent. Consequently, if $H \in \mathcal{E}_G$, then $U \in \mathcal{E}_G$.

(3) Let S be a subset of $G[X]$. Then there exists a finite subset S_0 of S such that $V_H(S) = V_H(S_0)$. We claim that this implies that $V_{H'}(S) = V_{H'}(S_0)$. For if this is not the case, there exists $v = (a_1, \ldots, a_n) \in H'^n$ such that S_0 vanishes at v but S does not vanish at v. So there exists an element $f \in S$ such that $f(v) \neq 1$. Choose now a G-homomorphism ϕ of H' into H so that $\phi(f(v)) = f((\phi(a_1), \ldots, \phi(a_n)) \neq 1$. But for each $f' \in S_0$, $f'(v) = 1$ and therefore $\phi(f'(v)) = f'(\phi(a_1), \ldots, \phi(a_n)) = 1$, i.e., $(\phi(a_1), \ldots, \phi(a_n))$ is a root of S_0. So $(\phi(a_1), \ldots, \phi(a_n))$ is a root of S and therefore $f(\phi(a_1), \ldots, \phi(a_n)) = 1$, a contradiction. □

The first major examples of equationally Noetherian groups are linear groups over commutative Noetherian rings. This was proved originally by R. Bryant [38] in the one variable case and then extended by V. Guba [123] to the case of free groups. The general result is the following.

Theorem 8.3.5. *Let H be a linear group over a commutative, Noetherian ring with unity and in particular a field. Then H is equationally Noetherian.*

Proof. Consider first the case where H is a subgroup of the general linear group $GL(n, K)$ over a field K. We can think of matrices from $GL(n, K)$ as elements of the K-vector space K^{n^2} of dimension n^2. The classical Zariski topology on K^{n^2}, which is Noetherian by Hilbert's basis theorem, induces the usual Zariski topology on $GL(n, K)$ and, consequently also on H, which is therefore Noetherian in this topology. Since multiplication and inversion in H are continuous functions in this induced Zariski topology, for every element $w(x_1, \ldots, x_n) \in G[x_1, \ldots, x_n]$, the set of all roots $V_H(w)$ of w is closed in this topology, since it is the pre-image of 1 under the continious map $H^n \longrightarrow H$ defined by w. This implies that every algebraic set over H is closed in this topology, because such sets are exactly intersections of sets of roots of single elements in $G[x_1, \ldots, x_n]$. In other words, the classical Zariski topology on H is a refinement of the non-commutative analogue of the Zariski topology on H that we have introduced here. It follows that this non-commutative Zariski topology is also Noetherian.

In general, the group $GL(n, R)$ over a commutative, Noetherian ring with unity R is a subgroup of a direct product of finitely many linear groups over a field; hence it is equationally Noetherian by Theorem 8.3.4. □

Recall that a free group of finite or countably infinite rank has a faithful representation in $SL(2, \mathbb{Z})$. It follows that any countable free group is linear and hence is equationally Noetherian.

Corollary 8.3.4. *Let F be a countable free group. Then F is equationally Noetherian.*

Not all equationally Noetherian groups are linear. We will prove that all Abelian groups are equationally Noetherian, thereby providing some additional examples. There are other examples of finitely generated equationally Noetherian non-linear groups, which are due to R. Bryant. He proved [38] that finitely generated Abelian-by-nilpotent groups are equationally Noetherian. Since the wreath product of a non-trivial, finitely generated Abelian group U by a finitely generated nilpotent group T is linear if and only if T is virtually Abelian [38] this provides us with more equationally Noetherian groups that are not linear. The paper [20] contains a further discussion about equationally Noetherian groups as well as additional examples of various kinds.

Theorem 8.3.6. *Every Abelian group A is equationally Noetherian.*

Proof. Every system $S = 1$ of equations over an Abelian group A is equivalent to a linear system, obtained by Abelianizing each of the elements in S. Each such linear system over A can be re-expressed in the form

$$S\colon m_{i1}x_1 + \cdots + m_{in}x_n = a_i \quad (i \in I, m_{ij} \in Z, a_i \in A)\,.$$

As usual (applying the Euclidean algorithm), this system is equivalent to a finite linear system $f_i = b_i$ ($b_i \in A$), $i = 1, \ldots, k$ in row-echelon form. So $k \leq n$. Notice, that all the equations $f_i = b_i$ are some integer linear combinations of a finite family S_0 of equations from the original system $S = 1$. Hence, $V_A(S_0) = V_A(S)$. This completes the proof. □

It is important to notice that the system of equations in row-echelon form obtained above consists of no more than n equations. However S_0 can contain an arbitrarily large number of equations. Consider the following example.

Example 8.3.2. Let $p_1, \ldots, p_n$ be distinct primes and suppose that a group H has elements of orders $p_1, \ldots, p_n$. Then the system

$$x^{p_1 \cdots p_{i-1} p_{i+1} \cdots p_n} = 1\,, \quad i = 1, \ldots, n\,,$$

is equivalent to the system

$$x = 1\,,$$

which is in row-echelon form. If we now re-express this latter system in terms of the original one, we find that all of the original equations are needed.

The example above suggests the following:

Problem. Let H be a G-group. Suppose that for every integer $n > 0$ and every subset S of $G[x_1, \dots, x_n]$ there exists a finite subset S_0 of $G[x_1, \dots, x_n]$ such that $V_H(S) = V_H(S_0)$. Is H G-equationally Noetherian?

The result on Abelian groups, Theorem 8.3.6 (see also Chapter 11) cannot be generalized to nilpotent groups. The following lemma shows that in general the restricted direct product of G-equationally Noetherian groups need not be G-equationally Noetherian. Hence this provides many examples of nilpotent groups that are not G-equationally Noetherian. On the other hand, direct powers of groups from $\mathcal{E}_1$ still belong to $\mathcal{E}_1$.

Lemma 8.3.8.
(1) *Let $\{H_i : i \in I\}$ be a family of G_i-equationally Noetherian G_i-groups. If there are infinitely many indices i for which G_i is not in the center of H_i, then the restricted direct product $D = \prod_{i\in I} H_i$ of the groups H_i, viewed as a $G = \prod_{i\in I} G_i$-group, is not G-equationally Noetherian.*
(2) *The class $\mathcal{E}_1$ is closed under unrestricted and restricted direct powers.*

Proof. (1) We choose an infinite subset J of I and elements $a_j \in G_j$ which are not in the center of H_j, $j \in J$. Consider now the subset $S = \{[x, a_j] : j \in J\}$ of $G[x]$. Then S is not equivalent to any of its finite subsets.

(2) Let $H \in \mathcal{E}_1$ and let I be a set of indices. Denote by H^I the unrestricted I-th power of H. Let $S \subseteq F(X)$, where $F(X)$ is the free group on X (i.e., $S = 1$ is a coefficient-free system of equations). Then $V_H(S) = V_H(S_0)$ for some finite subset S_0 of S. It is easy to see that $V_{H^I}(S) = V_{H^I}(S_0)$.

In view of the fact that $\mathcal{E}_1$ is closed under subgroups, this suffices for the proof of the lemma. □

The next result provides a description of which Baumslag–Solitar groups are equationally Noetherian.

Lemma 8.3.9. *Let*

$$B_{m,n} = \langle a, t; t^{-1}a^m t = a^n\rangle \quad (m > 0, n > 0) .$$

Then $B_{m,n}$ is equationally Noetherian provided either $m = 1$ or $n = 1$ or $m = n$; in all other cases $B_{m,n}$ does not belong to $\mathcal{E}_1$.

Proof. If either $m = 1$ or $n = 1$, then $B_{m,n}$ is metAbelian and linear. If $m = n$, observe that the normal closure N of t and a^m is the direct product of a free group of rank m and the infinite cyclic group on a^m. Moreover N is of index m in $B_{m,m}$. Now N is linear and hence so too is $B_{m,m}$. Thus in all the cases above $B_{m,n}$ is equationally Noetherian.

Suppose then that $m \neq 1, n \neq 1$ and $m \neq n$. On replacing t by t^{-1} if necessary, we can assume that m does not divide n. Observe that the elements

$$a^{m^k}, t^{-1}a^{m^k}t, \ldots, t^{-k}a^{m^k}t^k \tag{8.1}$$

all commute, but if j is chosen sufficiently large, a does not commute with $t^{-j}a^{m^k}t^j$. Let

$$S = \{[x_1, x_2^{-i}x_1x_2^i] : i = 1, 2, \ldots\} \subseteq F(x_1, x_2).$$

Suppose that $G \in \mathcal{E}_1$. Then there exists an integer $\ell > 0$ such that

$$V(S) = V(\{[x_1, x_2^{-i}x_1x_2^i] : i = 1, 2, \ldots, \ell\}).$$

But this implies that

$$[a^{m^\ell}, t^{-j}a^{m^\ell}t^j] \in V(S),$$

a contradiction.

It should be mentioned that $B_{m,n}$ with $m, n > 0$ is residually finite if and only if $m = 1$ or $n = 1$ or $m = n$. □

Notice that in the case of commutative rings, if R is a Noetherian ring then the ring of polynomials $R[x]$ is also Noetherian. We do not know whether the corresponding result, which we formulate here as a conjecture, also holds.

Conjecture 8.3.7. *Let G be an equationally Noetherian group. Then the free G-group $G[x_1, \ldots, x_n]$ is G-equationally Noetherian.*

We can prove this conjecture for various classes of equationally Noetherian groups, by making use of the specific properties of the groups involved. However the general conjecture remains unresolved. The best we can do is to prove the following theorem.

Theorem 8.3.8. *Let G be a linear group or a torsion-free hyperbolic equationally Noetherian group. Then the group $G[x_1, \ldots, x_n]$ is also equationally Noetherian.*

Proof. Suppose, first, that G is a linear group. The free product of G and a free group $F(x_1, \ldots, x_n)$ is again linear (see, for example,[252]); hence $G[x_1, \ldots, x_n]$ is equationally Noetherian.

If G is a torsion-free hyperbolic group, then the group $G[X]$ is G-separated by the G-equationally Noetherian group G (see [18]). Consequently $G[X]$ is G-equationally Noetherian by Theorem 8.3.4. Since $G[x_1, \ldots, x_n]$ is finitely generated and in $\mathcal{E}_1$, then by Lemma 8.3.7. $G[x_1, \ldots, x_n]$ is equationally Noetherian. □

The most general open questions in this direction are

Problem 8.3.9. Is the free product of two equationally Noetherian groups equationally Noetherian?

Problem 8.3.10. Is an arbitrary hyperbolic group equationally Noetherian?

8.3.3 Separation and discrimination

If $\mathcal{P}$ is a group property then the concepts of *residually-$\mathcal{P}$* and *fully-residually-$\mathcal{P}$* are fundamental. As we will see, the solution of the Tarski problems depends directly on the theory of fully residually free groups, a theory that we examined in great detail in the last two chapters.

In this section we look at residual and fully residual in a more general algebraic geometric setting where they are referred to as *separation* and *discrimination*. We will examine these again in Chapter 11. The results of this section hold in even more general algebraic settings but here we concentrate on groups.

Definition 8.3.5. Let H be a G-group. Then we say that a family

$$\mathcal{D} = \{D_i : i \in I\}$$

of G-groups, G-separates H, if for each non-trivial $h \in H$ there exists a group $D_i \in \mathcal{D}$ and a G-homomorphism $\phi: H \to D_i$ such that $\phi(h) \neq 1$.

We say that $\mathcal{D}$ G-discriminates H if for each finite subset $\{h_1, \dots, h_n\}$ of non-trivial elements of H there exists a $D_i \in \mathcal{D}$ and a G-homomorphism $\phi: H \to D_i$ such that $\phi(h_j) \neq 1,\ j = 1, \dots, n$.

If $\mathcal{D}$ consists of the singleton D, then we say that D G-separates H in the first instance and that D G-discriminates H in the second. If G is the trivial group, then we omit any mention of G and simply say that $\mathcal{D}$ separates H or $\mathcal{D}$ discriminates H.

Definition 8.3.6. Let $\mathcal{K}_G$ be the category of all G-groups, let $\mathcal{S}_G \subset \mathcal{K}_G$ be the subcategory of all G-groups G-separated by the singleton G and let $\mathcal{D}_G$ be the subcategory of $\mathcal{S}_G$ consisting of those G-groups which are G-discriminated by G.

If the group G has no G-zero divisors then according to [17] there exists a very simple criterion for a group in $\mathcal{S}_G$ to belong to $\mathcal{D}_G$. Benjamin Baumslag was the first to exploit this kind of argument in the case of free groups [22].

Theorem 8.3.11 ([17]). *Let G be a domain. Then a G-group H is G-discriminated by G if and only if H is a G-domain and H is G-separated by G.*

Proof. Let $H \in \mathcal{S}_G$ and suppose that H is a G-domain. Then for an arbitrary finite set $h_1, \dots, h_n$ of non-trivial elements of H there exist elements $z_2, \dots, z_n \in H$ such that the left-normed commutator $c = [h_1, h_2^{z_2}, \dots, h_n^{z_n}]$ is non-trivial. Hence we can separate c in G by a G-homomorphism $\phi: H \to G$ such that $\phi(c) \neq 1$. But this implies that $\phi(h_i) \neq 1$ for all i. This shows that G G-discriminates H.

Suppose now that $H \in \mathcal{D}_G$ and that f, h are two nontrivial elements of H. Then there exists a G-homomorphism $\phi: H \to G$ such that $\phi(f)$ and $\phi(h)$ are both non-trivial in G. Since G is a domain, it follows that $[\phi(f), \phi(h)^g] \neq 1$ for some $g \in G$. But then $[f, h^g] \neq 1$ in H, which shows that H is a G-domain.

It is not hard to see that if F is a non-Abelian free group, then $F \times F$ is separated by F, but it is not discriminated by F – this remark is due to [22]. □

Now we give an important characterization of finitely generated G-groups which are G-universally equivalent to G, when G is equationally Noetherian. This characterisation goes back to [214]in the case of free groups.

We say that the G-group H is *locally G-discriminated* by the G-group H' if every finitely generated G-subgroup of H is G-discriminated by H'.

Theorem 8.3.12. *Let H and H' be G-groups and suppose that at least one of them is G-equationally Noetherian. Then H is G-universally equivalent to H' if and only if H is locally G-discriminated by H' and H' is locally G-discriminated by H.*

Proof. Suppose that H is G-universally equivalent to H' and, furthermore, that one of them is G-equationally Noetherian. Then by Theorem 8.3.4 both H and H' are G-equationally Noetherian.

Let K be a finitely generated G-subgroup of H and let

$$K = \langle x_1, \dots, x_n; r_1 = 1, r_2 = 1, \dots \rangle$$

be a G-presentation of K with finitely many G-generators. The system of equations $\{r_i(x_1, \dots, x_n) = 1 : i = 1, 2, \dots\}$ is equivalent over H' to one of its finite subsets, say, $\{r_i(x_1, \dots, x_n) = 1 : i \leq m\}$. Let

$$u_1(x_1, \dots, x_n), \dots, u_k(x_1, \dots, x_n)$$

be arbitrary elements from K. We have to find a homomorphism from K to H' which separates the given elements $u_1(x_1, \dots, x_n), \dots, u_k(x_1, \dots, x_n)$ from the identity. The formula

$$\Phi = \exists x_1, \dots, x_n \Big(\bigwedge_1^m r_i(x_1, \dots, x_n) = 1 \bigwedge_1^k u_j(x_1, \dots, x_n) \neq 1 \Big)$$

holds in H and thus it holds in H'. Consequently, there exist elements $h_1, \dots, h_n$ in H' such that

$$r_1(h_1, \dots, h_n) = 1, \dots, r_m(h_1, \dots, h_n) = 1,$$

but

$$u_1(h_1, \dots, h_n) \neq 1, \dots, u_k(h_1, \dots, h_n) \neq 1.$$

It follows that $r_i(h_1, \dots, h_n) = 1,\ i = 1, 2, \dots$. Hence, the map

$$x_1 \longrightarrow h_1, \dots, x_n \longrightarrow h_n$$

can be extended to a G-homomorphism $\phi : K \longrightarrow H$ which separates the elements $u_1(x_1, \dots, x_n), \dots, u_k(x_1, \dots, x_n)$. This proves that K is G-discriminated by H'. Since

K was an arbitrary finitely generated G-subgroup of H it follows that H is locally G-discriminated by H'. Similarly, H' is locally G-discriminated by H.

Now suppose that the G-group H is locally G-discriminated by the G-group H'. We claim that any existential sentence with constants from G which holds in H holds also in H'. Let us consider a formula of the type

$$\Phi = \exists x_1,\ldots,x_n \Big(\bigwedge_1^s u_i(x_1,\ldots,x_n) = 1 \bigwedge_1^t v_j(x_1,\ldots,x_n) \neq 1 \Big),$$

where the words u_i and v_j may contain constants from G. Let the elements $a_1,\ldots,a_n \in H$ satisfy this sentence in H. Denote by K the G-subgroup generated by $\{a_1,\ldots,a_n\}$. By hypothesis, there exists a G-homomorphism $f\colon K \longrightarrow H'$ which separates the elements

$$v_1(a_1,\ldots,a_n),\ldots,v_t(a_1,\ldots,a_n)$$

in H'. This implies that the images $f(a_1),\ldots,f(a_n)$ satisfy the same equalities $u_i(f(a_1),\ldots,f(a_n)) = 1$, $i = 1,\ldots,s$ and the same inequalities

$$v_i(f(a_1),\ldots,f(a_n)) \neq 1,\ i = 1,\ldots,t$$

in H'. Therefore, the sentence Φ holds in H'. This shows that H is G-existentially equivalent to H', and hence H is G-universally equivalent to H'. □

A G-subgroup H of the G-group H' is said to be *G-existentially closed* in H', if any existential sentence with constants from G holds in H' if and only if it holds in the subgroup H.

Corollary 8.3.5. *Let H be a G-group which is locally G-discriminated by G. Then G is G-existentially closed in H and, in particular, G is universally equivalent to H.*

The proof follows along exactly the same lines as the proof of the second part of Theorem 8.3.10.

8.4 The affine geometry of G-groups

Crucial to algebraic geometry are affine algebraic sets and affine varieties. Here we translate these ideas to the context of G-groups.

Recall that *affine n-space over the G-group H* is defined as

$$H^n = \{(a_1,\ldots,a_n)\colon a_i \in H\}.$$

Let S be a subset of $G[X]$. Then the set

$$V_H(S) = \{v \in H^n\colon f(v) = 1, \text{ for all } f \in S\}.$$

is termed the *(affine) algebraic set* over H defined by S.

We sometimes denote $V_H(\{s_1,s_2,\ldots\})$ by $V_H(s_1,s_2,\ldots)$.

The union of two algebraic sets in H^n is not necessarily an algebraic set. We define a topology on H^n by taking as a sub-basis for the closed sets of this topology, the algebraic sets in H^n. We term this topology the *Zariski topology*. If H is a G-domain, then the union of two algebraic sets is again algebraic and so in this case the closed sets in the Zariski topology consist entirely of algebraic sets.

Then, fixing the G-group H, these algebraic sets can be viewed as the objects of a category, where morphisms are defined by polynomial maps, i.e., if $Y \subseteq H^n$, and $Z \subseteq H^p$ are algebraic sets then a map $\phi\colon Y \to Z$ is a *morphism* in this category (or a *polynomial map*) if there exist $f_1, \dots, f_p \in G[x_1, \dots, x_n]$ such that for any $(a_1, \dots, a_n) \in Y$

$$\phi(a_1, \dots, a_n) = (f_1(a_1, \dots, a_n), \dots, f_p(a_1, \dots, a_n)) .$$

It turns out that this category is isomorphic to a sub-category of the category of all G-groups. In order to explain more precisely what this sub-category consists of we need to turn our attention to the *ideals* of algebraic sets. In this section we make all these ideas precise.

8.4.1 Algebraic sets and the Zariski topology

From the previous discussion we have that an algebraic set is defined in the following way:

Definition 8.4.1. Let S be a subset of $G[X]$. Then the (affine) algebraic set over H defined by S is the set

$$V_H(S) = \{v \in H^n\colon\ f(v) = 1 \ \text{ for all } \ f \in S\}$$

We sometimes denote $V_H(S)$ simply by $V(S)$.

The following example provides us with algebraic sets.

Example 8.4.1. Let H be a G-group. Then the following subsets of H are algebraic:
(1) The G-singleton $\{a\}$ (here $a \in G$):

$$V_H(xa^{-1}) = \{a\} ;$$

(2) For any subset M of H, the centralizer $C_H(M)$:

$$V_H(\{[x, m]\colon m \in M\}) = C_H(M) .$$

The next lemma allows one to construct algebraic sets in higher dimensions.

Lemma 8.4.1. *Let H be a G-group and let U and W be affine algebraic sets in H^n and H^p, respectively. Then $U \times W$ is an algebraic set in H^{n+p}.*

The proof is analogous to that of the corresponding theorem in algebraic geometry. Indeed, if $U = V(S)$, where $S \subseteq G[x_1, \dots, x_n]$ and $W = V(T)$, where $T \subseteq G[y_1, \dots, y_p]$, then

$$U \times W = V(S \cup T),$$

where we view $S \cup T$ as a subset of $G[x_1, \dots, x_n, y_1, \dots, y_p]$.

The following lemma is useful.

Lemma 8.4.2. *Let H be a G-group and H^n an affine n-space over H. Then for arbitrary subsets S_i of $G[X]$, the following hold:*

(1) $V_H(1) = H^n$;
(2) $V_H(g) = \emptyset$ *for any non-trivial* $g \in G$;
(3) $S_1 \subseteq S_2 \Longrightarrow V_H(S_1) \supseteq V_H(S_2)$;
(4) $V_H(S) = V_H(gp_{G[X]}(S))$;
(5) $\bigcap_{i \in I} V_H(S_i) = V_H(\bigcup_{i \in I} S_i)$;
(6) *If H is a G-domain, then*

$$\begin{aligned} V_H(S_1) \cup V_H(S_2) &= V_H([s_1, s_2^g] \colon s_i \in S_i, g \in G) \\ &= V_H(s_1 \diamond s_2 \colon s_i \in S_i, i = 1, 2); \end{aligned}$$

(7) *If H is a G-domain, then*

$$V_H(Q_1) \cup V_H(Q_2) = V_H(Q_1 \cap Q_2),$$

for any ideals Q_1, Q_2 of $G[X]$.

Proof. The verification of the first 5 assertions is straightforward and will be left to the reader. In order to prove (6) we first prove that

$$V_H(\{[s_1, s_2^g] \colon s_i \in S_i, g \in G\}) \subseteq V_H(S_1) \cup V_H(S_2).$$

Suppose, that $v = (a_1, \dots, a_n) \in H^n$, $v \notin V_H(S_1) \cup V_H(S_2)$. So $v \notin V_H(S_1)$, $v \notin V_H(S_2)$. Hence there exist $s_1 \in S_1$ and $s_2 \in S_2$ such that

$$s_1(v) \neq 1, s_2(v) \neq 1.$$

Since H is a G-domain, there exists $g \in G$, such that

$$[s_1(v), s_2(v)^g] \neq 1.$$

Hence

$$v \notin V_H(\{[s_1, s_2^g] \colon s_i \in S_i, g \in G\}).$$

The reverse inclusion is immediate, which proves (6).

Now (7) follows from (6) since

$$Q_1 \diamond Q_2 \subseteq [Q_1, Q_2] \subset Q_1 \cap Q_2$$

and hence

$$V_H(Q_1) \cup V_H(Q_2) = V_H(Q_1 \diamond Q_2) \supseteq V_H([Q_1, Q_2]) \supseteq V_H(Q_1 \cap Q_2) .$$

The reverse inclusion is obvious. □

Notice, that the $\diamond$-product plays exactly the same role as multiplication of polynomials in the case of polynomial algebras.

The main consequence of the preceding lemma is that we can now define the *Zariski topology*.

Theorem 8.4.1. *Let H be a G-group. We define a subset of H^n to be closed if it is the intersection of an arbitrary number of finite unions of algebraic sets in H^n; this defines a topology on H^n, called the Zariski topology. The Zariski topology is a T_1-topology, i.e., the singletons are closed sets. Moreover if H is a G-domain, then the closed sets in the topology are the algebraic sets.*

This kind of topology was first introduced by R. Bryant [38] when $n = 1$.

Notice that continuity of maps in the Zariski topology depends only on the algebraic sets. The next lemma is an immediate consequence of this remark.

Lemma 8.4.3. *Let H and K be G-groups and let $f : H^n \longrightarrow K^m$ be a map. If the preimage of an algebraic set of K^m is an algebraic set of H^n, then the map f is continuous in the Zariski topology.*

We remark that, in general, the union of two algebraic sets need not be algebraic.

Example 8.4.2. Let A be an Abelian group, viewed as an A-group. Then any algebraic set in A^n is a coset with respect to some subgroup of A^n, where A^n is now an Abelian group, under the operation of coordinate-wise multiplication.

Indeed, suppose $S \subseteq A[x_1, \dots, x_n]$. If $s \in S$, then s can be written in the form $s = s^* s'$, where s' is in the derived group of $A[x_1, \dots, x_n]$ and

$$s^* = x_1^{m(s)_1} x_2^{m(s)_2} \dots x_n^{m(s)_n} a_s \ (a_s \in A) .$$

Put $S^* = \{s^* : s \in S\}$. Since A is Abelian $V_A(S) = V_A(S^*)$, which means that $V_A(S)$ consists exactly of all solutions in A^n of the multiplicatively written, linear system of equations

$$x_1^{m(s)_1} x_2^{m(s)_2} \dots x_n^{m(s)_n} = a_s^{-1} \ (s \in S). \qquad (\star)$$

The algebraic set $V_A(S_1)$ of the corresponding homogeneous system S_1:

$$x_1^{m(s)_1} x_2^{m(s)_2} \dots x_n^{m(s)_n} = 1 \ (s \in S) ,$$

is a subgroup of A^n and as usual

$$V_A(S) = bV_A(S_1)$$

where $b = (b_1, \ldots, b_n) \in A^n$ is an arbitrarily chosen solution of the system of equations given by Example 4 on Page 51 of [16].

Since the union of two such cosets need not be a coset, for example the union of two distinct cosets with respect to the same subgroup, it follows that the union of two algebraic sets need not be an algebraic set.

Note that the fact that $V_A(S) = bV_A(S_1)$ is essentially the gerelaization to ableian groups the well known argument that every solution of a system of linear equations is the sum of a particular solution ands the general solution of the associated homeogeneous system.

8.4.2 Ideals of algebraic sets

To understand the catgory of affine algebraic sets we must consider the ideals of algebraic sets. Recall that an ideal in a G-group is the kernel of a G-homomorphism.

As before, let H be a G-group, n a positive integer, H^n affine n-space over H and $G[X] = G[x_1, \ldots, x_n]$.

Definition 8.4.2. Let $Y \subseteq H^n$. Then

$$I_H(Y) = \{f \in G[X] \colon f(v) = 1 \text{ for all } v \in Y\}.$$

Notice, that $I_H(\emptyset) = G[X]$ and $1 \in I_H(Y)$ for every Y.

Lemma 8.4.4. *For any $Y \subseteq H^n$ the set $I_H(Y)$ is an ideal of the G-group $G[X]$.*

Proof. In order to prove the lemma, observe that every point $v = (a_1, \ldots, a_n) \in H^n$ can be used to define a G-homomorphism

$$\phi_v \colon G[X] \longrightarrow H$$

by *the evaluation map*:

$$\phi_v(f) = f(v), \; f \in G[X],$$

i.e. $\phi_v \colon x_i \mapsto a_i, g \mapsto g$, where $i = 1, \ldots, n$, $g \in G$. Now observe that

$$I_H(Y) = \bigcap_{v \in Y} \operatorname{Ker} \phi_v \, .$$

Since the intersection of ideals in a G-group is again an ideal, it follows that $I_H(Y)$ is an ideal of $G[X]$. □

Definition 8.4.3. An ideal in $G[X]$ is termed an H-closed ideal if it is of the form $I_H(Y)$ for some subset Y of H^n.

Sometimes we omit the subscript H and simply write $I(Y)$.

Definition 8.4.4. If Y is an affine algebraic set in H^n, then $I(Y)$ is termed the ideal of Y.

The various parts of the following lemma are either consequences of the foregoing discussion or can be proved directly from the definitions.

Lemma 8.4.5. *Let Y, Y_1, Y_2 be subsets of H^n. Then the following hold.*
(1) $Y_1 \subseteq Y_2 \Longrightarrow I(Y_1) \supseteq I(Y_2)$.
(2) $I(Y_1 \cup Y_2) = I(Y_1) \cap I(Y_2)$.
(3) $I(Y_1 \cap Y_2) \supseteq I(Y_1) \cup I(Y_2)$.
(4) $S \subseteq G[X] \Longrightarrow I(V(S)) \supseteq gp_{G[X]}(S)$.
(5) *If Y is an algebraic set in H^n, then $V(I(Y)) = Y$.*
(6) *If Q is an H-closed ideal of $G[X]$, then $I_H(V(Q)) = Q$.*

One consequence of the above lemma that is worth drawing attention to is the following.

Corollary 8.4.1.
(1) *If Y_1 and Y_2 are algebraic sets in H^n, then*

$$Y_1 = Y_2 \Longleftrightarrow I_H(Y_1) = I_H(Y_2)\,;$$

(2) *The functions I and V are inclusion reversing inverses when applied to the algebraic sets in H^n and the H-closed ideals in $G[X]$.*

Later we will need the following lemma.

Lemma 8.4.6. *Suppose that the G-group H is a G-domain and that Y is a subset of H^n. Then $V(I(Y)) = \overline{Y}$, the closure of Y in the Zariski topology on H^n.*

Proof. Let C be a closed subset in H^n containing Y. The set C is an algebraic set for H is a G-domain. Then $I(C) \subseteq I(Y)$, and hence $C = V(I(C)) \supseteq V(I(Y))$. Therefore $V(I(Y))$ is the minimal closed subset of H^n containing Y, i.e., $V(I(Y)) = \overline{Y}$. □

8.4.3 Morphisms of algebraic sets

Classical algebraic geometry depends on polynomial functions on affine n-space. In this section we extend the idea of a polynomial function to the group theoretic setting.

As before we shall assume that H is a G-group, that $G[X] = G[x_1, \dots, x_n]$ and that H^n is an affine n-space over H. We first define a polynomial function on H^n.

Definition 8.4.5. Let $f(x_1, \dots, x_n) \in G[X]$. The map $\mu_f \colon H^n \longrightarrow H$ defined by

$$\mu_f(a_1, \dots, a_n) = f(a_1, \dots, a_n)$$

is termed the polynomial function on H^n defined by f. Its restriction to an algebraic set Y in H^n is similarly termed a polynomial function on Y.

Lemma 8.4.7. *Any polynomial function $\mu_f \colon H^n \longrightarrow H$ is continuous in the Zariski topology.*

Proof. By Lemma 8.4.3 it is enough to prove that the pre-image of any algebraic set $Z \subseteq H$ is an algebraic set in H^n. Now, if $Z = V_H(S)$ then

$$\mu_f^{-1}(Z) = V_H(s(f(x_1, \ldots, x_n)) \mid s \in S) .$$

Hence $\mu_f^{-1}(Z)$ is an algebraic set in H^n. □

Lemma 8.4.8. *Group multiplication and inversion in H are continuous in the Zariski topology.*

To prove this it suffices to notice that multiplication is described by x_1x_2 and inversion by x_1^{-1}.

Corollary 8.4.2. *For any $a, b \in G$ the map $h \longrightarrow ahb$ is a homeomorphism of H in the Zariski topology.*

We now define a morphism of algebraic sets and the notion of isomorphism of algebraic sets.

Definition 8.4.6. Let H be a G-group and let $Y \subseteq H^n$, $Z \subseteq H^p$ be algebraic sets. Then a map

$$\phi\colon Y \longrightarrow Z$$

is termed a morphism of the algebraic set Y to the algebraic set Z if there exist $f_1, \ldots, f_p \in G[X]^n$ such that for any $(a_1, \ldots, a_n) \in Y$

$$\phi(a_1, \ldots, a_n) = (f_1(a_1, \ldots, a_n), \ldots, f_p(a_1, \ldots, a_n)).$$

Definition 8.4.7. Two algebraic sets Y and Z are said to be isomorphic if there exist morphisms

$$\phi\colon Y \longrightarrow Z,\ \ \theta\colon Z \longrightarrow Y$$

such that $\theta\phi = 1_Y, \phi\theta = 1_Z$.

We then have the following

Lemma 8.4.9. *Every morphism from an algebraic set $Y \subseteq H^n$ to an algebraic set $Z \subseteq H^p$ is a continuous map in the Zariski topology.*

The proof is similar to the proof of Lemma 8.4.7 and is therefore omitted.

Corollary 8.4.3. *The canonical projection $H^{n+m} \longrightarrow H^m$ is continuous in the Zariski topology.*

8.4.4 Coordinate groups

If H is a G-group, then the category of all algebraic sets over H is equivalent to the category of all coordinate groups defined over H, which is exactly the category of all finitely generated G-groups that are G-separated by H. We will show this in the next three sections and in Section 8.5 where we examine the abstract description of coordinate groups.

Let as before, H be a G-group, let $G[X] = G[x_1, \dots, x_n]$ and let $Y \subseteq H^n$ be an algebraic set defined over G.

We denote the set of all polynomial functions on Y by $\Gamma(Y)$. If $\mu, \nu \in \Gamma(Y)$, we define the product of μ and ν by

$$\mu\nu(y) = \mu(y)\nu(y) \quad (y \in Y)$$

and the inverse of μ by

$$\mu^{-1}(y) = \mu(y)^{-1} \quad (y \in Y).$$

Lemma 8.4.10. *The set $\Gamma(Y)$ of all polynomial functions on Y forms a G-group with respect to multiplication and inversion, as defined above, with G embedded in $\Gamma(Y)$ via the mapping $g \longrightarrow \mu_g$ of G into $\Gamma(Y)$ $(g \in G)$.*

The proof is straightforward.

Definition 8.4.8. $\Gamma(Y)$ is called the coordinate group of the algebraic set Y.

For each $i = 1, \dots, n$, define on $Y \subseteq H^n$ the polynomial function

$$t_i \colon (a_1, \dots, a_n) \mapsto a_i \quad ((a_1, \dots, a_n) \in Y).$$

These *coordinate functions* all belong to $\Gamma(Y)$. The following is an immediate consequence of the definitions.

Lemma 8.4.11. *The map $x_1 \mapsto t_1, \dots, x_n \mapsto t_n$ defines a G-epimorphism from $G[X]$ onto $\Gamma(Y)$ with kernel $I_H(Y)$. Hence*

$$G[x_1, \dots, x_n]/I_H(Y) \cong \Gamma(Y).$$

Denote

$$G_{R(Y)} = G[x_1, \dots, x_n]/I_H(Y).$$

In what follows we will refer to the group $G_{R(Y)}$ also as the coordinate group of the algebraic set Y.

Corollary 8.4.4. *For any algebraic set $Y \subseteq H^n$ the coordinate group $\Gamma(Y)$ is G-separated by H.*

Proof. Since $\Gamma(Y)$ is G-isomorphic to $G[x_1, \dots, x_n]/I_H(Y)$ it suffices to prove that the latter group is G-separated by H. The separation now follows from the description of $I_H(Y)$ as the intersection of the kernels of G-homomorphisms from $G[x_1, \dots, x_n]$ into H. □

In particular, if $G = H$, then the coordinate groups $\Gamma(Y)$ are all G-separated by G, i.e., they all are contained in the category $\mathcal{S}_G$.

8.4.5 Equivalence of the categories of affine algebraic sets and coordinate groups

We now show, as mentioned at in the last section, that if H is a G-group, then the category of all algebraic sets over H is equivalent to the category of all coordinate groups defined over H.

Let H be a G-group and let $\mathbf{AS}_H$ be the category of all algebraic sets over H defined by systems of equations with coefficients in G (morphisms in $\mathbf{AS}_H$ are the morphisms of algebraic sets defined above). Denote by $\mathbf{AG}_H$ the category of all coordinate groups of the algebraic sets in $\mathbf{AS}_H$ (morphisms in $\mathbf{AG}_H$ are G-homomorphisms). Notice that both categories are defined relative to a given G-group H.

Theorem 8.4.2. *Let H be a G-group. Then the categories $\mathbf{AS}_H$ and $\mathbf{AG}_H$ are equivalent to each other.*

We have to define two functors $\mathcal{F}\colon \mathbf{AS}_H \longrightarrow \mathbf{AG}_H$ and $\mathcal{G}\colon \mathbf{AG}_H \longrightarrow \mathbf{AS}_H$ such that

$$\mathcal{G}\mathcal{F} = 1_{\mathbf{AS}_H}, \quad \mathcal{F}\mathcal{G} = 1_{\mathbf{AG}_H}.$$

If Y is an algebraic set defined over H, then $\mathcal{F}(Y)$ is the coordinate group of Y:

$$\mathcal{F}(Y) = \Gamma(Y).$$

In order to define $\mathcal{F}$ on morphisms, suppose that $Y \subseteq H^n$ and $Z \subseteq H^p$ are algebraic sets and $\phi\colon Y \to Z$ is a morphism from Y to Z. We then define

$$\mathcal{F}(\phi)\ \Gamma(Z) \to \Gamma(Y)$$

as follows:

$$\mathcal{F}(\phi)(f) = f \circ \phi,$$

where f is a polynomial function on Z (i.e., an element from $\Gamma(Z)$) and $\circ$ denotes the composition of functions. It is not hard to see that $\mathcal{F}(\phi)$ is a G-homomorphism of G-groups and that

$$\mathcal{F}(\Psi\phi) = \mathcal{F}(\phi)\mathcal{F}(\Psi), \quad \mathcal{F}(1_Y) = 1_{\Gamma(Y)}.$$

Notice that $\mathcal{F}$ is a contravariant functor. We define next the functor $\mathcal{G}\colon \mathbf{AG}_H \longrightarrow \mathbf{AS}_H$ as follows. Since the objects in $\mathbf{AG}_H$ are simply the $\Gamma(Y)$, where Y is an algebraic set in some H^n, we define

$$\mathcal{G}(\Gamma(Y)) = Y.$$

Next we define $\mathcal{G}$ on G-homomorphisms. To this end, let

$$\vartheta\colon \Gamma(Y) \to \Gamma(Z)$$

be a G-homomorphism from one coordinate group $\Gamma(Y)$ to another coordinate group $\Gamma(Z)$. Now we use the G-isomorphisms

$$\Gamma(Y) \cong G[x_1,\dots,x_n]/I(Y), \quad \Gamma(Z) \cong G[x_1,\dots,x_k]/I(Z).$$

So

$$\vartheta(x_i I(Y)) = w_i(x_1, \ldots, x_k)I(Z),\ i = 1, \ldots, n\,.$$

Define now $\mathcal{G}(\vartheta)\colon Z \to Y$ by

$$\mathcal{G}(\vartheta)((a_1, \ldots, a_k)) = \left(w_1(a_1, \ldots, a_k), \ldots, w_n(a_1, \ldots, a_k) \right)\,.$$

It is not hard to verify then that $\mathcal{G}$ is a contravariant functor from $\mathbf{AG}_H$ to $\mathbf{AS}_H$. Moreover,

$$\mathcal{G}\mathcal{F} \cong 1_{\mathbf{AS}_H}, \quad \mathcal{F}\mathcal{G} \cong 1_{\mathbf{AG}_H}\,.$$

Corollary 8.4.5. *Let H be a G-group. Then algebraic sets $Y_1 \subset H^n$ and $Y_2 \subset H^p$ are isomorphic if and only if the coordinate groups $\Gamma(Y_1)$ and $\Gamma(Y_2)$ are G-isomorphic.*

8.4.6 The Zariski topology of equationally Noetherian groups

In this section we show that the equationally Noetherian property for a G-group H can be expressed in terms of the descending chain condition on algebraic sets over H. This of course impacts the structure of the corresponding Zariski topology.

Theorem 8.4.3. *Let H be a G-group. Then H is G-equationally Noetherian if and only if every properly descending chain of algebraic sets over H is finite.*

Proof. Suppose that H is G-equationally Noetherian. Every strictly descending chain of algebraic sets in H^n

$$V_1 \supset V_2 \supset \ldots$$

gives rise to a strictly ascending chain of ideals:

$$I(V_1) \subset I(V_2) \subset \ldots\,.$$

Put

$$S = \bigcup_i I(V_i)\,.$$

Then $V_H(S) = V_H(S_0)$ for some finite subset S_0 of S from the euationally Noetherian property. But then

$$S_0 \subseteq I(V_m)$$

for some m, which implies that

$$\bigcap_i V_i = V_H(S) = V_H(I(V_m)) = V_m\,,$$

Therefore the chain terminates in no more than m steps.

Conversely assume that the set of all algebraic sets over H satisfies the descending chain condition. Let S be a subset of $G[x_1, \ldots, x_n]$. If $V_H(S) \neq V_H(S_0)$ for any finite subset S_0 of S, then there exists an infinite sequence $s_1, s_2, \dots$ of elements of S such that

$$V_H(s_1) \supset V_H(s_1, s_2) \supset \dots$$

is an infinite, strictly descending chain of algebraic sets, a contradiction. Hence H is G-equationally Noetherian. □

We recall that a topological space is termed *Noetherian* if it satisfies the descending chain condition on closed subsets.

Theorem 8.4.4. *Let H be a G-group. Then for each integer $n > 0$, the Zariski topology on H^n is Noetherian if and only if H is G-equationally Noetherian.*

Proof. Suppose that the G-group H is G-equationally Noetherian. We need to prove that H^n is Noetherian for every $n > 0$ in the Zariski topology. It follows from Theorem 8.4.3 that the set A of all algebraic sets contained in H^n satisfies the descending chain condition. Let A_1 be the set of all finite unions of the sets in A and let A_2 be the set of all (possibly infinite) intersections of sets in A_1. By the definition of the Zariski topology, A_2 is the set of closed subsets of H^n. We first prove that A_1 satisfies the descending chain condition. Suppose that $M_1 = V_1 \cup \cdots \cup V_m$ and that $M_2 = W_1 \cup \cdots \cup W_k$ are sets in A_1 and that $M_1 \supset M_2$. Then for every $i \leq m$ we have $V_i \supset V_i \cap W_j$ which gives rise to a tree of subsets with root vertex V_i and with a unique edge from the root to every proper subset of the form $V_i \cap W_j$. A strictly descending chain of sets in A_1, say

$$M_1 \supset M_2 \supset \dots ,$$

gives rise to m trees of subsets such that each vertex of each tree is a finite intersection of sets in A, hence in A; moreover, for each such vertex there are only finitely many outgoing edges. In the resultant graph, every path corresponds to a strictly descending chain of algebraic sets and so is finite. By Koenig's lemma (see [16]) this implies that the whole graph is finite. Therefore, the above chain is also finite.

Since A_1 satisfies the descending chain condition and is closed under finite intersections, the intersection of an arbitrary collection of sets in A_1 is the intersection of some finite subcollection; hence it is also in A_1. Consequently, $A_2 = A_1$ and hence satisfies the descending chain condition. Consequently, H^n is Noetherian in the Zariski topology. Conversely, if H^n is Noetherian for every n, then H is G-equationally Noetherian by Theorem 8.4.3. This completes the proof. □

Corollary 8.4.6. *Let H be a G-equationally Noetherian group. Then every closed set in H^n is a finite union of algebraic sets.*

A non-empty subset Y of a topological space X is said to be *irreducible* if it can not be expressed as the union $Y = Y_1 \cup Y_2$ of two proper subsets, each one of which is closed in Y, in the induced topology.

Lemma 8.4.12. *In a Noetherian topological space X every non-empty closed subset Y can be expressed as a finite union $Y = Y_1 \cup \cdots \cup Y_n$ of irreducible closed subsets Y_i. If we require that $Y_i \nsubseteq Y_j$ for $i \neq j$, then the Y_i, the so-called irreducible irreducible components of Y, are unique (up to order).*

The proof is standard (see, for example [130].)

Corollary 8.4.7. *Let H be a G-equationally Noetherian group. Then every subset Y of H^n, which is closed in the Zariski topology, is a finite union of irreducible algebraic sets, each of which is uniquely determined by Y.*

Theorem 8.4.5. *Let H be a G-equationally Noetherian group and let Y be a subset of H^n, closed in the Zariski topology. If $Y = Y_1 \cup \cdots \cup Y_m$ is the decomposition of Y into its irreducible components, then the coordinate group $\Gamma(Y)$ is canonically embedded into the direct product of the coordinate groups $\Gamma(Y_i)$:*

$$\Gamma(Y) \hookrightarrow \Gamma(Y_1) \times \cdots \times \Gamma(Y_m)\,.$$

Proof. The irreducible decomposition $Y = Y_1 \cup \cdots \cup Y_m$ implies (by Lemma 8.4.5) the corresponding decomposition of the ideal $I(Y)$:

$$I(Y) = I(Y_1) \cap \cdots \cap I(Y_m).$$

The canonical homomorphisms

$$\lambda_i \colon \Gamma(Y) = G[X]/I(Y) \longrightarrow G[X]/I(Y_i) \cong \Gamma(Y_i)$$

give rise to an embedding

$$\lambda \colon \Gamma(Y) \longrightarrow \Gamma(Y_1) \times \cdots \times \Gamma(Y_m)$$

where $\lambda = \lambda_1 \times \cdots \times \lambda_m$. □

The following lemma establishes a very important property of coordinate groups of irreducible closed sets.

Lemma 8.4.13. *Let H be a G-equationally Noetherian group. If a closed set $Y \subseteq H^n$ is irreducible, then the coordinate group $\Gamma(Y)$ is G-discriminated by H.*

Proof. Let $S \subseteq G[X]$ and $Y = V_H(S)$. As we mentioned above $\Gamma(Y) \cong G[X]/I(S)$. Suppose that there exist finitely many non-trivial elements $u_1 I(S), \dots, u_n I(S)$ in $G[X]/I(S)$ that can not be discriminated in H by a G-homomorphism, i.e., for any G-homomorphism $\phi \colon G[X]/I(S) \longrightarrow H$ there exists an i such that $\phi(u_i I(S)) = 1$. It follows then that

$$Y = V_H(S) = V_H(S \cup \{u_1\}) \cup \cdots \cup V_H(S \cup \{u_n\}).$$

On the other hand, since $G[X]/I(S)$ is G-separated by H, (see Corollary 8.4.4), $V_H(S \cup \{u_i\})$ is a proper closed subset of $V_H(S) = Y$. This contradicts the irreducibility of Y.

In particular, if $G = H$ is equationally Noetherian, then the coordinate group $\Gamma(Y)$ of an irreducible closed set $Y \subseteq G^n$ is G-discriminated by G, that is $\Gamma(Y)$ belongs to the category $\mathcal{D}_G$. □

In the case when H is a G-equationally Noetherian G-domain we have the following important characterization of irreducible closed sets in terms of their coordinate groups.

Theorem 8.4.6. *Let H be a G-equationally Noetherian G-domain. Then a closed subset $Y \subseteq H^n$ is irreducible if and only if the coordinate group $\Gamma(Y)$ is G-discriminated by H.*

Proof. The only if statement has been already proved in Lemma 8.4.13. Now, let the coordinate group $\Gamma(Y)$ of a closed set $Y \subseteq H^n$ be G-discriminated by H. Suppose

$$Y = Y_1 \cup \cdots \cup Y_m$$

is the decomposition of Y into its irreducible components, then

$$I(Y) = I(Y_1) \cap \cdots \cap I(Y_m)$$

and by Theorem 8.4.5 there exists an embedding

$$\lambda\colon \Gamma(Y) \longrightarrow \Gamma(Y_1) \times \cdots \times \Gamma(Y_m)$$

which is induced by the canonical epimorphisms

$$\lambda_i\colon \Gamma(Y) = G[X]/I(Y) \longrightarrow G[X]/I(Y_i) \cong \Gamma(Y_i)\,.$$

We claim that at least one of these canonical epimorphisms is an isomorphism. Indeed, suppose that each epimorphism λ_i has a non-trivial kernel. Choose an arbitrary non-trivial element u_i from the kernel of λ_i, $i = 1, \ldots, m$. The group $\Gamma(Y)$ is G-discriminated by the G-domain H; hence $\Gamma(Y)$ is also a G-domain. Therefore, there are elements $g_2, \ldots, g_m \in G$ such that the commutator

$$u = [u_1, u_2^{g_2}, \ldots, u_m^{g_m}]$$

is non-trivial. Hence $\lambda(u) \neq 1$. But, for each $i = 1, \ldots, m$, $\lambda_i(u) = 1$, which implies that $\lambda(u) = 1$, contradicting the observation above. Hence, for a suitable choice of i, the homomorphism λ_i is an isomorphism. Consequently, $I(Y) = I(Y_i)$ and hence $Y = Y_i$, which implies that Y is irreducible. □

Over free groups this implies that the coordinate group of an irreducible equation must be fully residually free.

8.5 The theory of ideals

In standard algebraic geometry the theory and properties of ideals plays a major role. In particular, prime, irreducible and maximal ideals are all fundamental concepts. In this section we translate these to the algebraic geometry of G-groups and then prove several decomposition results.

8.5.1 Maximal and prime ideals

Throughout this section let G denote non-trivial group and let H be a G-group.

Ideals in G-groups enjoy many of the same properties as do ideals in commutative rings. We term an ideal M of the G-group H *a maximal ideal* of H if M is not contained in any other ideal of H. Notice that in view of the fact that $G \neq \{1\}$, every ideal of H is different from H.

Lemma 8.5.1. *If $G \neq \{1\}$ then every ideal in a G-group H is contained in a maximal ideal.*

The proof of the lemma is a standard application of Zorn's lemma.

Definition 8.5.1. A G-group H is termed G-simple, if the only proper ideal of H is the ideal $\{1\}$.

Notice, that any group G is G-simple. New examples of G-simple groups come surprisingly from nilpotent groups. Indeed, let H be a nontrivial nilpotent group and G be the center of H; then H is G-simple. Observe, that an ideal M of the G-group H is maximal if and only if H/M is G-simple.

We focus first on maximal ideals in $G[X]$. The following is an easy but important lemma.

Lemma 8.5.2. *If $a = (a_1, \dots, a_n) \in G^n$, then the ideal $I_G(a)$ is a maximal ideal of $G[X]$, corresponding to the point $a \in G^n$.*

Proof. Observe that

$$I_G(a) = gp_{G[X]}\{x_1 a_1^{-1}, \dots, x_n a_n^{-1}\}$$

and so the factor group

$$G[X]/I_G(a) \cong G$$

is G-simple; hence $I_G(a)$ is maximal. □

In general, if H is a G-group and $a \in H^n$, then $I_H(a)$ is not necessarilya maximal ideal of $G[X]$ since every ideal of $G[X]$ can be represented in this way – all we have to do is to choose the G-group H suitably.

Lemma 8.5.3. *Let Q be an ideal in $G[X]$. Then the following hold:*
(1) *$Q = I_H(a)$, where $a = (x_1Q, \dots, x_nQ) \in H^n$ with $H = G[X]/Q$;*
(2) *If M is a maximal ideal in $G[X]$ and a is a root of M in some G-group H, then a is the only root of M in H and $M = I_H(a)$.*

Proof. The first statement follows from the definition of the quotient $G[X]/Q$. To prove the second one, it suffices to notice that if $a \in V_H(M)$ then $M \subseteq I_H(a)$ and that $V_H(I_H(a)) = \{a\}$. □

It follows that every maximal ideal M in $G[X]$ is of the form $M = I_H(a)$ for some tuple $a \in H^n$, where H is a supergroup of G. However a need not lie in G^n. This leads to the idea of replacing G by some *universal completion*, say $\overline{G}$ of G which contains a root of every ideal in $G[X]$. In classical algebraic geometry such completions are, of course, just algebraically closed fields. For this chapter it suffices to only mention only the following result. An *algebraicially closed group* is defined in the same way as an algebraicially closed field, that is G is algebraicially closed if any equation over G which has a solution in an overgroup of G alreday has a solution within G.

Lemma 8.5.4. *Let G be an algebraically closed group. Then an ideal in $G[X]$, which is the normal closure of a finite set, is maximal if and only if it is of the form $I_G(a)$, where $a \in G^n$.*

As we remarked before, not all ideals of $G[X]$ of the type $I_H(a)$, $a \in H^n$, are maximal, since by Lemma 8.5.3, all ideals are of this form. Similarly, not all maximal ideals in $G[X]$ are of the form $I_G(a)$, $a \in G^n$. The following examples illustrate some of the possibilities that can occur.

Example 8.5.1. Let $H = \langle x_1, \dots, x_n; R\rangle$ be a presentation of a finitely generated simple group and let g be an element of H of infinite order. Let G be the subgroup generated by g. Then H is a G-group. Let ϕ be the obvious G-homomorphism of $G[X]$ onto H. Then the kernel M of ϕ is an ideal of $G[X]$ with $G[X]/M \cong H$. Since H is simple and hence G-simple, it follows that M is a maximal ideal in $G[X]$. But $G[X]/M$ is not isomorphic to G in the category of G-groups, which means that M is not of the form $I_G(a)$ with $a \in G^n$.

Another example of this kind is the following one.

Example 8.5.2. Let F be a free group freely generated by c and d and let $Q = gp_{F[x]}([c,d]x^{-2})$. Then Q is an ideal of $F[x]$, but no maximal ideal in $F[x]$ containing Q has any points in F and so it is never of the form $I_F(a)$, $a \in F$.

Proof. Put $H = F[x]/Q$. Then H is a free product with amalgamation of F and the infinite cyclic group generated by x:

$$H = F \underset{[c,d]=x^2}{*} \langle x\rangle.$$

Hence H is an F-group and therefore Q is an ideal in $F[x]$. Consequently, Q is contained in some maximal ideal M of $F[x]$. Since the equation $[c,d] = x^2$ has no solutions in F, M has no F-points. □

We have seen that there are two different types of maximal ideals in $G[X]$: ones that have G-points and those that do not. Following on the procedure in ring theory we call ideals of the first type *G-rational ideals* or *G-maximal ideals*. They will play an important role when we study radicals.

The following corollary is a consequence of the remarks above.

Corollary 8.5.1. *Let M be a maximal ideal in $G[X]$, where $G \neq 1$. Then the following conditions are equivalent:*
(1) *M is G-maximal;*
(2) *$M = I_G(a)$ for some $a \in G^n$;*
(3) *M has a G-point;*
(4) *$G[X]/M \cong G$.*

We now turn to irreducible and prime ideals.

Definition 8.5.2.
(1) The ideal Q of H is an *irreducible ideal* if $Q = Q_1 \cap Q_2$ implies that either $Q = Q_1$ or $Q = Q_2$, for any ideals Q_1 and Q_2 of H.
(2) An ideal Q of H is a *prime ideal* if H/Q is a G-domain.

The following result ties the irreducibility of algebraic sets with the irreducibility of their ideals.

Lemma 8.5.5. *Let H be a G-group. Then the following hold:*
(1) *If Q is an irreducible H-closed ideal in $G[X]$, then $V_H(Q)$ is an irreducible closed set in H^n;*
(2) *If H is a G-domain and Y is an irreducible closed set in H^n, then $I_H(Y)$ is an irreducible H-closed ideal in $G[X]$.*

Proof. (1) Suppose that $V_H(Q) = Y_1 \cup Y_2$, where Y_1 and Y_2 are closed sets in H^n. Then

$$Q = I_H(V_H(Q)) = I_H(Y_1 \cup Y_2) = I_H(Y_1) \cap I_H(Y_2).$$

Consequently, by the irreducibility of Q, either $Q = I_H(Y_1)$ or $Q = I_H(Y_2)$. It follows that either $V_H(Q) = V_H(I_H(Y_1)) = Y_1$ or else $V_H(Q) = V_H(I_H(Y_2)) = Y_2$.

(2) Suppose that $I_H(Y) = Q_1 \cap Q_2$ is the intersection of two ideals Q_1 and Q_2. Then by Lemma 8.4.2

$$Y = V_H(I_H(Y)) = V_H(Q_1 \cap Q_2) = V_H(Q_1) \cup V_H(Q_2)$$

(the last equality holds since H is a G-domain). Hence, by the irreducibility of Y, either $Y = V_H(Q_1)$ or else $Y = V_H(Q_2)$. It suffices to consider the first possibility. Then $I_H(Y) = I_H(V_H(Q_1)) \supseteq Q_1$ and therefore $Q_1 = I_H(Y)$. It follows that $I_H(Y)$ is irreducible. This completes the proof. □

8.5.2 Radicals and radical ideals

The concept of a *radical* appears in many different contexts in both commutative algebra and group theory. As we saw in Section 8.1, in classical algebraic geometry the concept of the radical of an ideal arises in the following way. If R is a commutative ring and I is an ideal in R then the *radical* of I, denoted $\sqrt{I}$ or $\mathrm{Rad}(I)$, is the set $\{a \in R; a^n \in I$ for some some integer $n > 0\}$. The radical of an ideal is an ideal and $I \subset \mathrm{rad}(I)$.

An ideal I is called a *radical ideal* if $I = \mathrm{rad}(I)$. In affine n-space k^n then the ideal of polynomials that vanish on a set X, that is $I(X)$ is always a radical ideal, that is $\mathrm{rad}(I(X)) = I(X)$.

Recall that *Hilbert's Nullstellensatz* says that if I is an ideal in $k[x_1, \ldots, x_n]$ where k is an algebraically close field then $I(V(I)) = \mathrm{rad}(I)$. This describes the full relationship between algebraic sets and their corresponding ideals.

In this section the notion of a radical and a radical ideal is extended to the category of G-groups. The concept of a radical is the counterpart of the notion of a closed ideal in a free G-group $G[X]$.

Let H be a G-group with $G \neq \{1\}$.

Definition 8.5.3. The Jacobson G-radical $J_G(H)$ of a G-group H is the intersection of all G-maximal ideals in H; if there are no such ideals, we define $J_G(H) = H$.

Similarly, we define the G-radical $\mathrm{Rad}_G(Q)$ of an arbitrary ideal Q in H as follows.

Definition 8.5.4. Let Q be an ideal in a G-group H. Then the G-radical $\mathrm{Rad}_G(Q)$ of Q is the intersection of all G-maximal ideals in H containing Q; if there are no such ideals, we define $\mathrm{Rad}_G(Q) = H$.

We term Q a G-radical ideal of H if $\mathrm{Rad}_G(Q) = Q$.

Lemma 8.5.6. *Let H be a G-group and let Q be an ideal in H. Then for an ideal P of H the following conditions are equivalent:*
(1) $P = \mathrm{Rad}_G(Q)$;
(2) *P is the pre-image in H of $J_G(H/Q)$;*
(3) *P is the smallest ideal in H such that P contains Q and the quotient group H/P is G-separated by G.*

The proof is straightforward.

In [18] it was proved that the free G-group $G[X]$ is G-discriminated by G provided G is non-Abelian and torsion-free hyperbolic. Hence we obtain the following lemma.

Lemma 8.5.7. *Let G be a non-Abelian torsion-free hyperbolic group. Then*

$$J_G(G[X]) = \{1\}.$$

Groups H with $J_G(H) = 1$ are very important because of their close relationship to coordinate groups of algebraic sets over G. We will say more about this in the section on coordinate groups.

We will need the following generalization of the notion of G-radical.

Definition 8.5.5. Let H and K be G-groups. Then the Jacobson K-radical $J_K(H)$ of H is the intersection of the kernels of all G-homomorphisms from H into K.

If Q is an ideal of H, then the K-radical $\mathrm{Rad}_K(Q)$ of Q in H is the pre-image in H of $J_K(H/Q)$.

We term an ideal Q of H *K-radical* if $Q = \mathrm{Rad}_K(Q)$.

Clearly, $\mathrm{Rad}_K(Q)$ is the smallest ideal P in H containing Q and such that H/P is G-separated by K.

Lemma 8.5.8. *The intersection of two K-radical ideals in a G-group H is a K-radical ideal.*

Proof. Let Q_1 and Q_2 be two arbitrary K-radical ideals in H. Clearly, every non-trivial element of the group $H/(Q_1 \cap Q_2)$ can be separated by the canonical epimorphism either onto the group H/Q_1 onto the group H/Q_2. Both of these groups are G-separated in K, therefore $H/(Q_1 \cap Q_2)$ is G-separated in K. Hence the ideal $Q_1 \cap Q_2$ is K-radical. □

In the following lemma we describe radical ideals of the free G-group $G[X]$. We recall that an ideal Q of $G[X]$ is K-closed if $Q = I_K(Y)$ for some subset Y of K^n (here K is an arbitrary G-group).

Lemma 8.5.9. *Let Q be an ideal in $G[X]$. Then for an arbitrary G-group K the following statements are equivalent:*

(1) *Q is K-closed in $G[X]$;*
(2) *$Q = \mathrm{Rad}_K(Q)$;*
(3) *G/Q is G-separated by K.*

Proof. The equivalence of 2) and 3) is a direct consequence of the definitions.

Now we prove that 1) implies 3). Let Q be a K-closed ideal in $G[X]$. Then $Q = I_K(Y)$ for some subset $Y \subseteq K^n$. Hence $G[X]/Q = \Gamma(Y)$ and the result follows from Corollary 8.4.4.

To finish the proof it suffices to show that 3) implies 1). Suppose that $G[X]/Q$ is G-separated by K. We claim that $I_K(V_K(Q)) = Q$, which shows that Q is K-closed. To this end, if ϕ is a G-homomorphism from $G[X]/Q$ into K, let $\phi(x_iQ) = a_i$ for each $i = 1, \dots, n$. If $h(x_1, \dots, x_n) \in Q$, then

$$1 = \phi(Q) = \phi(hQ) = h(a_1, \dots, a_n).$$

It follows that $V_K(Q) \neq \emptyset$. Now if $f \in G[X]$ and if $f \notin Q$, then we can find a homomorphism $\theta_f \colon G[X]/Q \longrightarrow K$ such that $\theta_f(fQ) \neq 1$. So if we put $b_i = \theta_f(x_iQ)$, then

$b = (b_1, \ldots, b_n) \in V_K(Q)$ but $f(b_1, \ldots, b_n) = \theta_f(fQ) \neq 1$. It follows that we have proved that $I_K(V_K(Q)) \subseteq Q$ and therefore that $I_K(V_K(Q)) = Q$, as needed. □

The next lemma provides another relationship between K-radical ideals of G-groups and K-closed ideals in $G[X]$.

Lemma 8.5.10. *Let H be a G-group and $\eta\colon G[X] \to H$ be a G-epimorphism from the free G-group $G[X]$ onto H. Then for any ideal Q in H and any G-group K the following conditions are equivalent:*
(1) *Q is K-radical in H;*
(2) *$\eta^{-1}(Q)$ is K-closed in $G[X]$.*

Proof. It follows from the appropriate isomorphism theorems that $\eta^{-1}(Q)$ is an ideal in $G[X]$ and
$$H/Q \cong G[X]/\eta^{-1}(Q)\,.$$
Therefore G-separability of one of the groups above implies G-separability of the other. The result then follows from Lemma 8.5.9. □

The next proposition details the connection between the equationally Noetherian condition and the ascending chain condition on radical ideals.

Lemma 8.5.11. *Let H be a finitely generated G-group. Then H satisfies the ascending chain condition on K-radical ideals for each G-equationally Noetherian group K.*

Proof. Let
$$Q_1 \subset Q_2 \subset \ldots$$
be a properly ascending chain of G-radical ideals in H. Let $\eta\colon G[X] \longrightarrow H$ be an epimorphism from a suitably chosen finitely generated free G-group $G[X]$ onto H. Observe, that if Q is a K-radical ideal in H then $\eta^{-1}(Q)$ is a K-radical ideal in $G[X]$. Hence
$$\eta^{-1}(Q_1) \subset \eta^{-1}(Q_2) \subset \ldots \tag{8.2}$$
is a properly ascending chain of K-radical ideals in $G[X]$. Therefore the algebraic sets defined by these ideals give rise to a properly descending chain of algebraic sets in K^n:
$$V_G(\eta^{-1}(Q_1)) \supset V_G(\eta^{-1}(Q_2)) \supset \ldots\,. \tag{8.3}$$
Since K is G-equationally Noetherian, this chain terminates, and therefore so do the other claims (see Theorem 8.4.3). □

Notice that it follows directly from the definitions, that if H and K are G-groups and Q is an ideal of H, then the following inclusions hold:
$$\mathrm{Rad}_G(Q) \supseteq \mathrm{Rad}_K(Q) \supseteq \mathrm{Rad}_{H/Q}(Q) = Q\,.$$
But if the group K is G-separated by G (i.e., if $K \in \mathcal{S}_G$) then $\mathrm{Rad}_K(Q) = \mathrm{Rad}_G(Q)$.

We next define the notion of a generic point.

Definition 8.5.6. Let H be a G-group and let Q be an ideal of $G[X]$. Then a point $h = (h_1, \ldots, h_n) \in H^n$ is termed a *generic point* of $V_H(Q)$ if

$$f \in I(V_H(Q)) \Longleftrightarrow f(h) = 1.$$

We have already seen that the G-group $G[X]/Q$ plays a special role here. The following lemma is an amplification of the definition above and this remark. It follows immediately from the definitions.

Lemma 8.5.12. *Let Q be an ideal of $G[X]$. For any G-group H containing the group $G[X]/Q$ the point $h = (x_1Q, \ldots, x_nQ) \in H^n$ is a generic point of $V_H(Q)$.*

Proof. Notice, that $h = (x_1Q, \ldots, x_nQ) \in V_H(Q)$. Now let $f \in G[x_1, \ldots, x_n]$. Then $f(x_1Q, \ldots, x_nQ) = 1$ in H if and only if $f(x_1Q, \ldots, x_nQ) = 1$ in $G[X]/Q$ which is equivalent to $f(x_1, \ldots, x_n) \in Q$. Therefore

$$f \in I(V_H(Q)) \Longleftrightarrow f(h) = 1,$$

as desired. □

Our next step is to define the *G-nilradical* of a G-group.

Definition 8.5.7. Let H be a G-group. The subgroup $L_G(H)$ of H generated by all locally nilpotent ideals in H is termed the G-nilradical of H.

In the case where $G = \{1\}$, $L_1(H)$ is usually refered to as the Hirsch–Plotkin radical of H.

Clearly every element in $L_G(H)$ is a G-nilpotent element. Hence, if H has no G-zero divisors then $L_G(H) = \{1\}$.

Lemma 8.5.13. *Suppose that the Hirsch–Plotkin radical $L_1(G)$ of G is trivial. Then for any G-group H, the G-nilradical $L_G(H)$ is an ideal in H which is contained in every maximal ideal of H and therefore*

$$J_G(H) \supseteq L_G(H)\,.$$

Proof. If $L_1(G) = \{1\}$, then the intersection $G \cap L_G(H)$ is a normal locally nilpotent subgroup of G, hence it is trivial. Consequently $L_G(H)$ is an ideal. Suppose now that $L_G(H) \nsubseteq M$, for some maximal ideal M in the G-group H. Then there exists an element $c \in L_G(H)$ such that $c \notin M$. Then $gp_H(c)$ is a locally nilpotent, normal subgroup of H. Now $Mgp_H(c)/M$ is a locally nilpotent, normal subgroup of the G-group H/M. Hence it meets GM/M (which is isomorphic to G) trivially. Therefore $Mgp_H(c)$ meets G trivially and so is an ideal of H which properly contains M, contradicting the maximality of M. This proves the lemma. □

Finally we have the

Lemma 8.5.14. *Let H be a G-equationally Noetherian G-group. Then $L_G(H)$ is a solvable subgroup of H.*

Proof. An equationally Noetherian group is a CZ-group, that is a group G which carries a T_1-topology satisfying the minimal condition for closed sets and for which multiplication, inversion and conjugation are conintuous (see [36] for definitions). Kaplanksy introduced CZ-groups to generalize linear groups. Since it is a CZ-group it follows that $L_G(H)$ is a locally nilpotent CZ-group. Hence it is solvable (See [38]). □

In the last section we saw that a prime ideal is always irreducible. The converse is not true in general however being a radical is sufficient to insure that an irreducible ideal is prime.

Lemma 8.5.15. *If Q is an irreducible ideal of H which is a K-radical for some G-domain K, then Q is a prime ideal.*

Proof. Since Q is K-closed in H,

$$Q = \bigcap_{\phi \in \mathrm{Hom}_G(H/Q,K)} \ker \phi,$$

where $\mathrm{Hom}_G(H/Q, K)$ is the set of all G-homomorphisms from H/Q into K. Suppose that Q is not prime. Then there exist $c_1, c_2 \in H - Q$ such that $[c_1, c_2^g] \in Q$, for every $g \in G$. Then $[c_1, c_2^g] \in \ker \phi$ for every $\phi \in \mathrm{Hom}_G(H/Q, K)$. But $H/\ker \phi$ is G-isomorphic to a G-subgroup of the G-domain K and so is itself a G-domain. Therefore, either $c_1 \in \mathrm{Ker}\ \phi$ or else $c_2 \in \mathrm{Ker}\ \phi$. Let Φ_i be the set of all those $\phi \in \mathrm{Hom}_G(H/Q, K)$ for which $c_i \in \mathrm{Ker}\ \phi$, $i = 1, 2$. Put

$$Q_i = \bigcap_{\phi \in \Phi_i} \mathrm{Ker} \phi , \quad (i = 1, 2) .$$

Then Q_1 and Q_2 are ideals in H and $Q = Q_1 \cap Q_2$. Since Q is irreducible, we find that either $Q = Q_1$ or $Q = Q_2$, which implies that either $c_1 \in Q$ or $c_2 \in Q$. It follows that Q is a prime ideal in H. □

Corollary 8.5.2. *If H is a G-domain, then every irreducible H-closed ideal Q of $G[X]$ is prime.*

Indeed, as we mentioned every H-closed ideal of $G[X]$ is an H-radical ideal of $G[X]$. The result now follows from the lemma above.

The next result is useful.

Lemma 8.5.16. *Let H be a G-group. If Q is an ideal of H, then Q is prime and K-radical for some G-domain K if and only if H/Q is G-discriminated by K.*

Proof. By definition an ideal Q in H is K-radical if and only if G/Q is separated by K. In this event if K is a G-domain then by Theorem 8.3.9 H/Q is G-discriminated by K.

Now suppose that H/Q is G-discriminated by K and K is a G-domain. If Q is not prime in H then H/Q has a pair of non-trivial elements c_1 and c_2 such that $[c_1, c_2^g] = 1$ for every $g \in G$. Since H/Q is G-discriminated by K there exists a G-homomorphism

$\phi: H/Q \to K$ such that $\phi(c_1) \neq 1$ and $\phi(c_2) \neq 1$. Since $[\phi(c_1), \phi(c_2)^g] = 1$ for every $g \in G$, it follows that $\phi(c_1)$ is a G-zero divisor in K, which is impossible. This completes the proof. □

8.5.3 Some decomposition theorems for ideals

The prominent result in algebraic number theory is that ideals in algebraic number rings have unique decompositions into prime ideals (see [90]). In this section we prove some similar decomposition results for ideals over G-domains where G is an equationally Noetherian group.

Theorem 8.5.1. *Let H be a G-equationally Noetherian G-domain. Then each H-closed ideal in $G[X]$ is the intersection of finitely many prime H-closed ideals, none of which is contained in any one of the others and this representation is unique up to order. Consequently, distinct irredundant intersections of prime H-closed ideals define distinct H-closed ideals.*

Proof. Let Q be an H-closed ideal in $G[X]$. Let $Y = V_H(Q)$. Since H is a G-equationally Noetherian G-domain, Y can be decomposed uniquely into a finite union of irreducible algebraic sets:

$$Y = Y_1 \cup \cdots \cup Y_m.$$

This gives rise to the decomposition

$$Q = I(Y) = I(Y_1) \cap \cdots \cap I(Y_m).$$

By Proposition 18 in [16], each of the $I(Y_j)$ is irreducible. Now by Corollary 8.5.2 each of the ideals $I(Y_j)$ is prime. That proves the existence of the irredundant decomposition of Q. Moreover, if $Q = Q_1 \cap \cdots \cap Q_k$ is a decomposition of Q as an intersection of irreducible H-closed ideals, none of which is contained in any one of the other, then

$$Y = V_H(Q) = V_H(Q_1) \cup \cdots \cup V_H(Q_k)$$

is an irredundant decomposition of Y into irreducible closed sets. It follows that $k = m$ and that the $V_H(Q_j)$ are simply a rearrangement of the Y_ℓ and hence that the Q_j are simply the $I(Y_\ell)$, in a possibly different order. To finish the proof it suffices to notice that the intersection of two (or finitely many) H-radical ideals is again H-radical – this was proved in the last section Lemma 8.5.8. □

The above theorem has a counterpart for the ideals of finitely generated G-groups.

Theorem 8.5.2. *Let H be a finitely generated G-group and let K be a G-equationally Noetherian G-domain. Then each K-radical ideal in H is a finite irredundant intersection of prime K-radical ideals. Moreover, this representation is unique up to order. Furthermore, distinct irredundant intersections of prime K-radical ideals define distinct K-radical ideals.*

Proof. Let Q be a K-radical ideal in H. Then $J_K(H/Q) = 1$. The group H/Q is finitely generated as a G-group, so we can express H/Q as a factor group of a finitely generated free G-group $G[X]$, say,

$$H/Q = G[X]/P\,.$$

Since $J_K(G[X]/P) = J_K(H/Q) = 1$, the ideal P is K-radical. By Lemma 8.5.9, the ideal P is K-closed in $G[X]$. Hence, by Theorem 8.5.1, P can be expressed as a finite intersection of prime K-closed ideals in $G[X]$:

$$P = P_1 \cap \cdots \cap P_m\,.$$

We claim that each such prime decomposition of P in $G[X]$ gives rise to a prime decomposition of Q in H. Indeed, let Q_i be the pre-image of the ideal P_i/P in H with respect to the canonical G-epimorphism

$$H \to H/Q \cong G[X]/P\,.$$

Then Q_i is K-radical and prime in H. Observe, that

$$Q = Q_1 \cap \cdots \cap Q_m\,.$$

Similarly, each prime decomposition of Q in H gives rise to a prime decomposition of P in $G[X]$. The uniqueness of the decomposition follows from Theorem 8.5.1. □

In the case $K = G$ we have the following corollary.

Corollary 8.5.3. *Let G be an equationally Noetherian domain and let H be a finitely generated G-group. Then each G-radical ideal in H is a finite irredundant intersection of prime G-radical ideals. Moreover, this representation is unique up to order. Furthermore, distinct irredundant intersections of prime G-radical ideals define distinct G-radical ideals.*

8.6 Coordinate groups

In this section we describe coordinate groups in purely group-theoretic terms. First we provide an abstract characterization of coordinate groups.

Lemma 8.6.1. *Let H be a finitely generated G-group. If $J_K(H) = \{1\}$ for some G-group K, then H is G-isomorphic to the coordinate group $\Gamma(Y)$ of some algebraic set $Y \subseteq K^n$ defined over G. Conversely, every coordinate group $\Gamma(Y)$ of an algebraic set $Y \subseteq K^n$, defined over G, is a finitely generated G-group with $J_K(\Gamma(Y)) = \{1\}$.*

Proof. Suppose that H is a finitely generated G-group with $J_K(H) = \{1\}$, where K is a given G-group. We express H as a factor group of a finitely G-generated G-free group $G[X]$:

$$H \cong G[X]/Q.$$

Thus $J_K(G[X]/Q) = \{1\}$, i.e., $\mathrm{Rad}_K(Q) = Q$. Consequently from Lemma 8.5.9, Q is K-closed in $G[X]$ and so H is isomorphic to the coordinate group of the algebraic set $V_K(Q)$ defined over G.

Let $\Gamma(Y)$ be the coordinate group of some algebraic set $Y \subseteq K^n$. So $\Gamma(Y) \cong H = G[X]/Q$, where $Q = I_K(Y)$. Again by Lemma 8.5.9, $Q = \mathrm{Rad}_K(Q)$, hence

$$J_K(\Gamma(Y)) = J_K(G[X]/Q) = \{1\},$$

as desired. □

Lemma 8.6.1 demonstrates once more the role of the category $\mathcal{S}_G$ in algebraic geometry over groups.

Corollary 8.6.1. *Let H be a G-group. Then the coordinate groups of algebraic sets in H^n defined over G are exactly the n-generator G-groups which are G-separated by H.*

In the event that $H = G$ this corollary shows that the coordinate groups of algebraic sets over G are exactly the finitely generated G-groups in the category $\mathcal{S}_G$; in particular, they are exactly the finitely generated G-subgroups of the unrestricted direct power $G_\omega(I)$ of I copies of G provided that the set of indices I is sufficiently large.

We now give an example of a finitely generated group in $\mathcal{S}_G$.

Corollary 8.6.2. *If G is a non-Abelian torsion-free hyperbolic group, then $G[x_1, \ldots, x_n]$ is the coordinate group of the algebraic set G^n.*

Proof. By Lemma 8.5.7 $J_G(G[X]) = 1$ for any non-Abelian torsion-free hyperbolic group G. So the desired result follows from Corollary 8.6.1. □

It follows also from Corollary 8.6.1 that if G satisfies a non-trivial identity, or, more generally, a G-identity, then $G[X]$ is never the coordinate group of an algebraic set in G^n.

Finally, on combining Theorem 8.3.4 with Lemma 8.6.1 we obtain the following result.

Lemma 8.6.2. *Let H be a G-equationally Noetherian G-group. Then for every algebraic set $Y \subseteq H^n$ the coordinate group $\Gamma(Y)$ is G-equationally Noetherian.*

8.6.1 Coordinate groups of irreducible varieties

We now present several useful characterizations of irreducible algebraic sets in terms of their ideals and also their coordinate groups.

Theorem 8.6.1. *Let H be a G-equationally Noetherian G-domain and Y be an algebraic set in H^n. Then the following conditions are equivalent:*

(1) *Y is irreducible;*
(2) *$I_H(Y)$ is a prime ideal in $G[X]$;*

(3) $\Gamma(Y)$ *is a G-equationally Noetherian G-domain;*
(4) $\Gamma(Y)$ *is G-discriminated by H.*

Proof. The equivalence of (1) and (2) follows from Lemma 8.5.5 and Proposiiton 19 of [16].

The equivalence of (2) and (3) is a consequence of the definition of a prime ideal and Lemma 8.6.2.

(4) is equivalent to (1) by Lemma 8.6.1, Theorem 8.3.9 and Theorem 8.3.4. □

Theorem 8.6.1 provides a useful way of proving irreducibility of algebraic sets, as we see from he following lemma.

Lemma 8.6.3. *Let G be an equationally Noetherian non-Abelian torsion-free hyperbolic group. If a finitely generated group H given by the presentation*

$$H = \langle x_1, \dots, x_n; S\rangle$$

(so S is a subset of the free group, freely generated by X) is discriminated by G, then $V_G(S)$ *is an irreducible algebraic set.*

Proof. Let $Q = gp_{G[X]}(S)$ be the normal closure of S in $G[X]$. Then

$$G[X]/Q \cong G * H\,.$$

It was proved in [17] that, under the above assumptions, the free product $G * H$ is G-discriminated by G. It follows that $Q = \mathrm{Rad}_G(Q)$ and hence that $G[X]/Q$ is the coordinate group of the algebraic set $V_G(S)$. So, by Theorem 8.6.1 this algebraic set $V_G(S)$ is irreducible. □

There are numerous examples of groups that are discriminated by free groups. Here we mention a few typical examples. The orientable surface groups

$$\langle x_1, y_1, \dots, x_n, y_n; [x_1, y_1]\dots[x_n, y_n] = 1\rangle$$

are discriminated by a free group, whenever $n \geq 1$, while the non-orientable surface groups

$$\langle x_1, \dots, x_n; x_1^2 \dots x_n^2 = 1\rangle$$

are discriminated by a free group whenever $n > 3$. It follows that all of these groups are separated by every non-Abelian torsion-free hyperbolic group G, since every non-Abelian, torsion-free hyperbolic group contains a non-Abelian free subgroup. These results together with Lemma 8.6.3 imply the following

Lemma 8.6.4. *Let* $S = \{[x_1, y_1]\dots[x_n, y_n]\}, n \geq 1$ *or let* $S = \{x_1^2 \dots x_n^2\}$, $n \neq 3$. *Then for any equationally Noetherian non-Abelian torsion-free hyperbolic group G the algebraic set* $V_G(S)$ *is irreducible.*

Proof. The orientable case and the non-orientable case for $n > 3$ both follow immediately from the discussion above. So we left to consider only the cases when S is either $\{x_1^2\}$, or $\{x_1^2x_2^2\}$.

If $S = \{x_1^2\}$, it is easy to see that

$$\mathrm{Rad}_G(S) = gp_{G[X]}(x_1)$$

and consequently $G[X]/\mathrm{Rad}_G(S) \cong G$. Hence $V_G(S)$ is irreducible by Theorem 8.6.1.

Similarly, if $S = \{x_1^2x_2^2\}$, then

$$\mathrm{Rad}_G(S) = gp_{G[X]}(x_1x_2^{-1}).$$

Indeed, G is commutative transitive, therefore for any solution $x_1 = u, x_2 = v$ of $S = 1$ in G the elements u and v commute, therefore $(uv)^2 = 1$ and hence $uv = 1$. In this event,

$$G[X]/\mathrm{Rad}_G(S) \cong G * \langle x_1 \rangle$$

is G-discriminated by G and the desired conclusion follows from Theorem 8.6.1. □

Notice, that if the group G in Lemma 8.6.4 is free, then the conlusion also holds in the non-orientable case with $n = 3$. This follows from a result of Schutzenberger [230] which states that if $x_1 = u, x_2 = v, x_3 = w$ is a solution of the equation $x_1^2x_2^2x_3^2 = 1$ in a free group F, then the elements u, v, w all commute. Hence

$$\mathrm{Rad}_F(x_1^2x_2^2x_3^2) = gp_{F[X]}(x_1x_2x_3)$$

and the group

$$F[X]/\mathrm{Rad}_F(x_1^2x_2^2x_3^2) \cong F * F(x_1, x_2)$$

is discriminated by F.

In the event that $G = H$ we can add one more equivalent condition to Theorem 8.6.1, which establishes a surprising relationship between coordinate groups of irreducible algebraic sets over G and finitely generated models of the universal theory of the group G.

Theorem 8.6.2. *Let G be an equationally Noetherian domain and Y be an algebraic set in G^n. Then the following conditions are equivalent:*

(1) *Y is irreducible;*
(2) *$\Gamma(Y)$ is G-universally equivalent to G.*
Moreover, any finitely generated G-group which is G-universally equivalent to G is the coordinate group of some irreducible algebraic set over G.

Proof. The equivalence of (1) and (2) follows from Theorem 8.6.1 and Theorem 8.3.10. □

To prove, under either of these conditions, that any finitely generated G-group which is G-universally equivalent to G, is the coordinate group of some irreducible algebraic set over G, consider a finitely generated G-group H which is G-universally equivalent to G. Then by Theorem 8.3.10 H is G-discriminated by G. Hence by Corollary 8.6.1 the group H is the coordinate group of some algebraic set over G. So the desired conclusion follows from the equivalence of 1) and 2) above.

8.6.2 Decomposition theorems

Here we prove some decomposition results for coordinate groups.

Theorem 8.6.3. *Let H be a G-equationally Noetherian G-domain. If Y is any algebraic set in H^n, then the coordinate group $\Gamma(Y)$ is a subgroup of a direct product of finitely many G-groups, each of which is G-discriminated by H.*

The proof follows from Lemma 8.4.1, Theorem 8.4.5, and Theorem 8.4.6.

We now have the following result extending Theorem 8.6.2 to non-Abelian equationally Noetherian torsion-free hyperbolic groups.

Theorem 8.6.4. *Let H be a non-Abelian equationally Noetherian torsion-free hyperbolic group. Then every finitely generated group E which is separated by H is a subdirect product of finitely many finitely generated groups, each of which is discriminated by H.*

Proof. Let E be a finitely generated group that is separated by a non-Abelian equationally Noetherian torsion-free hyperbolic group H. Let

$$E = \langle x_1, \dots, x_n; f_1 = 1, f_2 = 1, \dots \rangle$$

be a presentation of E on finitely many generators $x_1, \dots, x_n$ and possibly infinitely many defining relators $f_1, f_2, \dots$. Then we can think of the set $S = \{f_1, f_2, \dots\}$ of defining relators of E as a subset of the free H-group $H[X]$. Observe that if $gp_F(S)$ is the normal closure of S in $F = \langle x_1, \dots, x_n \rangle$, then

$$I(V_H(S)) \cap F = gp_F(S).$$

Indeed, if $f \in F$ and $f \notin gp_F(S)$, then there is a homomorphism ϕ from F into H, which factors through E (i.e., $gp_F(S)$ is in the kernel of ϕ) such that $\phi(f) \neq 1$. This homomorphism ϕ can be extended to a H-homomorphism, again denoted by ϕ, from $H[X]$ into the H-group H. If we put $\phi(x_i) = a_i$ $(i = 1, \dots, n)$, then f does not vanish at $a = (a_1, \dots, a_n)$ although $a = (a_1, \dots, a_n) \in V_H(S)$. Hence $f \notin I(V_H(S))$. Therefore E embeds in $\Gamma(Y)$, where $Y = V_H(S)$. But H is an equationally Noetherian domain. So, by Theorem 8.6.3, we find that E is embeddable in a direct product of finitely many finitely generated H-groups, each of which is H-discriminated by H. Since every hyperbolic group, in particular H, is finitely generated it follows that every finitely generated H-group is finitely generated as an abstract group. □

Corollary 8.6.3. *Every finitely generated group that is separated by a free group is a subdirect product of finitely many finitely generated groups which are discriminated by free groups.*

8.7 The Nullstellensatz

One of the cornerstones of classical algebraic geometry is Hilbert's Nullstellenstaz (see Section 8.1). In this section we look at versions of this result that hold in the algebraic geometry of groups.

Recall that *Hilbert's Nullstellensatz* describes the full relationship between algebraic sets and their corresponding ideals. In partciular it is the following result.

Theorem 8.7.1 (Hilbert's Nullstellensatz). *Let k be an algebraically closed field and ler I be an ideal in $k[x_1, \ldots, x_n]$. Then $I(V(I)) = \mathrm{rad}(I)$.*

In particular this implies the following. If $f_1, \ldots, f_m \in k[x_1, \ldots, x_n]$ and g is another polynomial that vanishes whenever $f_1, \ldots, f_m$ vanish then there is an integer $r > 0$ and polynomials $h_1, \ldots, h_m \in k[x_1, \ldots, x_n]$ such that

$$g^r = h_1 f_1 + \cdots + h_m f_m .$$

The Nullstellensatz provides a correspondence between algebraic sets and radical ideals and between prime ideals and afffine varieties. We have looked at various group theoretic versions of these correspondences.

Hilbert's Nullstellensatz has various equivalent formulations. One of them asserts that every proper ideal in the polynomial ring $K[X]$ has a root in K. A similar result holds for G-groups. A G- algebraically closed G group is a group closed under solutions of equations which allow constants from G.

Theorem 8.7.2. *Let H be a G-algebraically closed G-group. Then every ideal in $G[X]$ which is the normal closure of a finite set, has a root in H^n.*

Proof. Let Q be an ideal in $G[X]$ which is the normal closure of the finite set S. Then $G[X]/Q$ is a G-group and S has a root in H. Since G is algebraically closed, S has a root in G. □

One of the consequences of Theorem 8.7.2 is the following result which, as in the classical case, shows that there exists a one-to-one correspondence between finitely generated ideals in $G[X]$ and points in the affine space G^n.

Theorem 8.7.3. *Let G be an algebraically closed group. Then an ideal Q in $G[X]$, which is the normal closure in $G[X]$ of a finite set, is maximal if and only if it is of the form $I(a)$, where $a \in G^n$.*

Proof. Suppose that Q is a maximal ideal in $G[X]$ and that Q is the normal closure of a finite set. Since Q has a root a in G^n, $Q = I(a)$. The converse follows on appealing to Lemma 8.5.3. □

Another classical form of the Nullstellensatz describes the closed ideals I in the ring of polynomials $k[X]$ over an algebraically closed field k as radical ideals. We recall that an ideal I of $K[X]$ is a radical ideal if $f \in I$ whenever $f^n \in I$ $(f \in K[X])$. A similar result holds also for G-algebraically closed groups. In order to formulate it, we introduce the following definition.

Definition 8.7.1. Let H be a G-group and let S be a subset of $G[X]$. Then we say that S satisfies the Nullstellensatz over H if

$$I(V_H(S)) = gp_{G[X]}(S).$$

In particular, an ideal Q in $G[X]$ satisfies the Nullstellensatz if and only if Q is H-radical.

Theorem 8.7.4. *Let H be a G-group and suppose that H is G-algebraically closed. Then every finite subset S of $G[x_1, \ldots, x_n]$ with $V_H(S) \neq \emptyset$, satisfies the Nullstellensatz.*

Proof. Observe that if $Q = gp_{G[X]}(S)$ in $G[X]$, then Q is an ideal of $G[X]$ since, by hypothesis, $V_H(S) \neq \emptyset$. Consequently the quotient group $K = G[X]/Q$ is a G-group.

Now suppose that $S = \{f_1, \ldots, f_k\}$ and that $f \notin Q$. Let $H' = H *_G K$ be the free product of H and K amalgamating G. Then H' is a G-group and if $b_i = x_i Q$, we find that $f_j(b_1, \ldots, b_n) = 1$ for $j = 1, \ldots, k$ but that $f(b_1, \ldots, b_n) \neq 1$. Since H is G-algebraically closed and is a G-subgroup of H', there exist elements $a_1, \ldots, a_n$ in H such that $f_j(a_1, \ldots, a_n) = 1$ for $j = 1, \ldots, k$ but that $f(a_1, \ldots, a_n) \neq 1$. Put $v = (a_1, \ldots, a_n) \in H^n$. Then $v \in V_H(S)$ and $f(v) \neq 1$. Therefore $f \notin I(V_H(S))$. It follows that $I(V_H(S)) \subseteq Q$ and hence that $I(V_H(S)) = Q$. □

The next simple but useful result follows immediately from Lemma 8.5.9.

Lemma 8.7.1. *Suppose that H is a G-group and that $V_H(S) \neq \emptyset$, where S is a subset of $G[x_1, \ldots, x_n]$ Then S satisfies the Nullstellensatz over H if and only if $G[X]/gp_{G[X]}(S)$ is G-separated by H.*

The problem of the description of systems of equations which satisfy the Nullstellensatz, for example, over a non-Abelian free group is, in general, a difficult one. Here we discuss only the standard quadratic equations, without coefficients. The orientable one of genus n takes the form

$$[x_1, y_1] \ldots [x_n, y_n] = 1$$

while the non-orientable one of genus n takes the form

$$x_1^2 \ldots x_n^2 = 1\,.$$

Theorem 8.7.5. *Let $S = [x_1, y_1] \dots [x_n, y_n]$ or $S = x_1^2 \dots x_n^2$ (in the latter case $n > 3$). Then S satisfies the Nullstellensatz over any torsion-free non-Abelian hyperbolic group G.*

Proof. Let G be a torsion-free non-Abelian hyperbolic group. Let Q be the normal closure $gp_{G[X]}(S)$ of S in $G[X]$. Then

$$G[X]/Q \cong G * F/gp_F(S),$$

where here F is the free group on X. From Section 8.6.2 we know that that the group $G * F/gp_F(S)$ is G-discriminated by G. Consequently, S satisfies the Nullstellensatz over G, as desired. □

In the case of a non-Abelian free group F, Kharlampovich and Myasnikov [152] have described the radical of an arbitrary quadratic equation (with coefficients) over F; in particular, they give a description of the quadratic equations over F that satisfy the Nullstellensatz. This description plays an important role in the solution of the Tarski problems and we will discuss it further in the next chapter.

9 The solution of the Tarski problems

9.1 The Tarski problems

With the development of algebraic geometry over free groups, coupled with the extensive structure theory of fully residually free groups, the stage was set to give a proof of the Tarski problems. This was done independently by Kharlampovich and Myasnikov in [152, 153, 154, 155, 156] and by Z.Sela in [233, 234, 235, 236, 237]. Kharlampovich and Myasnikov's proof used the algebraic geometry over free groups and their theory of fully residually free groups (see Chapters 6 and 8). Sela's proof used a similar technique that he called diophantine geometry over free groups, coupled with his treatment of fully residually free groups. He approached this class of groups as limiting groups of homomorphisms into free groups and called them limit groups. This latter name has now become standard (see Chaper 7). In this chapter we give an overview outline of the approach of Kharlampovich and Myasnikov. First we restate the problems and solutions. We use the notation and definitions from Chapters 4 and 5 and in this chapter we take the liberty of repeating a great deal of material from previous chapters.

Conjecture (Tarski Conjecture 1). *Any two non-Abelian free groups are elementarily equivalent. That is any two non-Abelian free groups satisfy exactly the same first-order theory.*

Conjecture (Tarski Conjecture 2). *If the non-Abelian free group H is a free factor in the free group G then the inclusion map $H \to G$ is an elementary embedding.*

Clearly the second conjecture is stronger than the first and implies the first. If true then the theory of the non-Abelian free groups would be complete, that is given a sentence ϕ of L_0 then either ϕ is true in every non-Abelian free group or ϕ is false in every non-Abelian free group.

In addition to the completeness of the theory of the non-Abelian free groups the question of its decidability also arises. Recall that this means the question of whether there exists a recursive algorithm which, given a sentence ϕ of L_0 decides whether or not ϕ is true in every non-Abelian free group. Tarski further conjectured that the theory of the non-Abelian free groups is decidable.

Conjecture (Tarski Conjecture 3). *The elementary theory of the countable non-Abelian free groups is decidable.*

Kharlampovich and Myasnikov [152, 153, 154, 155, 156] and Sela [233, 234, 235, 236, 237] proved that the first two Tarski conjectures described above are true. Using an extension of results of Makanin and Razborov, Kharlampovich and Myasnikov further settled the third Tarski conjecture and showed that the elementary theory of free groups is decidable. We summarize the results in the next theorem.

Theorem 9.1.1. *If the non-Abelian free group H is a free factor in the free group G then the inclusion map $H \to G$ is an elementary embedding. Hence any two non-Abelian free groups are elementarily equivalent. Moreover the theory of the non-Abelian free groups is decidable.*

Tarski presented his conjectures about 1945 and the solutions came after a long and fairly extensive set of partial results. It was known that free Abelian groups of finite rank are elementarily equivalent if and only if their ranks coincide. The analogous result is true for free nilpotent groups of finite rank and fixed class. Further it was known that the elementary theory of any free Abelian group of finite rank is decidable but that the elementary theory of a free non-Abelian nilpotent group of finite rank is undecidable (see Chapter 5).

We now review some of the material in Chapter 5. The first progress on the Tarski problems was due to Vaught who showed that the Tarski conjectures 1,2 are true if G and H are both free groups of infinite rank.

Theorem 9.1.2. *If the infinite rank free group H is a free factor in the free group G then the inclusion map is an elementary embedding. In particular if G and H are free groups of infinite rank then $Th(G) = Th(H)$.*

The basic idea in Vaught's proof is to use the following criteria for elementary embeddings; if H_0 is a subgroup of H and that to every finite subset $\{a_1, \ldots, a_n\}$ of H_0 and every element $b \in H$ there exists an automorphism σ of H fixing $a_1, \ldots, a_n$ and mapping b into H_0, then the inclusion map from H_0 into H is an elementary embedding. This was the content of Theorem 5.1. Applying this criterion to free groups of infinite rank, suppose that F is free on an infinite subset S and that G is free on an infinite subset S_0 of S. Then permutations of S will induce enough automorphisms to guarantee that the inclusion map of G into F is an elementary embedding We note that Vaught's proof extends to infinite rank free algebras in any variety of algebras of any type. The precise details of Vaught's proof are given in Chapter 5. We will say more about Vaught's criterion in Section 9.3.

The elementary chain theorem (Theorem 4.2.3) gives that the direct union of elementary embeddings is elementary. Therefore Vaught's proof that the Tarski conjectures 1,2 are true for free groups of infinite rank reduced the conjectures to free groups of finite rank.

The next significant progress was due to Merzljakov. Recall that a *positive sentence* is a first-order sentence which is logically equivalent to a sentence constructed using (at most) the connectives $\vee, \wedge, \forall, \exists$. The *positive theory* of a group G consists of all the positive sentences true in G. Merzljakov [185] showed that the non-Abelian free groups have the same positive theory.

Theorem 9.1.3. *Two non-Abelian free groups have the same positive theory.*

A proof of Merzljakov [185] can be given involving generalized equations and a quantifier elimination process. This was a precursor to the methods used in the eventual solution of the overall Tarski problems.

Since positive sentences are preserved by homomorphic images and since every group is a homomorphic image of a free group this theorem has the following curious corollary. If the positive sentence ϕ is true in some non-Abelian free group and hence in all non-Abelian free groups then it is true in any group whatsoever.

Further work following Merzljakov centered on restricted theories of free groups. We have seen that any two non-Abelian free groups satisfy the same universal theory. Sacerdote [226] proved that this could be extended to *universal-existential sentences* (see Section 2). This is called the π_2-theory.

Theorem 9.1.4. *Any two non-Abelian free groups satisfy the same universal-existential theory, that is they satisfy exactly the same universal-existential sentences.*

Advances in a different direction were given by Makanin and Razborov. Makanin [179] proved that there exists an algorithm to determine, given a finite system of equations over a free group, whether the system possesses at least one solution. In other words, the Diophantine problem is decidable for free groups (compare this to undecidability of the Diophantine problems for integers).

Razborov [213] working with the Makanin algorithm determined an algorithm to effectively describe the solution sets of a finite system of equations over a free group. Kharlampovich and Myasnikov further developed the Makanin–Razborov method. Their technique allows one to transform arbitrary finite systems of equations over free groups to some "canonical forms" and describe precisely the irreducible components of algebraic sets in free groups. These canonical forms consist of finitely many quadratic equations in a triangular form (see Definition 6.3 and Theorem 6.2). Observe that quadratic equations over free groups (Comerford and Edmonds [60]), and hyperbolic groups (see Grigorchuk and Kurchanov [120]) are well-studied, and there are good methods for solving them.

The following result is a corollary of the decidability of the Diophantine problem

Theorem 9.1.5 ([Mak 2]).

(1) *The existential (and hence the universal) theory of a free group is decidable.*
(2) *The positive theory of a free group is decidable.*

In addition to the Tarski problems themselves, the quest for a solution has inspired many other results in group theory. This is especially true concerning the theory of solutions of equations within groups. In 1959 Vaught asked the question whether the sentence

$$\forall x \forall y \forall z (x^2 y^2 z^2 = 1 \rightarrow xy = yx \wedge xz = xz \wedge yz = zy)$$

holds in all free groups. R.Lyndon [172] then proved that for each solution of $x^2y^2z^2 = 1$ in a free group the elements commute pairwise. This result launched the theory of equations over free groups. The first general results in this were due to Lyndon [172], Lorenc [166] and Appel [5] where they described the solution set of an arbitrary one variable equation over a free group. In 1966 Malcev described the solution set of the equation $[x, y] = [a, b]$ over the free group $F(a, b)$, a problem considered earlier by Nielsen [203]. A version of this was also solved by Csorgo, Fine and Rosenberger [61]. Malcev's solution has the following nontrivial implication for the elementary theory of a free group of rank 2; the set of all free bases of $F(a, b)$ can be defined by a first order formula in the language of group theory. That is the elements $u, v \in F(a, b)$ are a free basis if and only if they satisfy the following formula with constants a, b;

$$\exists z([u, v] = z^{-1}[a, b]z \vee [u, v] = z^{-1}[b, a]z) .$$

The focus of study eventually turned to strictly quadratic equations over free groups, that is equations in which every variable x occurs exactly twice as either x or x^{-1}. Group relators that are quadratic have always been essential in combinatorial group theory due to their close connection with surface group relators (see Chapters 2 and 3). Comerford and Edmonds [60] and Grigorchuk and Kurchanov [120, 121] described the solution sets of quadratic equations over arbitrary free groups. Further work of Hoare, Karrass and Solitar shows that every quadratic equation over a free group is automorphically equiavlent to a standard one.

Makanin in 1982 [179] proved that if a given equation over a free group F has a solution in F then this equation has a solution of bounded length and this bound can be effectively computed from the equation itself. Makanin's work allowed Razborov [213] to describe the solution set of a system of equations over F. Makanin further extended his results [180], proving that that the universal theory of a non-Abelian free group F is algorithmically decidable.

9.2 Components of the solution

In order to proceed from the partial results to a full solution of the Tarski problems four new ingredients were necessary: algebraic geometry over groups, the theory of fully residually free groups including the relationship to free exponential groups, an implicit function theorem (over free groups) and a description of irreducible algebraic varieties (over free groups) in terms of triangular quasi-quadratic systems. The last two, the implicit function theorem and the decription of irreducible algebraic varieties, work extensively with the methods of Makanin and Razborov.

It was clear from the beginning that to deal with the Tarski problems it was necessary to give a precise definition of solution sets of equations and inequations over free groups. In analogy with the classical theory of equations over number fields, algebraic geometry over groups was developed by G. Baumslag, A. Myasnikov and V.

Remeslennikov [16], [193]. The theory of algebraic geometry over groups translates the basic notions of classical algebraic geometry; algebraic sets, the Zariski topology, Noetherian domains, irreducible varieties, radicals and coordinate groups to the setting of equations over groups. This provides the necessary machinery to transcribe important geometric ideas into pure group theory. Some of these ideas go back to R. Bryant [38], V. Guba [123], B. Plotkin [211] and E. Rips [217]. The fundamental ideas of algebraic geometry over free groups were discussed in Chapter 8.

The next basic ingredient in the approach to the Tarski problems is the idea of a group being discriminated by a free group. This led to a systematic study of fully residually free groups. A group G is *fully residually free* provided to every finite set $S \subset G \setminus \{1\}$ of non-trivial elements of G there is a free group F_S and an epimorphism $h_S\colon G \to F_S$ such that $h_S(g) \neq 1$ for all $g \in S$. In other language we say that the group G is *discriminated* by a free group. Remeslennikov [214] and independently Gaglione and Spellman [108] established a surprising connection between fully residually free groups and the universal theory of free groups. In particular a finitely generated non-Abelian group G is fully residually free if and only if it is universally free, that is has the same universal theory as the non-Abelian free groups. Since elementary free groups must be universally free, fully residually free groups must play a prominent role in the study of the elementary theory. We discussed the extensive research on fully residually free groups in Chapters 6 and 7. We note that in Sela's work, finitely generated fully residually free groups are called limit groups. In Chapters 6 and 7 the structure theory of fully residually free groups was studied and the relationship between the Kharlampovich–Mysasnikov approach and the Sela approach via limiting groups of homomorphisms into free groups was described.

R. Lyndon [173] introduced the concept of an exponential group, that is a group which allows parametric exponents in an associative unitary ring A. In particular he studied the free exponential group $F^{\mathbb{Z}[t]}$ where exponents are allowed from the polynomial ring $\mathbb{Z}[t]$ over the integers $\mathbb{Z}$ (see Chapter 6). Lyndon established that the free exponential group $F^{\mathbb{Z}[t]}$ and hence any finitely generated subgroup of it, is fully residually free and hence, if it is non-Abelian, universally free. Kharlampovich and Myasnikov [151, 152, 153, 154, 155, 156] established the converse; therefore a finitely generated group is fully residually free if and only if it is embeddable in $F^{\mathbb{Z}[t]}$.

Lyndon's original motivation for introducing exponential groups was from the solution sets of equations over free groups. In [173] he found that the solution set of any equation with one variable over a free group F can be obtained from finitely many parametric words by specializing their parameters in the integers. A parametric word with parameters in $\mathbb{Z}[t_1, \ldots, t_n]$ is a formal expression obtained from a basis for F by finitely many concatenations and exponentiations from $\mathbb{Z}[t_1, \ldots, t_n]$. If one specializes the parameters $t_1, \ldots, t_n$ in $\mathbb{Z}$ one obtains an element of F. Lyndon proved that for any equation with one variable over a free group F one can effectively find a finite set of parametric words with parameters from the ring $\mathbb{Z}[t_1, \ldots, t_n]$ such that any solution of this equation can be obtained from some specialization of these words. K.

Appel refined Lyndon's result and showed that the solution set of a one variable equation over a free group can be parametrized by a finite set of words of the form fg^th where $f, g, h \in F$ and $t \in \mathbb{Z}[t]$.

A further detailed study of the structure of exponential groups was carried out by A. Myasnikov and V. Remeslennikov [196, 197]. They proved that the group $F^{\mathbb{Z}[t]}$ can be obtained starting from F by an infinite chain of a special type of HNN extensions called *extensions of centralizers*. If G is a group and C is the centralizer of a nontrivial element in G then

$$G(C, s) = \langle G, s; s^{-1}cs = c; c \in C\rangle$$

is a free extension of the centralizer C by s. From the work of Myasnikov and Remeslennikov, to construct $F^{\mathbb{Z}[t]}$, one needs to extend each centralizer sufficiently many times so that every proper centralizer is isomorphic to a free Abelian group of infinite rank – the additive group of $\mathbb{Z}[t]$. This further implies that any finitely generated subgroup of $F^{\mathbb{Z}[t]}$ and hence any fully residually free group, is actually a subgroup of a group that can be obtained from F by finitely many extensions of centralizers. For such groups, Bass–Serre theory (see Chapter 2) can be used to determine the structure.

Very important in this direction is the description of the coordinate groups $F_{R(S)}$ of an algebraic set $V_F(S)$ defined by a quadratic equation $S = 1$ with coefficients in S. By $R(S)$ we mean the coordinate group of the radical of S (see Chapter 8). This, by examining the relationship between the coordinate groups and groups embeddable by a sequence of extensions of centralizers in the free exponential group $F^{\mathbb{Z}[t]}$, shows that the coordinate groups of irreducible algebraic varieties are precisely the finitely generated fully residually free groups (limit groups).

The free exponential groups and their ties to the fully residually free groups were discussed in Chapter 6.

The final ingredients of the proof of the Tarski problems is a reduction of systems of equations over free groups to certain normal forms or systems of quadratic equations and then an implicit function theorem. We will describe these in Sections 9.5 and 9.6.

9.3 The Tarski–Vaught test and the overall strategy

As we have seen, Vaught proved Tarski Conjecture 2 for all free groups of infinite rank and hence reduced the problem to non-Abelian free groups of finite rank. His main result was that if the infinite rank free group F_1 is a free factor of the infinite rank free group F_2 then F_1 is an elementary subgroup (see below) of F_2, that is the identity map embedding F_1 into F_2 is an elementary embedding. Sacerdote went on to prove that all free groups of finite rank have the same π_2-theory, that is they satisfy exactly the same $\forall\exists$ (and equivalently $\exists\forall$) sentences. As we will see, Sacerdote's results pinpoints the main strategy in solving the whole problem and provides the first step in an induction.

Recall that $H \subset G$ is an *elementary subgroup* or equivalently that the identity map provides an *elementary embedding* if for any formula $\phi(\overline{x})$ and for any tuple $\overline{h}$ coming from elements of H then $\phi(\overline{h})$ is satisfied in G implies that $\phi(\overline{h})$ is satisfied in H. The main technique Vaught used in proving the Tarski conjecture for infinite rank and Sacerdote used for the π_2-theory is the following, that is known as the *Tarski–Vaught Test.*

Test (Tarski–Vaught Test). If H is a subgroup of G then H is an elementary subgroup of G if and only if for any formula $\phi(x, \overline{z})$ and for any tuple $(\overline{h})$ of elements from H there exists a $c \in G$ such that $\phi(c, \overline{h})$ is satisfied in G implies that there exists $c \in H$ such that $\phi(c, \overline{h})$ is satisfied in H.

Roughly the Tarski–Vaught Test says that a subgroup H of G is an elementary subgroup if and only if H is *algebraically closed* in G. In analogy with commutative algebra if we consider first order sentences with variables as our equations then any equation with constants from H which has a solution in G already has a solution within H.

If we wish to apply the Tarski–Vaught Test to the case of a free factor in a free group of finite rank we must then understand the nature of solving equations in free groups and over free groups. Here is where the work of Makanin and Razborov became crucial. Their work provided first a method to determine if an equation over a free group was solvable and hence provided a technique for Kharlampovich and Myasnikov to show that the elementary theory of non-Abelian free groups was decidable. However it was the introduction of algebraic geometry over free groups (see Chapter 8) that led to the necessary understanding of groups that have the same elementary theory as a non-Abelian free group of finite rank.

As in commutative algebra, to understand the solution set of an equation or a set of equations, we need its algebraic variety and in turn the coordinate group of its radical. Over free groups, first it was shown that the algebraic variety for a system of equations can be reduced to the variety of a finite subsystem. This is the group theoretic analog of Hilbert's Nullstellensatz and is called the *equationally Noetherian property* (see Chapter 8). Next it was shown by both Kharlampovich–Myasnikov and by Sela that the coordinate groups of irreducible algebraic varieties (over the non-Abelian free groups) are precisely the class of finitely generated fully residually free groups, that is the class of limit groups. Hence the full structure of this extensive class of groups, that we examined in Chapters 6 and 7, can be brought to bear on the study of equational systems over free groups. From this point there are several very strong structural results. In the proofs of Kharlampovich–Myasnikov and Sela it is shown that a general system of equations with a few special cases that must be handled separately can be shown to be equivalent to what is called a *quasi-trianglular system* of quadratic equations. The coordinate groups of such systems are called QT-groups and a special subclass of them, called special NTQ-groups, are precisely the groups that can be shown to have the same elementary theory as the non-Abelian free groups. Further the structure of the algebraic variety of a system of equations can be broken down by the Makanin–Razborov method and is tied to the group theoretic breakdown of the coordinate groups. Since the coordinate groups are limit groups this breakdown is well-understood as the JSJ decomposition of limit groups. This decomposition combined with a type of implicit function theorem provides for a quantifier elimination process that permits an induction starting with Sacerdote's π_2-result.

Before describing in detail these various aspects of the solution we describe a type of general strategy to use the Tarski–Vaught Test to prove Tarski Conjecture 2.

Let F_ω be a free group with basis $A_\omega = \{a_1, a_2, \ldots\}$, a countably infinite set. For each positive integer r let F_r be the free factor $\langle a_1, \ldots, a_r\rangle$ of F_ω. Let L_0 be as before the usual first order language with equality appropriate for group theory and for each $r \geq 2$ let $L_0[F_r]$ be the extension of L_0 formed by adjoining the nontrivial elements of F_r as new constant symbols.

Without assuming the Tarski conjectures for each integer $r \geq 2$ let

$$T_r = \bigcap_{r \leq n < \infty} Th_{L_0[F_r]}(F_n)$$

be the set of those sentences of $L_0[F_r]$ true in every free group containing F_r as a free factor. Sacerdote's Theorem is that if $2 \leq r < s < \infty$ then F_r and F_s satisfy precisely the same existential-universal and universal-existential sentences of $L_0[F_r]$. Sacerdote's result is then the starting point although both Kharlampovich–Myasnikov reprove it. The basic idea is then to use quantifier elimination (see Section 9.6) to reduce everything back to Sacerdote's result.

To this end let $T_r \cap (\forall\exists \cup \exists\forall)$ be the set of all universal-existential or existential-universal sentences of $L_0[F_r]$ true in F_r and hence true in every free group containing F_r as a free factor. Let B_r be the set of all Boolean combinations of $T_r \cap (\forall\exists \cup \exists\forall)$. That is B_r consists of all those sentences of $L_0[F_r]$ which are obtained from $T_r \cap (\forall\exists \cup \exists\forall)$ by conjunctions, disjunctions and negations. Suppose that we have a sentence Φ of $L_0[F_r]$ which is more complicated than a universal-existential or existential-universal sentence. What is done is to use quantifier elimination to show that there is a sentence $\Phi^* \in B_r$ such that Φ holds in F_s for all $r \leq s \leq \omega$ if and only if Φ^* does. This reduces the whole theorem to Sacerdote's case and proves that the embeddings $F_r \to F_s$ are elementary.

The quantifier elimination is handled by the *implicit function theorem* that we will discussed in Section 9.5. Basically consider a universal-existential sentence of $L_0[F_r]$,

$$\forall x_1, \ldots, x_m \exists y_1, \ldots, y_n \phi(x_1, \ldots, x_m, y_1, \ldots, y_n)\,. \tag{1}$$

We wish to show that for $r < s < \omega$ the sentence (1) holds in F_s if and only if it holds in F_r. One direction is relatively simple. Suppose that (1) holds in F_s. Then for an arbitrary element $\overline{u} = (u_1, \ldots, u_m) \in F_r^m$ the sentence

$$\exists y_1, \ldots, y_m \phi(u_1, \ldots, u_m, y_1, \ldots, y_n)$$

is an existential sentence of $L_0[F_r]$ true in F_s. Since F_s is discriminated by retractions $F_s \to F_r$ it follows that

$$\exists y_1, \ldots, y_m \phi(u_1, \ldots, u_m, y_1, \ldots, y_n)$$

must hold in F_r. Since $\overline{u} \in F_r^m$ was arbitrary the original sentence (1) must hold in F_r.

The other way, that (1) holding in F_r implies that it holds in F_s is the real work. Let $F(X) = \langle x_1, \ldots, x_m; \ \rangle$. From the implicit function theorem it is the case that whenever (1) holds in F_r there is an ordered n-tuple $(Y_1, \ldots, Y_n)$ from $F_r * X$ such that

$$\forall x_1, \ldots, x_m \phi(x_1, \ldots, x_m, Y_1(x_1, \ldots, x_m), \ldots, Y_n(x_1, \ldots, x_m))$$

also holds in F_r. However

$$\forall x_1, \ldots, x_m \phi(x_1, \ldots, x_m, Y_1(x_1, \ldots, x_m), \ldots, Y_n(x_1, \ldots, x_m))$$

is a universal sentence of $L_0[F_r]$. Again from the fact that F_s is discriminated by retractions $F_s \to F_r$ it follows that F_r and F_s satisfy the same universal sentences of $L_0[F_r]$.

Therefore if (1) holds in F_r then

$$\forall x_1, \ldots, x_m \phi(x_1, \ldots, x_m, Y_1(x_1, \ldots, x_m), \ldots, Y_n(x_1, \ldots, x_m))$$

holds in F_s and so (1) must hold in F_s. The cases where the hypotheses of the implicit function theorem are violated must be treated separately.

In the next sections we go over more precisely the basic ingredients as well as the implicit function theorem and then in Section 9.6 give a brief outline of the proof of the elementary embedding and in Section 9.8 a proof of the decidability. As part of the proof we get a complete description of the elementary free groups, that is those nonfree groups that have the same elementary theory as the non-Abelian free groups. This has some very nice consequences that we describe in Chapter 10.

9.4 Algebraic geometry and fully residually free groups

The proof of the Tarski conjectures depends on the algebraic geometry of free groups. In particular it depends on the description of a fully residually free group as the coordinate group of an irreducible algebraic variety. Here we review some of the basic definitions from the algebraic geometry over groups that was described in the last chapter.

Let G be a group generated by a finite set A, $F(X)$ be a free group with basis $X = \{x_1, x_2, \ldots, x_n\}$, and let $G[X] = G * F(X)$ be the free product of G and $F(X)$. A *system of equations* over G is an expression $S = 1$ where $S \subset G[X]$. As an element of the free product, the left side of every equation in $S = 1$ can be written as a product of some elements from the basis $X = \{x_1, \ldots, x_n\}$ and some elements from A. The elements from X are called *variables* while the elements from A are called *constants*. To emphasize this we sometimes write $S = 1$ as $S(X, A) = 1$ or just $S(X) = 1$.

A *solution* of the system $S(X) = 1$ over a group G is a tuple of elements $g_1, \ldots, g_n \in G$ such that after replacing each x_i by g_i the left hand side of every equation in $S = 1$ becomes the trivial element of G. Equivalently, a solution of the system $S = 1$ over G can be described as a G-homomorphism (see Chapter 8) $\Phi\colon G[X] \to G$ such that $\Phi(S) = 1$. Denote by $N(S)$ the normal closure of S in $G[X]$, and by G_S the quotient group $G[X]/N(S)$. Then every solution of $S(X) = 1$ in G gives rise to a G-homomorphism $G_S \to G$ and vice versa. By $V_G(S)$ we denote the set of all solutions in G of the system $S = 1$, it is called the *algebraic set* defined by S. This algebraic set $V_G(S)$ uniquely corresponds to the normal subgroup

$$R(S) = \{T(X) \in G[X]\colon (S(X) = 1 \to T(X) = 1)\}$$

of the group $G[X]$. Notice that if $V_G(S) = \emptyset$, then $R(S) = G[X]$. The subgroup $R(S)$ contains S, and it is called the *radical* of S, denoted by $\mathrm{Rad}(S) = R(S)$. The quotient group

$$G_{R(S)} = G[X]/R(S)$$

is the *coordinate group* of the algebraic set $V(S)$. Again, every solution of $S(X) = 1$ in G can be described as a G-homomorphism $G_{R(S)} \to G$.

To study the coordinate groups of equations in a given fixed group G it is convenient to consider the category of G-groups, that is groups which contain the group G as a distinguished subgroup. If H and K are G-groups then a homomorphism $\phi\colon H \to K$ is a G-homomorphism if $\phi(g) = g$ for every $g \in G$. In the category of G-groups morphisms are G-homomorphisms; subgroups are G-subgroups, etc. For most of our applications we consider the group G to be a CSA-group (see Chapter 6).

By $\mathrm{Hom}_G(H, K)$ we denote the set of all G-homomorphisms from H into K. It is not hard to see that $G[X] = G * F(X)$ is a free object in the category of G-groups. This group is called a free G-group with basis X. A G-group H is a finitely generated G-group if there exists a finite subset $B \subset H$ such that the set $G \cup B$ generates H.

A Zariski topology is defined on G^n by taking algebraic sets in G^n as a sub-basis for the closed sets of this topology. If G is a non-Abelian CSA group, in particular, a non-Abelian freely discriminated group, then the union of two algebraic sets is again algebraic. Therefore the closed sets in the Zariski topology over G are precisely the algebraic sets.

A G-group H is called *equationally Noetherian* if every system $S(X) = 1$ with coefficients from G is equivalent over G to a finite subsystem $S_0 = 1$, where $S_0 \subset S$. Then $V_G(S) = V_G(S_0)$. If G is G-equationally Noetherian, then we say that G is equationally Noetherian. It is known that linear groups (in particular, freely discriminated groups) are equationally Noetherian (see Chapter 8). If G is equationally Noetherian then the Zariski topology over G^n is Noetherian for every n, that is, every proper descending chain of closed sets in G^n is finite. This implies that every algebraic set V in G^n is a finite union of irreducible subsets called the *irreducible components* of V. Further the decomposition of V is unique. Recall that a closed set V is irreducible if it is not the union of two proper closed subsets. As in classical algebraic geometry we call an irreducible algebraic set an *algebraic variety* or just a *variety*.

What ties the algebraic geometry over groups and the solution of the Tarski problem to the class of fully residually free groups is the important result that the fully residually free groups are precisely the coordinate groups of irreducible algebraic sets in the category of F-groups where F is a non-Abelian free group. The breakdown via JSJ decomposition of a fully residually free group (see Chapter 7) then corresponds to a breakdown of a system of equations via the Makanin–Razborov process. Further the fully residually free groups embed into the coordinate groups of a special system of quadratic equations called NTQ-systems and hence we can concentrate on such systems.

We next summarize all the important properties of fully residually free groups including their relation to the coordinate groups of algebraic varieties. The proofs of these results can be found in Chapters 6, 7 and 8. The first result gives the important group theoretic properties of fully residually free groups. Most of the proofs of these can be found in Chapter 6.

Theorem 9.4.1. *Let G be a fully residually free group. Then G satisfies the following properties:*
(1) *G is torsion-free and each subgroup is also fully resiudally free.*
(2) *G is CSA and hence CT. Further if G is finitley generated each Abelian subgroup of G is finitely generated and contained in a unique maximal Abelian subgroup.*
(3) *If G is finitely generated then G is finitely presented and has only finitely many conjugacy classes of maximal Abelian subgroups.*
(4) *G is linear and has a solvable word problem.*
(5) *Every 2-generator subgroup of G is either free or Abelian*
(6) *If rank(G) $= 3$ then either G is free of rank 3, or free Abelian of rank 3, or a free rank one extension of centralizers of a free group of rank 2 (see Chapter 6).*

The next result summarizes the important equivalences for finitely generated fully residually free groups. These characterize this class of groups in several different ways and then play a crucial role in the proof of the Tarski problems. Number (8) in the theorem is especially important since it ties the class of limit groups to solutions of equations over free groups. We note that fully residually free groups in general need not be finitely generated. For example an infinitely generated free group is certainly fully residually free. However for our applications and for many of the important properties finite generation is crucial. The proofs of the next theorem are in Chapters 6 and 7, except for (8), which is in Chapter 8.

Theorem 9.4.2. *Let G be a finitely generated group. Then the following are equivalent:*
(1) *G is fully residually free,*
(2) *G is universally free if G is non-Abelian,*
(3) *G is a limit group,*
(4) *G is a constructible limit group,*
(5) *G is a limit of free groups in the Gromov–Hausdorf Topology,*
(6) *G embeds into a nonstandard free group, that is an ultrapower of a free group,*
(7) *G embeds into the free Lyndon completion* $F^{\mathbb{Z}[t]}$*,*
(8) *G is the coordinate group of an irreducible variety over a free group.*

We note that this theorem is a general result and can be considered as without constants or coefficients from any particular free group. Each statement has a correspsonding result if we allow coefficients. If F is a particular non-Abelian free group then for example in (1) we can say that a finitely generated F-group G is F-discriminated by F is equivalent to G being universally equivalent to F in the language $L_0(F)$ where this language allows constants from F.

The applications to equations over free groups and hence the solution to the Tarski problems especially uses (7) and (8) in the theorem and we look at these a bit more deeply. Kharlampovich and Myasnikov prove the following result that they call the embedding theorem. This result combined with the fact that the finitely generated subgroups of $F^{\mathbb{Z}[t]}$ are precisely the finitely generated fully residually free groups shows that the finitely generated fully residually free groups are exactly the coordinate groups of irreducible algebraic varities.

Theorem 9.4.3 (The Embedding Theorem). *Given an irreducible system of equations* $S = 1$ *over a free group F then one can effectively embed the coordinate group* $F_{R(S)}$ *into* $F^{\mathbb{Z}[t]}$.

Corollary 9.4.1. *The coordinate groups of irreducible algebraic varieties over a free group F are precisely the finitely generated fully residually free groups.*

Recall that the finitely generated subgroups (see Chapter 6) of $F^{\mathbb{Z}[t]}$ are precisely subgroups of groups built from finitely generated free groups by iterated extensions of centralizers. From Bass–Serre theory then we get that each finitely generated fully residually free group has graph of groups decomposition with cyclic edges. From this we get that each finitely generated fully residually free group has a JSJ decomposition. This breakup of the coordinate group will lead to a breakup of the algebraic variety.

Finally the following result was proved that provided a tremendous amount of information on a fully residually free group and its subgroups especially in regard to various algorithmic problems (see Chapter 7).

Theorem 9.4.4. *Let G be a finitely generated fully residually free group. Then G acts freely on some* $\mathbb{Z}^n$*-tree with lexicographic order for a suitable n.*

9.5 Quadratic equations and quasitriangular systems

The big breakthrough in doing applications of the algebraic geometry and particularly in the solution of the Tarski problem came with the discovery that to study the solution sets and hence the varieties of general equations over free groups it was only necessary to study *quadratic equations*. That is it was proved that not only are the fully residually free groups the coordinate groups of irreducible algebraic varieties but were embedded into the coordinate groups of special systems called NTQ-systems of quadratic equations. The solutions of quadratic equations over free groups were already extensively studied. Further implicit function theorems were developed for such NTQ-systems. Here we introduce the necessary material about quadratic equations and quadratic systems. Complete proofs can be found in [152, 153, 154, 155, 156].

Definition 9.5.1. An equation $S = 1$ in variables from $X = \{x_1, \ldots, x_n\}$ is *quadratic* if every variable x from X occurs in S no more than twice each time as x or x^{-1}.

A quadratic equation $S(X) = 1$ need not contain all the variables from X, it can be empty in some variable, linear in some variables or strictly quadratic on some subset of X.

We now write X instead of $\overline{x}$ for a finite system $\{x_1, \ldots, x_m\}$. The reason is that we consider X also as a generating system of a group $F(X)$ and sometimes split X into pieces $X_1, X_2, \ldots$.

We extend this to G-groups.

Let $S \subset G[X]$. Denote by $var(S)$ the set of variables that occur in S.

Definition 9.5.2. A set $S \subset G[X]$ is called *quadratic* if every variable from $var(S)$ occurs in each element S not more than twice as either x or x^{-1}. The set S is *strictly quadratic* if every variable x from $var(S)$ occurs in each element of S exactly twice each time as either x or x^{-1}.

A system $S = 1$ over G is *quadratic (strictly quadratic)*, if the corresponding set S is quadratic (strictly quadratic). An element of a quadratic (strictly quadratic) set S is called a *quadratic* or *strictly quadratic* word respectively. If S is a singelton we just write S for the set.

Quadratic equations can be placed into several *standard forms*.

Definition 9.5.3. A *standard quadratic equation* over the group G is an equation of one of the following forms (below d, c_i are nontrivial elements from G):

$$\prod_{i=1}^{n}[x_i, y_i] = 1, \quad n > 0; \tag{st1}$$

$$\prod_{i=1}^{n}[x_i, y_i]\prod_{i=1}^{m}(z_i^{-1}c_iz_i)d = 1, \quad n, m \geq 0, m + n \geq 1; \tag{st2}$$

$$\prod_{i=1}^{n} x_i^2 = 1, \quad n > 0; \tag{st3}$$

$$\prod_{i=1}^{n} x_i^2 \prod_{i=1}^{m} (z_i^{-1} c_i z_i) d = 1, \quad n, m \geq 0, n + m \geq 1; \tag{st4}$$

Lemma 9.5.1. *Let S be a strictly quadratic word over G. Then there is a G-automorphism $f \in \mathrm{Aut}_G(G[X])$ such that $f(S)$ is a standard quadratic word over G.*

The proof of this is in [60].

Definition 9.5.4. Strictly quadratic words of the type $[x, y], x^2, z^{-1}cz$, where $c \in G$, are called *atomic quadratic words* or simply *atoms*.

By Definition 9.5.4 a standard quadratic equation $S = 1$ over G has the form

$$r_1 r_2 \dots r_k d = 1,$$

where r_i are atoms, $d \in G$. This number k is called the *atomic rank* of this equation, we denote it by $r(S)$.

Definition 9.5.5. Let $S = 1$ be a standard quadratic equation written in the atomic form $r_1 r_2 \dots r_k d = 1$ with $k \geq 2$. A solution $\phi\colon G_{R(S)} \to G$ of $S = 1$ is called:

(1) commutative if $[\phi(r_i), \phi(r_{i+1})] = 1$ for all $i = 1, \dots, k-1$, and noncommutative otherwise;

(2) in a general position if $[\phi(r_i), \phi(r_{i+1})] \neq 1$ for all $i = 1, \dots, k-1$.

Theorem 9.5.1 ([KhM 1]). *In the following cases a standard quadratic equation $S = 1$ always has a solution in a general position:*

(1) *$S = 1$ is of type (st1), $n > 2$;*

(2) *$S = 1$ is of type (st2), $n > 0, n + m > 1$;*

(3) *$S = 1$ is of type (st3), $n > 3$;*

(4) *$S = 1$ is of type (st4), $n > 2$.*

The following theorem describes the radical of a standard quadratic equation which has at least one solution in a freely discriminated group G. For a element $g \in G$, $N(g)$ stands for the normal closure of this element in G.

Theorem 9.5.2 ([KhM 1]). *Let G be a freely discriminated group and let $S = 1$ be a standard quadratic equation over G which has a solution in G. Then*

(1) *If $S = [x, y]d$ or $S = [x_1, y_1][x_2, y_2]$ then $R(S) = N(S)$;*

(2) *If $S = x^2 d$ then $R(S) = N(xb)$ where $b^2 = d$;*

(3) *If $S = c^z d$ then $R(S) = N([zb^{-1}, c])$ where $d^{-1} = c^b$;*

(4) *If $S = x_1^2 x_2^2$ then $R(S) = N([x_1, x_2])$;*

(5) *If $S = x_1^2 x_2^2 x_3^2$ then $R(S) = N([x_1, x_2], [x_1, x_3], [x_2, x_3])$;*

(6) *If $r(S) \geq 2$ and $S = 1$ has a noncommutative solution then $R(S) = N(S)$;*

(7) *If $S = 1$ is of the type (st4) and all solutions of $S = 1$ are commutative, then $R(S)$ is the normal closure of the following system:*

$$\begin{aligned}\{x_1 \dots x_n = s_1 \dots s_n, [x_k, x_l] = 1, \quad &[a_i^{-1} z_i, x_k] = 1, \; [x_k, C] = 1,\\ &[a_i^{-1} z_i, C] = 1,\\ &[a_i^{-1} z_i, a_j^{-1} z_j] = 1 \quad (k, l = 1, \dots, n; i, j = 1, \dots, m)\},\end{aligned}$$

where $x_k \to s_k, z_i \to a_i$ is a solution of $S = 1$ and

$$C = C_G(c_1^{a_1}, \dots, c_m^{a_m}, s_1, \dots, s_n)$$

is the corresponding centralizer. The group $G_{R(S)}$ is an extension of the centralizer C.

Definition 9.5.6. A standard quadratic equation $S = 1$ over F is called *regular* if either it is an equation of the type $[x, y] = d$ $(d \neq 1)$, or the equation $[x_1, y_1][x_2, y_2] = 1$, or it has a non-commutative solution and it is not an equation of the type $c_1 z_1 c_2^{z_2} = c_1 c_2$, $x^2 c^z = a^2 c$, $x_1^2 x_2^2 = a_1^2 a_2^2$.

The following theorem is one of the main technical tools in the solution of the Tarski problems.

Theorem 9.5.3. *Let $S(X, A) = 1$ be a regular standard quadratic equation over $F(A)$. Then every equation $T(X, Y, A) = 1$ compatible with $S(X, A) = 1$ admits a S-lift.*

See later in this section for a formal definition of an S-lift.

In what follows we usually write just $S(X) = 1$ or $T(X, Y) = 1$. Now we formulate the implicit function theorem over free groups in its simplest form.

Theorem 9.5.4 (Implicit function theorem). *Let $S(X) = 1$ be a regular standard quadratic equation over a non- Abelian free group F and let $T(X, Y) = 1$ be an equation over F, $|X| = m$, $|Y| = n$. Suppose that for any solution $U \in V_F(S)$ there exists a tuple of elements $W \in F^n$ such that $T(U, W) = 1$. Then there exists a tuple of words $P = (p_1(X), \dots, p_n(X))$, with constants from F, such that $T(U, P(U)) = 1$ for any $U \in V_F(S)$.*

The implicit function theorem hence allows for a type of *quantifier elimination*.

What is next of importance are special types of systems of quadratic equations. First we define *quasi-triangular systems*.

Definition 9.5.7. Let $X_1, \dots, X_m$ be disjoint tuples of variables. A system of equations $U(X_1, \dots, X_m) = 1$ with coefficients from a free group F of the following form

$$\begin{aligned}S_1(X_1, \dots, X_m) &= 1\\ S_2(X_2, \dots, X_m) &= 1\\ &\dots\\ S_m(X_m) &= 1\end{aligned}$$

is said to be *triangular quasi-quadratic* if for every i the equation

$$S_i(X_i, \dots, X_m) = 1$$

is quadratic in the variables from X_i.

Denote by G_i the coordinate group of the subsystem $S_i = 1, \dots, S_m = 1$ of the system $U = 1$. The system $U = 1$ is said to be *non-degenerate* (NTQ) if for each i the equation $S_i(X_i, \dots, X_m) = 1$ has a solution in G_{i+1}.

The coordinate group of an NTQ-system is called an *NTQ-group*. In Sela's terminology this is called an *ω-residually free tower*. Further if the only Abelian subgroups of an NTQ-group G are cyclic then G is hyperbolic and called a *special NTQ-group*. Sela calls these *hyperbolic-ω-residually free towers*. The proofs of Kharlampovich and Myasnikov and Sela of the elementary embedding of one free group into another have the byproduct that the special NTQ-groups are precisely the finitely generated groups that have the same elementary theory as the class of non-Abelian free groups. Further from the JSJ structure of these groups it follows that an orientable surface group of genus $g \geq 2$ and a nonorientable surface group of genus $g \geq 4$ are special NTQ-groups and hence have precisely the same first-order theory as the class of non-Abelian free groups. We call such groups *elementary free groups* and will discuss them in general in Chapter 10.

Kharlampovich and Myasnikov [152] use these ideas of algebraic geometry to prove that every finitely generated fully residually free group embeds into Lyndon's group $F^{\mathbb{Z}[t]}$. As remarked earlier it follows from this that finitely generated fully residually free groups must be finitely presented. The proof of the embedding result follows from a sequence of theorems. We mention two of these from the tail end of the sequence (see [152]).

Theorem 9.5.5. *For every finite system $S(X) = 1$ of equations over a free group F one can find effectively a finite family of nondegenerate triangular quasi-quadratic systems $U_1, \dots, U_k$ and word mappings*

$$\rho_i \colon V_F(U_i) \to V_F(S), \quad i = 1, \dots, k$$

such that for every $b \in V_F(S)$ there exists an i and $c \in V_F(U_i)$ for which $b = \rho(c)$, that is

$$V_F(S) = \rho_1(V_F(U_1)) \cup \cdots \cup \rho_k(V_F(U_k))$$

and all sets $\rho_i(V_F(U_i))$ are irreducible. Moreover every irreducible component of $V_F(S)$ can be obtained as the closure of $\rho_i(V_F(U_i))$ in the Zariski topology.

Theorem 9.5.6. *For a system $S = 1$ over a free group the set $V(S)$ is irreducible if and only if $F_{R(S)} \subset F_{R(S_1)}$ for a nondegenerate triangular quasi-quadratic system S_1.*

From the previous two theorems it follows that to consider the coordinate group of a general equation over a free group it can be reduced to looking at the coordinate group of an NTQ-system.

One can relax the condition on exponential groups that the system of exponents must be a ring. Instead one can take a torsion-free additive Abelian group containing a pure infinite cyclic subgroup (to be identified with $\mathbb{Z}$). As another pretty application ofalgebraic geometry over groups we have the following result due also to Kharlampovich and Myasnikov [152].

Theorem 9.5.7. *Let $S(X) = 1$ be a system of equations over a free group F. Then there exists a finite set of n-tuples of parametric words*

$$U = (u_1, \ldots, u_n) \in (F^{\mathbb{Z}^k})^n$$

such that the set of all their specializations $^*U$ is a dense subset of the variety $V_F(S)$ in the Zariski topology.

It follows from this theorem that solution sets of consistent systems of equations over free groups can *almost* be parametrized.

We now consider lifting equations into generic points. These correspond to lifting solutions in classical algebraic geometry.

Definition 9.5.8. Let $S(X) = 1$ be a system of equations over a group G which has a solution in G. We say that a system of equations $T(X, Y) = 1$ is compatible with $S(X) = 1$ over G if for every solution U of $S(X) = 1$ in G the equation $T(U, Y) = 1$ also has a solution in G, i.e., the algebraic set $V_G(S)$ is a projection of the algebraic set $V_G(S \cup T)$.

Let $S(X) = 1$ be a system of equations over G and suppose $V_G(S) \neq \emptyset$. The canonical embedding $X \to G[X]$ induces the canonical map

$$\mu\colon X \to G_{R(S)}\,.$$

We are ready to formulate the main definition.

Definition 9.5.9. Let $S(X) = 1$ be a system of equations over G with $V_G(S) \neq \emptyset$ and let $\mu\colon X \to G_{R(S)}$ be the canonical map. Let a system $T(X, Y) = 1$ be compatible with $S(X) = 1$ over G. We say that $T(X, Y) = 1$ admits a lift to a generic point of $S = 1$ over G (or, shortly, S-lift over G) if $T(\mu(X), Y) = 1$ has a solution in $G_{R(S)}$ (here Y are variables and $\mu(X)$ are constants from $G_{R(S)}$).

The next result characterizes lifts in terms of the coordinate groups of the corresponding equations.

Theorem 9.5.8. *Let $S(X) = 1$ be an equation over G which has a solution in G. Then for an arbitrary equation $T(X,Y) = 1$ over G the following conditions are equivalent:*
(1) *$T(X,Y) = 1$ is compatible with $S(X) = 1$ and $T(X,Y) = 1$ admits S-lift over G;*
(2) *$G_{R(S)}$ is a retract of $G_{R(S \cup T)}$, i.e., $G_{R(S)}$ is a subgroup of $G_{R(S \cup T)}$ and there exists a $G_{R(S)}$-homomorphism $G_{R(S \cup T)} \to G_{R(S)}$.*

In the notation of the theorem every solution U of an equation $T(\mu(X), Y) = 1$ in $G_{R(S)}$ gives rise to a retraction, i.e., $G_{R(S)}$-homomorphism, $\phi_U : G_{R(S \cup T)} \to G_{R(S)}$, and vice versa. This allows one to consider solutions of $T(\mu(X), Y) = 1$ in $G_{R(S)}$ as retractions $G_{R(S \cup T)} \to G_{R(S)}$.

Definition 9.5.10. Let $T(X,Y) = 1$ be compatible with $S(X) = 1$ over G. We say that $T(X,Y) = 1$ admits a complete S-lift (or a separating S-lift) if the set of all solutions of $T(\mu(X), Y) = 1$ in $G_{R(S)}$ (viewed as retractions) separates the group $G_{R(S \cup T)}$.

The next result characterizes complete lifts in terms of the coordinate groups of the corresponding equations.

Theorem 9.5.9. *Let $S(X) = 1$ be an equation over G which has a solution in G. Then for an arbitrary equation $T(X,Y) = 1$ over G the following conditions are equivalent:*
(1) *$T(X,Y) = 1$ is compatible with $S(X) = 1$ over G and it admits a complete S-lift;*
(2) *$G_{R(S \cup T)} \cong (G_{R(S)})_{R(T(\mu(X),Y))}$.*

9.6 Quantifier elimination and the elimination process

The final necessary component of the proof of the Tarski problems is the *elimination process*, abbreviated EP by Kharlampovich and Myasnikov. This is the most technical and difficult portion of the program and is based on initial work of Makanin and then improved upon by Razborov.

In general a *quantifier elimination* is a concept of simplification used in mathematical logic and model theory. First order formulas with fewer quantifiers are considered simpler with quantifier-free formulas as the simplest. A theory has quantifier elimination if for every formula in the theory there is another formula with fewer quantifiers logically equivalent to it relative to the theory.

Quantifier elimination permits an induction on the number of quantifiers. A first order theory L has quantifier elimination if and only if for any two models B and C of the theory with a common substructure A, B and C are elementarily equivalent in the language of L augmented with constants from A. To prove the elementary equivalence of B and C under quantifier elimination it suffices to prove the equivalence of the existential theory.

The proofs of the Tarski theorems by Kharlampovich and Myasnikov use an elimination process originally introduced by Makanin. This elimination process, that is abbreviated EP, is a symbolic rewriting process that transforms formal systems of equations in groups. Makanin in 1982 introduced the initial version of the EP. His method provides a decision algorithm to verify consistency of a given system of equations, that is he handles the decidability of the Diophantine problem over free groups. To accomplish this, Makanin estimates the length of the minimal solution (if it exists). As part of this EP, Makanin introduced the fundamental notions of generalized equations and elementary and entire transformations. In 1987, Razborov (1987) extended the EP much further. Razborov's version of the EP produces all solutions of a given system in a free group F. He used special groups of automorphisms, and fundamental sequences to encode solutions.

In 1996 Kharlampovich and Myasnikov, building on the above, found an effective description of solutions of equations over free and fully residually free groups in terms of NTQ systems (see Section 9.5). In particular they represented a solution set of a system of equations canonically as a union of solutions of a finite family of NTQ groups.

Theorem 9.6.1 ([KhM 5]). *One can effectively construct the EP that starts with an arbitrary system $S(X, A) = 1$ and results in finitely many NTQ systems*

$$U_1(Y) = 1, \ldots, U_m(Y) = 1$$

such that

$$V_F(S) = P_1(V(U_1)) \cup \cdots \cup P_m(V(U_m))$$

for some word mappings $P_1, \ldots, P_m$.

The word mapping P_i maps a tuple Y_i to a tuple $X \in V(S)$. P_i may be pictured as an A-automorphism from $F_{R(S)}$ into $F_{R(U_i)}$, then any solution $\psi\colon F_{R(U_i)} \to F$ precomposed with P_i provides a solution $\phi\colon F_{R(S)} \to F$.

The elimination process in this case can be viewed as a non-commutative analog of the classical elimination process in algebraic geometry. Hence, going from the bottom to the top, every solution of the subsystem $S_n = 1, \ldots, S_i = 1$ then can be extended to a solution of the next equation $S_{i-1} = 1$. The effectiveness of the EP allows for the determination of the decidability of the first order theory of free groups (see Section 9.8).

Theorem 9.6.2. *All solutions of the system of equations $S = 1$ in $F(A)$ can be effectively represented as homomorphisms from $F_{R(S)}$ into $F(A)$ encoded into the following canonical Hom-diagram. Here all groups, except, maybe, the one in the root, are fully residually free (given by a finite presentation), arrows pointing down correspond to epimorphisms (defined effectively in terms of generators) with non-trivial kernels, and loops correspond to automorphisms of the coordinate groups.*

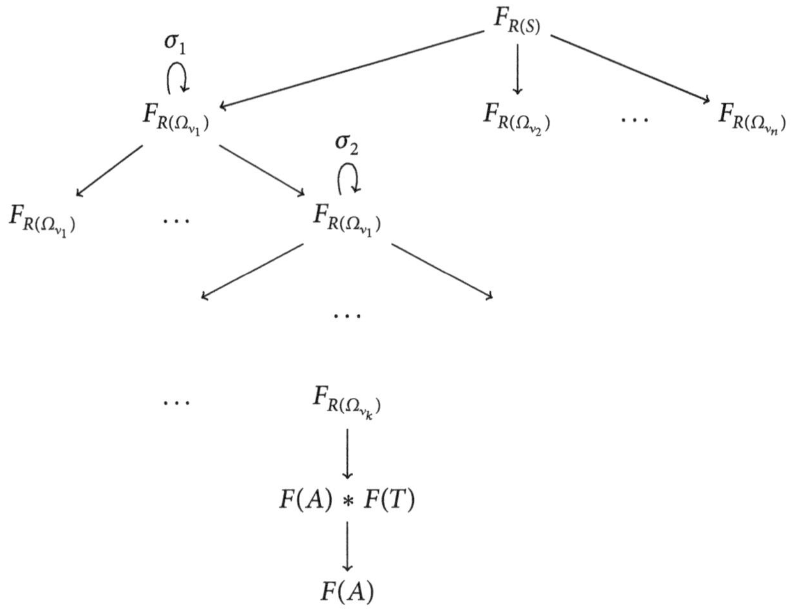

A family of homomorphisms encoded in a path from the root to a leaf of this tree is called a *fundamental sequence* or fundamental set of solutions (because each homomorphism in the family is a composition of a sequence of automorphisms and epimorphisms). Later Sela called such family a *resolution*. Therefore the solution set of the system $S = 1$ consists of a finite number of fundamental sets. Further each fundamental set factors through one of the NTQ systems from Theorem 9.6.1. If $S = 1$, or equivalently the coordinate group $G = F_{R(S)}$, is fully residually free, then one of the fundamental sets discriminates G. From this we get the following.

Theorem 9.6.3. *Finitely generated fully residually free groups are subgroups of coordinate groups of NTQ systems. Further there is an algorithm to construct the embedding.*

The crux of the elimination process as applied to systems of equations over a free group F is the following chain of ideas. The precise details can be found in [156]. First a generalized equation is a system of equations of a particular type that we will describe below. Then:

Given a system $S(X) = 1$ of equations in a free group $F(A)$ one can effectively construct a finite set of generalized equations

$$\Omega_1, \ldots, \Omega_k$$

such that:
(1) given a solution of $S(X) = 1$ in $F(A)$ one can effectively construct a reduced solution of one of Ω_i in the free semigroup with basis $A \cup A^{-1}$

(2) given a solution of some Ω_i in the free semigroup with basis $A \cup A^{-1}$ one can effectively construct a solution of $S(X) = 1$ in $F(A)$.
(3) Given a generalized equation Ω there are finitely many elementary transformations that can be applied to get a new generalized equation Ω' such that if σ is a solution of Ω then σ' is a solution of Ω'.
(4) The elimination process is a sequence of elementary transformations, applied according to some precise rules to an initial pair Ω_0, σ_0:

$$\Omega_0, \sigma_0 \to \Omega_1, \sigma_1 \to \cdots \to \Omega_k, \sigma_k .$$

(5) The EP is a branching process such that on each step one of the finite number of elementary transformations is applied according to some precise set of rules to form the sequence above.
(6) From a group theoretic point of view the elimination process provides informtaion about the coordinate groups of the systems involved. This allows the transformation of the pure combinatorial and algorithmic results obtained in the elimination process into statements about the coordinate groups.

We close this section by describing the generalized equations and allowable elementary transformations, which are quite technical. In the subsequent two sections we show briefly how this material is put together into the proofs of both elementary embedding and the decidability.

Definition 9.6.1. A *combinatorial generalized equation* consists of the following components:
(1) A finite set of bases $BS = BS(\Omega)$. The set of bases $\mathcal{M}$ consists of $2n$ elements

$$\mathcal{M} = \{\mu_1, \ldots, \mu_{2n}\}.$$

The set $\mathcal{M}$ comes equipped with two functions: a function $\epsilon\colon \mathcal{M} \to \{1, -1\}$ and an involution $\Delta\colon \mathcal{M} \to \mathcal{M}$ (that is, Δ is a bijection such that Δ^2 is an identity on $\mathcal{M}$). Bases μ and $\overline{\mu} = \Delta(\mu)$ are called dual bases. We denote bases by letters μ, λ, and so on.
(2) A set of boundaries $BD = BD(\Omega)$. BD is a finite initial segment of the set of positive integers $BD = 1, 2, \ldots, \rho + 1 + m$, where m is the cardinality of the basis $A = \{a_1, \ldots, a_m\}$ of the free group $F = F(A)$. We use letters i, j, etc. for boundaries.
(3) Two functions $\alpha\colon BS \to BD$ and $\beta\colon BS \to BD$. We call $\alpha(\mu)$ and $\beta(\mu)$ the initial and terminal boundaries of the base μ (or endpoints of μ). These functions satisfy the following conditions for every base $\mu \in BS$: $\alpha(\mu) \langle \beta(\mu)$ if $\epsilon(\mu) = 1$ and $\alpha(\mu) \rangle \beta(\mu)$ if $\epsilon(\mu) = -1$.
(4) The set of boundary connections (p, λ, q), where p is a boundary on λ (between $\alpha(\lambda)$ and $\beta(\lambda)$) and q is a boundary on $\overline{\lambda}$

For a combinatorial generalized equation, one can canonically associate a system of equations in variables $h_1, \ldots, h_\rho$ over $F(A)$. The variables h_i are sometimes called items. This system is called a generalized equation, and (slightly abusing the language) is denoted by the same symbol Ω. The generalized equation consists of the following three types of equations.

(1) Each pair of dual bases $\lambda, \overline{\lambda}$ provides an equation

$$(h_{\alpha(\lambda)}h_{\alpha(\lambda)+1}\cdots h_{\beta(\lambda)-1})^{\epsilon(\lambda)} = (h_{\alpha(\overline{\lambda})}h_{\alpha(\overline{\lambda})+1}\cdots h_{\beta(\overline{\lambda})-1})^{\epsilon(\overline{\lambda})}.$$

These equations are called *basic equations*.

(2) Every boundary connection (p, λ, q) gives rise to a boundary equation

$$(h_{\alpha(\lambda)}h_{\alpha(\lambda)+1}\cdots h_{\beta(\lambda)-1}) = (h_{\alpha(\overline{\lambda})}h_{\alpha(\overline{\lambda})+1}\cdots h_{q-1})$$

if $\epsilon(\lambda) = \epsilon(\overline{\lambda})$ and

$$(h_{\alpha(\lambda)}h_{\alpha(\lambda)+1}\cdots h_{\beta(\lambda)-1}) = (h_{\alpha(\overline{\lambda})}h_{\alpha(\overline{\lambda})+1}\cdots h_{\beta(\overline{\lambda})-1})^{-1}$$

if $\epsilon(\lambda) = -\epsilon(\overline{\lambda})$.

(3) Constant equations: $h_{\rho+1} = a_1, \ldots, h_{\rho+1+m} = a_m$.

We assume that every generalized equation comes associated with a combinatorial one. Denote by $F_{R(S)}$ the coordinate group of the generalized equation.

Definition 9.6.2. Let $\Omega(h) = \{L_1(h) = R_1(h), \ldots, L_s(h) = R_s(h)\}$ be a generalized equation in variables $h = (h_1, \ldots, h_\rho)$. A sequence of reduced nonempty words $U = (U_1(Z), \ldots, U_\rho(Z))$ in the alphabet $(A \cup Z)^{\pm 1}$ is a *solution* of Ω if:
(1) all words $L_i(U), R_i(U)$ are reduced as written,
(2) $L_i(U) = R_i(U)$ if $i \in [1, s]$.

If we specify a particular solution δ of a generalized equation Ω then we use a pair (Ω, δ).

The basic idea in using the elimination process is to move from a system of equations to a system of generalized equations, then to a family of generalized equations in a manner that is tied to homomorphisms of coordinate groups and tied to solutions, homomorphisms into free groups. This is done by *elementary transformations*. In particular if Ω is a generalized equation then an *elementary transformation* associates to Ω a family of generalized equations $ET(\Omega) = (\Omega_1, \ldots, \Omega_k)$ and surjective homomorphisms

$$\pi_i \colon F_{R(\Omega)} \to F_{R(\Omega_i)}, i = 1, \ldots, k$$

and to a solution δ_i of Ω_i such that the following diagram commutes.

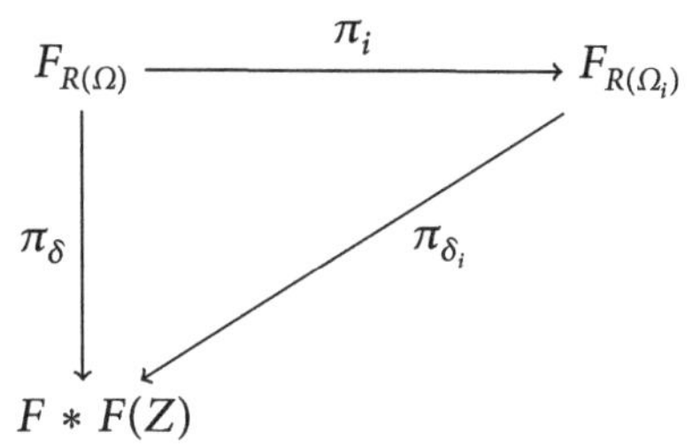

The actual transformations, as well as the generalized equations used, are very technical and can be found in [156].

9.7 Proof of the elementary embedding

In [156] these various ingredients, the reduction to NTQ systems, the description of the breakup of the coordinate groups of equations in terms of the JSJ decomposition of the groups and the corresponding breakup of the algebraic varieties and finally the implicit funciton theorem and quantifier elimination given by the elimination process, was put together to give the final proof that if a free group F_r is a free factor of the free group F_s then it is an elementary subgroup. The techniques of this proof also prove the decidability. Sela's proof proceeds in much the same way but with different terminology.

As indicated in Section 9.3 the basic strategy is an induction on the number of quantifiers. The starting off point for the induction is Sacerdote's theorem which says that if F_r is a free factor of F_s then they satisfy exactly the same π_2-theories, that is exactly the same $\forall\exists$ sentences. Sacerdote's proof is complicated and not enirely clear so both Kharlampovich–Myasnikov and Sela reprove it using their own machinery. However in the case of Kharlampovich–Myasnikov this also provides a technique for the induction step.

Recall that Merzlyakov proved that all non-Abelian free groups satisfy exactly the positive sentences. Sacerdote, in his proof for the π_2-theory, used Merzlyakov's ideas and the small cancellation technique in Van-Kampen diagrams for group presentations (see Chapter 3). Sacerdote shows that an arbitrary π_2-sentence is either positive, in which case Merzlyakov's result can be used, or by using the topology of the Van-Kampen diagrams can be written as a boolean combination of positive sentences and sentences with only one quantifier. It follows from Sacerdote's theorem that free non-Abelian groups of finite rank satisfy the same boolean combinations of $\forall\exists$-sentences. In Myasnikov–Kharlampovich's handling of Sacerdote's result they follow the same idea as Sacerdote to reduce to either positive sentences and apply Merzlyakov's result or to a boolean combination of positive sentences and sentences with only one quantifier. To get the quantifier elimination they rely on their implicit funcion theorem (see

Section 9.6) coupled with their extension of the Makanin–Razborov diagrams. This replaces Sacerdote's use of van-Kampen diagrams. There are several special cases where the implicit function theorem does not apply directly and these are handled separately. The details are intricate and can be found in [156].

Thus Sacerdote's theorem on π_2-sentences is the first step in an induction. For the inductive step they consider a general sentence

$$\Phi = \exists X_1 \forall Y_1 \ldots \exists X_k \forall Y_k (U(X_1, Y_1, \ldots, X_k, Y_k) = 1 \to V(X_1, Y_1, \ldots, X_k, Y_k) = 1)$$

where U and V are nontrivial elements in the free grup $F(X_1, Y_1, \ldots, X_k, Y_k)$. To prove the Tarski Elementary Embedding Theorem it is shown that this sentence is true if and only if some boolean combination of sentences with less alterations of quantifiers is true and further the reduction does not depend on the free group F nor even on the coordinate group $F_{R(S)}$. Ultimately this reduction leads to Boolean combinations of π_2-sentences and hence Sacerdote's theorem can be used completing the induction. As for Sacerdote's result the details are complicated and can be found in [156].

In order to arrive at this reduction, the following ideas, that we have introduced in the previous sections, must be interwoven.

(1) The solution set of a systems of equations $S = 1$ over a free group depends on its coordinate group $F_{R(S)}$ where $R(S)$ is the radical of the system. Further general varieties break up into irreducible algebraic varieties and further the finitely generated fully residually free groups, that is the limit groups, are precisely the coordinate groups of irreducible algebraic varieties.
(2) Finitely generated fully residually free groups can be embedded into the coordinate groups of NTQ-systems. This allows us in trying to solve our general system to concentrate on NTQ-systems. As we have seen these systems are constructed inductively from quadratic equations and we can then apply the techniques developed for quadratic equations to the study of these systems.
(3) The implicit function theorem for algebraic varieties corresponding to regular quadratic and NTQ-systems over free groups. As a by-product of placing these ideas together in the proof it will follow that the coordinate groups or special NTQ-systems turn out to form the class of finitely generated elementary free groups, that is the class of finitely generated groups elementarily equivalent to a non-Abelian free group. We note that any nonstandard free group, that is a proper ultrapower $^*F$ of a non-Abelian free group, is elementary free but not finitely generated.
(4) The variation and extension of the Makanin–Razborov process for solving equations over free groups. This extension provides a description of the solution set of a system of equations in a free group as a diagram of homomorphisms tied together with a decomposition of the coordinate group. This leads to what are called *generalized equations* and an *elimination process*.

Now consider the general sentence above, and we assume that there are more than two alterations of quantifiers so that it is not a π_2-sentence:

$$\Phi = \exists X_1 \forall Y_1 \dots \exists X_k \forall Y_k (U(X_1, Y_1, \dots, X_k, Y_k) = 1 \rightarrow (X_1, Y_1, \dots, X_k, Y_k) = 1) .$$

If this sentence is positive then it is true or false in every non-Abelian free group independent of rank by Merzljakov's result. If not consider the algebraic variety of this equation and the corresponding system given by the irreducible algebraic components. Call this resulting system S and the corresponding coordinate group $F_{R(S)}$. This is a fully residually free group and from the embdedding of these into coordinate groups of NTQ-systems we can look at the equivalent NTQ-system. Now using the extended Makanin–Razborov elimination process combined with the implicit function theorem we can reduce the NTQ-system to a system with a fewer number of quantifiers. The elimination process is a branching process and hence grows in size. A difficult portion of the analysis is to show that this process is finite, that is, it will terminate in a finite number of steps (see [156]).

Ultimately we get that our original system is equivalent to a system that can be written as a boolean combination of sentences with less alternations of quantifiers. Further this reduction does not depend on the particular coordinate group $F_{R(S)}$ only the general properties (1) through (4) above.

This argument proves the Tarski conjecture in the strong second form and hence also proves the first Tarski conjecture. The argument also proves the following three results.

Theorem 9.7.1. *Every formula in the language of a free group is equivalent to a boolean combination of* $\forall\exists$*-formulas.*

Theorem 9.7.1 was originally formulated and proved by Sela [236].

Theorem 9.7.2. *A coordinate group of a special regular NTQ-system has the same elementary theory as a non-Abelian free group.*

In [155] it was proved that any finitely generated group which is π_2-equivalent to a non-Abelian free group is isomorphic to the coordinate group of a regular NTQ-system. Combining this with Theorem 9.7.2 gives the complete characterization of finitely generated groups with exactly the same first order theory as the non-Abelian free groups. Such groups are called *elementary free groups* or *elementarily free groups*.

Theorem 9.7.3. *A finitely generated group G is an elementary free group if and only G is isomorphic to the coordinate group of a regular NTQ-system. We call such groups NTQ-groups.*

We note that Sela of course comes up with the same characterization. In his language the NTQ-groups are called *ω-residually free towers*. Therefore from Sela the finitely generated elementary free groups are the hyperbolic-ω-residually free towers.

In the next chapter we will look in more detail at elementary free groups.

9.8 Proof of decidability

The proof outlined in the last section proves the first two Tarski conjectures in the strongest possible form. That is, the free group $F(a_1, \ldots, a_n)$ freely generated by $a_1, \ldots, a_n$ is an elementary subgroup of $F(a_1, \ldots, a_n, \ldots, a_{n+p})$ for every $n \geq 2$ and $p \geq 0$. This leaves us the third Tarski conjecture, that the elementary theory $\mathrm{Th}(F)$ of a non-Abelian free group is decidable. This was proved by Kharlampovich and Myasnikov along the same lines using induction as the proof in the last section by showing that each reduction is in fact effective. By effective we mean algorithmic in a finite number of steps.

Recall that a theory of a group G, $\mathrm{Th}(G)$, is *decidable* if there exists a recursive algorithm which, given a sentence ϕ of L_0, decides whether or not ϕ is true. For the class of non-Abelian free groups the decidability means that there exists a recursive algorithm which, given a sentence ϕ of L_0, decides whether or not ϕ is true in every non-Abelian free group. The proof of Kharlampovich and Myasnikov is even a bit stronger. There exists a recursive algorithm which, given a sentence ϕ of $L_0[F]$, decides whether or not ϕ is true, where $L_0[F]$ allows constants from the free group F.

The starting off point for the proof of the decidability of the theory of non-Abelian free groups was the work of Makanin. Makanin [179] proved the algorithmic decidability of the Diophantine problem over free groups. He combined this with Merzjlakov's proof that all non-Abelian free groups share the same positive theory to prove the algorithmic decidability of the positive theory of non-Abelian free groups. He further proved the algorithmic decidability of the universal theory. This then implies, although it was not relevant at the time, the algorithmic decidability of the universal theory of limit groups. Makanin developed a powerful machinery, which is now called the Makanin elimination process, to deal with equations over free groups. Razborov extended Makanin's method and described the solution set of an arbitary system of equations over a free group in terms of Makanin–Razborov diagrams. As we have seen the Makanin–Razborov process for describing the solution sets of arbitrary systems of equations can be reduced to examining the solution sets of NTQ systems and hence the solution of quadratic equations.

Kharlampovich and Myasnikov combined the same type of induction process used in the proof in Section 9.7 together with Makanin's technique to obtain the decidability of the elementary theory of a non-Abelian free group.

What is done, although as in the previous section the details are complicated, is to show the decidability of the π_2-sentences. Then in the elimination process, based on generalized equations and the extended Makanin–Razborov process, each reduction is algorithmically effective, that is can be done algorithmically. It follows that the decidability question also fits into the induction scheme and carries through. Some of the proofs involving effectiveness have been simplified by using the description of limit groups in terms of infinite words in $F^{\mathbb{Z}[t]}$ (see Chapter 7). Further some of the reductions are handled by the extended implicit function theorem (Theorem 9.5.).

10 On elementary free groups and extensions

10.1 Elementary free groups

As a by-product of the positive solution of the *Tarski conjectures* by Kharlampovich and Myasnikov [152, 153, 154, 155, 156] and Sela [233, 234, 235, 236, 237] it was proved that the class of non-free groups that have exactly the same first order theory as the class of non-Abelian free groups was nonempty. Such groups are called *elementary free groups* (or *elementarily free groups*) and both sets of authors provide complete characterizations of the finitely generated instances of them. As we saw at the end of the previous chapter, in the Kharlampovich–Myasnikov approach these are the *special NTQ-groups*. The primary examples of such groups are the orientable surface groups S_g of genus $g \geq 2$ and the nonorientable surface groups N_g of genus $g \geq 4$.

Surface groups have always played a fundamental role in combinatorial group theory, most prominently as motivating examples for many important concepts. The paper by Ackermann, Fine and Rosenberger [1] provides an overall examination of these. That surface groups are elementary free provides a powerful tool to prove some results in surface groups that are otherwise quite difficult to obtain. For example, J. Howie [135] and independently O. Bogopolski and O. Bogopoloski and S. Sviridvov [32], [33] proved that a theorem of Magnus concerning the normal closures of elements in free groups also holds in surface groups of appropriate genus (see Section 10.2). Their proofs were nontrivial. However it was proved (see [87] and [114]) that this result is first order and hence automatically true in any elementary free group. In [88] a large collection of such results was given, where they were called *something for nothing results*. We examine a collection of these theorems in Sections 10.2 and 10.3. Any first order result, true in a non-Abelian free group, must hold in any elementary free group. However in elementary free groups other properties must hold beyond first order results. In the present chapter we consider these results also.

A finitely generated elementary free group G must be a limit group (see Chapters 6 and 7) and many of its properties follow from the structure theory of limit groups. Hence such a group must be CSA and any 2-generator subgroup is either free or Abelian. Further we prove that a finitely generated elementary free group has cyclic centralizers. This is not a first order statement, however from this we obtain that if two elements commute in a finitely generated elementary free group then they are both powers of a single element. This is not true in a general elementary free group and we give an example. From the cyclic centralizer property we can obtain that a finitely generated elementary free group must be hyperbolic, stably hyperbolic and a Turner group, that is the test elements, if there are any, in any finitely generated elementary free group, are precisely those elements that do not lie in any proper retract. Further we can prove that any finitely generated elementary free group is conjugacy separable and hence has a solvable conjugacy problem. The automorphism group of a finitely generated elementary free group is tame, a result we mentioned in Chapter 7.

The starting off point for this chapter are the following two results that were mentioned at the end of the last chapter. Recall that a *special NTQ-group* is the coordinate group of a special NTQ-system.

Theorem 10.1.1. *The class of finitely generated elementary free groups coincides with the class of special NTQ-groups. Further this class extends beyond the class of non-Abelian free groups.*

From the JSJ-decomposition of limit groups (see Chapters 6 and 7) it follows that orientable surface groups S_g of genus $g \geq 2$ and non-orientable surface groups N_g of genus $g \geq 4$ are special NTQ-groups and hence elementary free.

Theorem 10.1.2. *An orientable surface group S_g of genus $g \geq 2$ and a non-orientable surface group N_g of genus $g \geq 4$ are special NTQ-groups and hence are elementary free.*

This theorem allows us to apply to surface groups of appropriate genus any first order theorem holding in a non-Abelian free group.

10.2 Surface groups and Magnus' theorem

If g is an element in a group G we let $N(g)$ denote the normal closure of the element g in G. Magnus proved the following theorem about the normal closures of elements in non-Abelian free groups:

Theorem 10.2.1 (Magnus). *Let F be a non-Abelian free group and $R, S \in F$. Then if $N(R) = N(S)$, it follows that R is conjugate to either S or S^{-1}.*

J. Howie [135] and independently O. Bogopolski [32] and Bogopolski and V.Sviridov [33] gave a proof of this for surface groups. Howie's proof was for orientable surface groups while Bogopolski and Sviridov also handled the nonorientable case. Their proofs were nontrivial and Howie's proof used the topological properties of surface groups. Howie further developed, as part of his proof of Magnus's theorem for surface groups, a theory of one-relator surface groups. These are surface groups modulo a single additional relator. Bogopolski and Bogopolski–Sviridov proved in addition, that Magnus's Theorem holds in even a wider class of groups.

With some work it can be determined that Magnus's result is actually a first-order theorem on non-Abelian free groups, and hence from the theorems concerning the solution of the Tarski problems, it holds automatically in all elementary free groups. In particular, Magnus's theorem will hold in surface groups, both orientable and nonorientable of appropriate genus. As remarked above, if G is a group and $g \in G$ then $N(g)$, as in the statement of Magnus's Theorem, will denote the normal closure in G of the element g.

Theorem 10.2.2. *Let G be an elementary free group and $R, S \in G$. Then if $N(R) = N(S)$ it follows that R is conjugate to either S or S^{-1}.*

Before exhibiting the proof of this theorem we mention the following two corollaries which extend Magnus's Theorem to surface groups and recover the results of Howie [135], Bogopolski [32] and Bogopolski–Sviridov [33].

Corollary 10.2.1 ([Ho], [Bo]). *Let S_g be an orientable surface group of genus $g \geq 2$. Then S_g satisfies Magnus's theorem, that is if $u, v \in S_g$ and $N(u) = N(v)$ it follows that u is conjugate to either v or v^{-1}.*

Corollary 10.2.2 ([BoS]). *Let N_g be a nonorientable surface group of genus $g \geq 4$. Then N_g satisfies Magnus's theorem, that is if $u, v \in N_g$ and $N(u) = N(v)$ it follows that u is conjugate to either v or v^{-1}. The genus $g \geq 4$ is essential here.*

We now present a proof of Theorem 10.2.2. From Theorem 10.2.2 the two corollaries, describing this result in surface groups, follow easily based on the solution to the Tarski problems coupled with the facts that orientable surface groups of genus $g \geq 2$ and nonorientable surface groups of genus $g \geq 4$ are elementary free.

Proof of Theorem 10.2.2. We show that Magnus's theorem is actually a first-order result in non-Abelian free groups. Since it is known to be true in non-Abelian free groups, then from the solution to the Tarski problems it will be true in any elementary free group.

Magnus's theorem can be given by a sequence of elementary sentences of the form (see also [GLS]).

$$\{\forall R, S \in G, \forall g \in G \exists g_1, \ldots, g_t, h_1, \ldots, h_k\}$$
$$(g^{-1}Rg = g_1^{-1}S^{\pm 1}g_1 \ldots g_t^{-1}S^{\pm 1}g_t) \wedge (g^{-1}Sg = h_1^{-1}R^{\pm 1}h_1 \ldots h_k^{-1}R^{\pm 1}h_k)\}$$
$$\Longrightarrow \{\exists x \in G(x^{-1}Rx = S \vee x^{-1}Rx = S^{-1})\}$$

Magnus's theorem is therefore a first-order result and the theorem follows. □

As described prior to the proof, it follows that any elementary free group and hence surface groups of the appropriate genus, satisfy Magnus's theorem. This recovers the results in [129], [33] and [32]. Actually more is true. An examination of the sentences capturing that Magnus's theorem (Theorem 10.2.2) is first-order, shows that the sentences are universal-existential. Hence the theorem holds in the almost locally free groups of Gaglione and Spellman [109].

Corollary 10.2.3. *Let G be an almost locally free group and $R, S \in G$. Then if $N(R) = N(S)$ it follows that R is conjugate to either S or S^{-1}.*

10.3 Questions and something for nothing

The proof of Magnus's theorem for surface groups that was presented in the last section is a type of something for nothing result. By this we mean that nontrivial proofs, such as those of Howie and Bogopolski, of results in certain classes of groups, fall out directly from the solution to the Tarski problems. These types of proofs and results lead to several different types of questions. In particular we consider the following two:

(1) First: Which additional nontrivial free group results are true in surface groups but difficult to obtain directly?
(2) Second: What first-order properties of non-Abelian free groups are true beyond the class of elementary free groups.

In this scetion we consider the question about surface groups and present a series of results true in all elementary free groups and in particular surface groups of the appropriate genus. In subsequent sections we will consider aspects of the second question.

As a simple example of the first type of question we consider the well known property concerning commutativity in free groups. It is well known (see [178]) that non-Abelian free groups have cyclic centralizers of nontrivial elements. This is a consequence of the following somewhat weaker result.

Theorem 10.3.1 ([MKS]). *Let F be a non-Abelian free group. If $x, y \in F$ and x, y commute then both x and y are powers of a single element $w \in F$.*

This result is given by the sentence

$$\forall\{x, y \in F\}([x, y] = 1) \to \exists\{w \in F\}\exists\{m, n \in \mathbb{Z}\}(x = w^m \wedge y = w^n)\,.$$

This is not first-order in the language of group theory since we must quantify over the integers which are not included in the language L_0. Hence this result is not necessarily true in elementary free groups. As an example, let D be a nonprincipal ultrafilter on $\mathbb{Z}$ (see Chapter 4). Let $F = \langle a_1, a_2; \;\rangle$ the free group of rank 2 on a_1, a_2 and let ${}^*F = F^{\mathbb{Z}}/D$ be the corresponding ultrapower so that ${}^*F$ is elementary free (see Chapter 4). Consider the elements

$$[(a_1)_{k\in\mathbb{Z}}]_D = [(\ldots, a_1, a_1, \ldots, a_1, \ldots)]_D$$

and

$$[(a_1^k)_{k\in\mathbb{Z}}]_D = [(\ldots, a_1^{-2}, a_1^{-1}, 1, a_1, a_1^2, \ldots)]_D\,.$$

These commute but there is no fixed element B of which they are both powers.

However the following result can be proved directly.

Theorem 10.3.2. *Let G be a finitely generated elementary free group. Then G has cyclic centralizers of nontrivial elements. It follows that if $x, y \in G$ and x, y commute then both x and y are powers of a single element $w \in G$.*

Proof. Let G be a finitely generated elementary free group. Then G is finitely generated and fully residually free. It follows from the fact that finitely generated fully residually free groups are commutative transitive that G has Abelian centralizers. Applying Szmielew's criteria for elementary equivalence of Abelian groups(see [245]) it follows that in any elementary free group the centralizer of any nontrivial element is elementarily equivalent to the infinite cyclic group. In particular such centralizers must satisfy the sentences:

$$\forall x_1, x_2(x_1x_2 = x_2x_1);$$

$$\exists x(x \neq 1);$$

for each integer $n \geq 2$ the sentence

$$\forall x((x^n = 1) \to (x = 1));$$

and the sentence

$$\forall x_1, x_2, x_3 \exists y((x_1x_2^{-1} = y^2) \vee (x_1x_3^{-1} = y^2) \vee (x_2x_3^{-1} = y^2));$$

asserting that, modulo 2, there are at most 2 distinct elements.

A result of Gaglione, Lipschutz and Spellman (Lemma 3.6 in [GLS]) shows that up to isomorphism the only finitely generated group M which can satisfy these properties simultaneously is the infinite cyclic group. Here we will repeat the proof given there.

Suppose not and M is a finitely generated Abelian group satisfying the above sentences. Then M contains a rank 2 free Abelian direct factor A and suppose that $M = A \times B$.

Now let $(a_1, a_2, a_3) \in A^3$. Then there is $a \in A, b \in B$ such that

$$a_1a_2^{-1} = a^2b^2 \vee a_1a_3^{-1} = a^2b^2 \vee a_2a_3^{-1} = a^2b^2.$$

Since the product is direct $b^2 = 1$ is the only possibility. Then, writing $A(X^2)$ for the subgroup of A generated by the squares, $a_1 \equiv a_2 \bmod A(X^2)$ or $a_1 \equiv a_3 \bmod A(X^2)$ or $a_2 \equiv a_3 \bmod A(X^2)$. Since $(a_1, a_2, a_2) \in A^3$ was arbitrary, the index $[A\colon A(X^2)] \leq 2$. However if A has rank 2 it follows that $[A\colon A(X^2)] = 4$. This contradiction shows that M is cyclic. □

As a corollary we get that the result must be true in surface groups, a fact that can also be obtained directly from the amalgam structure of such groups, or from their faithful representations in $PSL(2, \mathbb{C})$.

Corollary 10.3.1. *Let G be either an orientable surface group of genus $g \geq 2$ or a nonorientable surface group of genus $g \geq 4$. If $x, y \in G$ and x, y commute then both x and y are powers of a single element $w \in F$.*

We now present an example that is less trivial, and not obvious in a surface group. First we need the next theorem, that can be easily proved in non-Abelian free groups.

Theorem 10.3.3. *Let F be a free group and n, k nonzero integers. For all $x, y \in F$ if $[x^n, y] = [x, y^k]$ then either $n = k = 1$ or x, y commute and both are powers of a single element.*

The first part of the result that either $n = k = 1$ or $[x, y] = 1$ is first-order given by a sequence of elementary sentences, one for each $(n, k) \in \mathbb{Z}^2 \setminus \{(1, 1)\}$ with neither n nor k zero;

$$\forall x, y \in F([x^n, y] = [x, y^k]) \to ([x, y] = 1) .$$

Therefore this part of the result must hold in any elementary free group. Further if the elementary free group is finitely generated the second part must also hold.

Corollary 10.3.2. *Let G be an elementary free group and n, k nonzero integers. If $x, y \in G$ and if $[x^n, y] = [x, y^k]$ then either $n = k = 1$ or x, y commute. If G is finitely generated then both x and y are powers of a single element $w \in G$.*

Since nonfree surface groups are finitely generated we have the following.

Corollary 10.3.3. *Let G be either an orientable surface group of genus $g \geq 2$ or a nonorientable surface group of genus $g \geq 4$ and n, k nonzero integers. If $x, y \in G$ and if $[x^n, y] = [x, y^k]$ then either $n = k = 1$ or x, y commute and then both x and y are powers of a single element $w \in G$.*

Csorgo, Fine and Rosenberger [61] proved the following extension of this.

Theorem 10.3.4 ([CsFR]). *Suppose F is a non-Abelian free group and x, y, u, v are elements of F with $[x, y] \neq 1$ and u, v in the subgroup generated by x, y. Then if $[x, y]$ is conjugate to a power of $[u, v]$ within $\langle x, y\rangle$, that is there exists an integer k with $[x, y] = g([u, v]^k)g^{-1}$, for some $g \in \langle x, y\rangle$ and $[x, y^m] = [u, v^n]$ for natural numbers m, n it follows that $m = n$. Further if $m = n \geq 2$ then y is conjugate within $\langle x, y\rangle$ to v or v^{-1}.*

As with Magnus's theorem this can be shown to be given by a sequence of first-order sentences and is hence a first-order result. Therefore this holds in any elementary free group.

Theorem 10.3.5. *Let G be an elementary free group and $x, y, u, v \in G$ with $[x, y] \neq 1$ and u, v in the subgroup generated by x, y. Then if $[x, y]$ is conjugate to a power of $[u, v]$ within $\langle x, y\rangle$, that is there exists an integer k, with $[x, y] = g([u, v]^k)g^{-1}$ for some $g \in \langle x, y\rangle$, and $[x, y^m] = [u, v^n]$ for natural numbers m, n it follows that $m = n$. Further if $m = n \geq 2$ then y is conjugate within $\langle x, y\rangle$ to v or v^{-1}.*

In particular we get the extension to surface groups.

Corollary 10.3.4. *Let G be either an orientable surface group of genus $g \geq 2$ or a nonorientable surface group of genus $g \geq 4$ and suppose that $x, y, u, v \in G$ with $[x, y] \neq 1$ and u, v in the subgroup generated by x, y. Then if $[x, y]$ is conjugate to a power of*

$[u, v]$ within $\langle x, y\rangle$, that is there exists an integer k with $[x, y] = g([u, v]^k)g^{-1}$ for some $g \in \langle x, y\rangle$, and $[x, y^m] = [u, v^n]$ for natural numbers m, n it follows that $m = n$. Further if $m = n \geq 2$ then y is conjugate within $\langle x, y\rangle$ to v or v^{-1}.

10.4 Results on elementary free groups

In the previous section it was proved that in any finitely generated elementary free group centralizers must be cyclic. It follows that in any finitely generated elementary free group if two elements commute then they must both be powers of a single element. This mirrors the situation in non-Abelian free groups. However it is not a first order result and an example was given of an elementary free group, necessarily non-finitely generated in which it did not hold. In this section we give a series of results that are true in elementary free groups, some require finite generation, but which are not first order. Some are true for all finitely generated residually free groups. Since these are precisely the finitely generated universally free groups and every elementary free group is universally free it follows that these results will be true for finitely generated elementary free groups.

10.4.1 Hyperbolicity and stable hyperbolicity

Hyperbolic groups are finitely presented groups that have become a fundamental structure in geometric group theory. Finitely generated free groups are hyperbolic. Roughly a group G is hyperbolic if the geometry of the Cayley graph of G is hyperbolic in the sense that it satisfies the δ-hyperbolic property (see Chapter 3). Torsion-free hyperbolic groups have cyclic centralizers for nontrivial elements and it has been proved that a limit group G (finitely generated fully residually free group) is hyperbolic if and only if in G the centralizer of a nontrivial element is cyclic. A finitely generated elementary free group, being also universally free, is also fully residually free (by the Gaglione–Spellman–Remeslennikov result) and from Theorem 10.3.1 such a group also has the cyclic centralizer property. Thus we have proved that any finitely generated elementary free group must be hyperbolic.

Theorem 10.4.1. *Let G be a finitely generated elementary free group. Then G is hyperbolic.*

Sela [232] proved that the isomorphism problem is solvable within the class of torsion-free hyperbolic groups. A finitely generated elementary free group is then within the class torsion-free hyperbolic groups. This provides a proof of the the following corollary which was proved directly by Bumagin, Kharlampovich and Myasnikov [39] (see Section 7.6)

Corollary 10.4.1. *The isomorphism problem is solvable within the class of finitely generated elementary free groups.*

A group G is *stably hyperbolic* if G is hyperbolic and for any endomorphism $\phi\colon G \to G$ for all $n \in \mathbb{N}$ there is an $m \geq n$ such that $\phi^m(G)$ is hyperbolic. Using the cyclic centralizer result we obtain.

Theorem 10.4.2. *Let G be a finitely generated elementary free group. Then G is stably hyperbolic.*

Proof. Let G be a finitely generated elementary free group. As in the proof of the last theorem G is finitely generated and fully residually free. Since G is finitely generated and elementary free it has cyclic centralizers and is therefore hyperbolic. Now let $\phi\colon G \to G$ be an endomorphism. Since G is finitely generated then $\phi^n(G)$ is also a finitely generated fully residually free group for any natural number n. The property of having cyclic centralizers holds in any subgroup and therefore $\phi^n(G)$ is hyperbolic for any n. Therefore G is stably hyperbolic. □

10.4.2 The retract theorem and Turner groups

An element g in a group G is a *test element* if whenever $f(g) = g$ for some endomorphism of G then f must be an automorphism. This concept dates back to Nielsen who showed that $[x, y]$ is a test element in the free group on $\{x, y\}$. Test elements in a free group are called *test words* (see [106]).

Turner [249] gave the following characterization of test words in finitely generated free groups. This is now referred to as either the *Retract Theorem* or *Turner's Theorem*.

Theorem 10.4.3. *Let F be a finitely generated non-Abelian free group. Then an element $g \in F$ is a test word if and only if g lies in no proper retract.*

The question of whether Turner's theorem is first-order or not, was considered in [84]. We call an element g in a group G *nonprojectible* if it lies in no proper retract of G. We then call a group G a *Turner group* if for $g \in G$ being nonprojectible in G implies that g is a test element. Equivalently G is a Turner group if and only if the Retract Theorem holds. Hence Turner's theorem says that non-Abelian free groups are Turner groups.

A result of O'Neill and Turner (see [208]) shows that stably hyperbolic groups are Turner groups. From Theorem 10.4.2 it then follows that finitely generated elementary free groups are Turner groups, that is they satisfy the Retract Theorem.

Theorem 10.4.4. *Let G be a finitely generated elementary free group. Then G is a Turner group, that is G satisfies the Retract Theorem and hence the test elements in G are precisely those elements that avoid any proper retract.*

Proof. Let G be a finitely generated elementary free group. From Theorem 10.4.2 G is stably hyperbolic. Therefore G satisfies the Retract Theorem from the result of O'Neill and Turner. □

As in the previous cases this then extends to surface groups of appropriate genus.

Corollary 10.4.2. *Let G be either an orientable surface group of genus $g \geq 2$ or a nonorientable surface group of genus $g \geq 4$. Then G is a Turner group.*

In [208] it was proved directly that there are test elements in surface groups. However this also follows directly from the previous corollary since not every element in either S_g or N_g falls in a proper retract.

Corollary 10.4.3. *Let G be either an orientable surface group of genus $g \geq 2$ or a nonorientable surface group of genus $g \geq 4$. Then G has test elements.*

In [84] the following results were proved showing that Turner's Theorem is not first-order and not the model class of any set of sentences of L_0.

Theorem 10.4.5 (Nondefinability Theorem).
(1) *There is no set $P(x)$ of formulas of L_0 such that, for an arbitrary group G and arbitrary element $g \in G$, $P(g)$ holds if and only if g is nonprojectible.*
(2) *There is no set $T(x)$ of formulas of L_0 such that, for an arbitrary group G and arbitrary element $g \in G$, $T(g)$ holds if and only if g is a test element.*

Theorem 10.4.6 (Nonaxiomatizability Theorem). *The class of Turner groups is not the model class of any set of sentences of L_0.*

10.4.3 Conjugacy separability of elementary free groups

A group G is *conjugacy separable* if given any two elements $g_1, g_2 \in G$ either g_1 is conjugate to g_2 or there exists a homomorphism $\rho\colon G \to H$ where H is a finite group and in which $\rho(g_1)$ is not conjugate to $\rho(g_2)$. It is known that all free groups are conjugacy separable. Here we next prove that all finitely generated elementary free groups are conjugacy separable.

Theorem 10.4.7. *Let G be a finitely generated elementary free group. Then G is conjugacy separable.*

Proof. Suppose G is an finitely generated elementary free group and g_1, g_2 are two nonconjugate elements of G. Since free groups are conjugacy separable to show that G is conjugacy separable it suffices to show that there is a free homomorphic image of G in which the images of g_1 and g_2 are nonconjugate.

Suppose there is no free homomorphic image of G in which g_1 is not conjugate to g_2. Note that a finitely generated elementary free group, in fact more generally a finitely generated fully residually free group, must be finitely presented (see [153]). Fix a finite presentation for G,

$$\langle a_1, \ldots, a_n; R_1(a_1, \ldots, a_n) = \cdots = R_m(a_1, \ldots, a_n) = 1\rangle$$

and suppose that $g_i = w_i(a_1, \ldots, a_n)$ for $i = 1, 2$. Then since there are no free homomorphic images of G in which g_1 and g_2 are not conjugate the following universal-existential sentence, which we denote by $\langle 1\rangle$, of L_0 would be true in every non-Abelian free group

$$\forall x_1, \ldots, x_n \exists y (\wedge_{i=1}^{m} (R_i(x_1, \ldots, x_n) = 1)) \rightarrow (w_2(x_1, \ldots, x_n) = y^{-1} w_1(x_1, \ldots, x_n) y).$$

It follows that $\langle 1\rangle$ would have to be true in G. But this contradicts the fact that g_1 is not congugate to g_2 in G. Therefore there must exist a free homomorphic image in which g_1 and g_2 are not conjugate and hence G is conjugacy separable. □

10.4.4 Tame automorphisms of elementary free groups

In Chapter 7 we showed that the automorphism groups of limit groups are tame. Since elementary free groups must be limit groups it carries over that the automorphism group of an elementary free group is also tame. Here we repeat some of the discussion found in Sections 7.2.2 and 7.2.3.

A minimal finite presentation of a finitely presented group G is a presentation that is minimal with respect to the number of generators. Hence a presentation

$$G = \langle x_1, \ldots, x_n; R_1 = \cdots = R_m = 1\rangle$$

is a minimal finite presentation for G if $n = \text{rank}(G)$, the minimal number of generators necessary to present G. Now suppose that

$$G = \langle x_1, \ldots, x_n; R_1 = \cdots = R_m = 1\rangle$$

with $1 \leq n, m < \infty$ is minimal finite presentation of G. Let $F = \langle x_1, \ldots, x_n; \ \rangle$ be the free group of rank n on $\{x_1, \ldots, x_n\}$. An automorphism $\alpha\colon G \to G$ is *tame* if it is induced by or lifts to an automorphism on F (considered as free on the generators of G). If each automorphism of G is tame we say that the automorphism group $\text{Aut}(G)$ is *tame*. In [125] Gupta and Shpilrain gave a survey of some of the known general results on tame automorphisms and tame automorphism groups. If G is a surface group a result of Zieschang [258] and improved upon by Rosenberger [221] shows that G has only one Nielsen class of minimal generating systems. An easy consequence of this is that $\text{Aut}(G)$ is tame. Rosenberger (see [176] or [220]) uses the term *almost quasifree* for a finitely presented group which has a tame automorphism group. If G is almost quasifree,

$$G = \langle x_1, \ldots, x_n; R_1 = \cdots = R_m = 1\rangle, \quad 1 \leq n, \quad m < \infty$$

a minimal finite presentation of G and, in addition, each automorphism of $F = \langle x_1, \ldots, x_n; \ \rangle$ induces an automorphism of G, G is called *quasifree*. Rosenberger observed that a non-cyclic, non-free one-relator group is quasifree only if it has a presentation $\langle a, b; [a, b]^n = 1\rangle$ for $n \geq 1$. This is a Fuchsian group if $n \geq 2$ and isomorphic to a free Abelian group of rank 2 if $n = 1$.

We have seen that any finitely generated fully residually free group and hence any elementary free group has a JSJ decomposition with cyclic edge groups and vertex groups of specific types if it is not Abelian or a surface group. We mention that Bumagin, Kharlampovich and Myasnikov [39] used the JSJ decomposition to describe the automorphism group of a limit group. Further specifics can be found in [39].

As a consequence of the JSJ decomposition of a fully residually free group and the work of Bumagin, Kharlampovich and Mysasnikov [39] on the structure of the automorphism groups of limit groups, the tameness of Aut(G) for a limit group was proved by Fine, Kharlmapovich, Myasnikov, Remeslennikov and Rosenberger [99].

Theorem 10.4.8 ([FKMRR]). *The automorphism group* Aut(G) *of a finitely generated freely indecomposable fully residually free group G is tame with respect to a presentation for the JSJ decomposition for G.*

Since each finitely generated elementary free group is universally free and hence fully residually free the proof of the corollary is immediate.

Corollary 10.4.4. *The automorphism group of a finitely generated freely indecomposable elementary free group G is tame.*

We note that the converse of this corollary is false. That is there do exist groups (in fact hyperbolic groups) where every automorphism is tame but which are not fully residually free. As an example the groups

$$G = \langle a_1, \ldots, a_n; a_1^{\alpha_1} \ldots a_n^{\alpha_n} = 1\rangle$$

with $n \geq 3, 2 \leq \alpha_1, \ldots, \alpha_n$ and

$$H = \langle s_1, \ldots, s_n; s_1^2 = \cdots = s_{n-1}^2 = s_n^{2k+1} = s_1 \ldots s_n = 1\rangle$$

with $n = 2\ell, n \geq 4$ even and $k \geq 1$, are all hyperbolic. Further every automorhphism is tame (see [219] and [105]). However not all of these groups are fully residually free.

10.4.5 The isomorphism problem for elementary free groups

The proof of Theorem 10.4.8 used the structure theory of automorphisms of limit groups developed by Bumagin, Kharlampovich and Myasnikov [39]. As a further application of the study of these automorphisms of fully residually free groups, Bumagin, Kharlampovich and Myasnikov [39] were further able to prove that the isomorphism problem is solvable for finitely generated fully residually free groups. We saw a special case of these earlier in the chapter for hyperbolic limit groups using a result of Sela on the isomorphism problem for torsion-free hyperbolic groups. This is actually part of the algorithmic study of this class of groups. In particular Kharlampovich, Myasnikov, Remeslennikov and Serbin [158] were able to translate the method of Stallings foldings to fully residually free groups by considering *infinite words* in $F^{\mathbb{Z}[t]}$. Doing this they were able to algorithmically solve many problems in fully residually free groups mirroring the algorithmic solutions in absolutely free groups,

Theorem 10.4.9. *The isomorphism problem is solvable in the class of finitely generated fully residually free groups. That is given two finite presentations that are known to define fully residually free groups there is an effective algorithm to determine if the defined groups are isomorphic.*

10.4.6 Faithful representations in $PSL(2, \mathbb{C})$

We now take a brief relook at material that was given in Section 7.3. It is well-known that the surface groups have faithful representations as discrete subgroups of $PSL(2, \mathbb{C})$ in fact also within $PSL(2, \mathbb{R})$. Fine and Rosenberger [94, 95] proved that any finitely generated fully residually free group has a faithful representation within $PSL(2, \mathbb{C})$. This follows from the fact that the limit groups are precisely the finitely generated subgroups of non-standard free groups. In the case where the limit group is hyperbolic, that is has cyclic centralizers, this representation can be effectively constructed using the JSJ decomposition.

Theorem 10.4.10. *Let G be a hyperbolic limit group and in particular any finitely generated elementary free group.Then a faithful representation of*

$$\rho\colon G \to PSL(2, \mathbb{C})$$

can be effectively constructed from the JSJ decomposition of G.

Using the embedding of a limit group in a nonstandard free group the restriction to hyperbolicity can be removed.

Theorem 10.4.11. *Any limit group and in particular any finitely generated elementary free group has a faithful representation in $PSL(2, \mathbb{C})$*

The details of the proofs of these two result can be found in Section 7.3.

Theorem 10.4.10 was proved first using faithful representations of amalgams. The explicit constructible representation then depends on the JSJ decomposition of a limit group. This in turn depends upon faithful $PSL(2, \mathbb{C})$ representations of two extremely important amalgam constructions: *cyclically pinched one-relator groups* and *conjugacy pinched one-relator groups* (see Chapter 3). Again we presented the details of the proofs in Section 7.3.

10.4.7 Elementary free groups and the Howson property

A group G satisfies the *Howson property* or is a *Howson group* if the intersection of two finitely generated subgroups is again finitely generated. Free groups satisfy the Howson property as do Abelian groups and orientable surface groups of genus $g \geq 2$. However Kapovich [144] gave an example of a torsion-free hyperbolic one-relator group that is not Howson.

Kapovich [144] (and reproved in a different way in [158]) proved that all fully residually free groups satisfy the Howson property. Therefore all llimit groups and all elementary free groups are Howson. However as we will prove the Howson property is not first order.

Theorem 10.4.12. *Let G be an elementary free group, then G satisfies the Howson property.*

We first need the following.

Lemma 10.4.1. *Let $G = F_2 \times \mathbb{Z}$ be the direct product of a free group of rank 2 and an infinite cyclic group. Then G is not Howson.*

Proof. Let $F_2 = \langle a_1, a_2 \rangle$ and $\mathbb{Z} = \langle b \rangle$ and $G = F_2 \times \mathbb{Z}$. Consider the finitely generated subgroups

$$H_1 = \langle (a_1, b), (a_2, 1) \rangle$$
$$H_2 = \langle (a_1, 1), (a_2, b) \rangle$$

of G. We show that $H_1 \cap H_2$ is not finitely generated and hence G is not Howson.

Notice that $(w(a_1, a_2), b^n) \in H_1$ if and only if the exponent sum on a_1 in $w(a_1, a_2)$ is n. Similarly $(w(a_1, a_2), b^n) \in H_2$ if and only if the exponent sum on a_2 in $w(a_1, a_2)$ is n.

Writing e_i for the exponent sum on a_i, $i = 1, 2$ respectively we see that $(w(a_1, a_2), b^n) \in H_1 \cap H_2$ if and only if

$$e_1(w(a_1, a_2)) = e_2 * (w(a_1, a_2)) = n.$$

Now since $\mathbb{Z}$ is Abelian $e_1 - e_2 \colon F \to \mathbb{Z}$, $i = 1, 2$ is a well-defined surjective homomorphism.

Letting

$$P = \{w(a_1, a_2) \colon \exists n \in \mathbb{Z}, (w(a_1, a_2), b^n) \in H_1 \cap H_2\}$$

be the projection onto the first coordinate in $H_1 \cap H_2$ (so a subgroup of F_2) we have that $\mathrm{Ker}(e_1 - e_2) \subset P$. So P contains the nontrivial normal subgroup $\mathrm{Ker}(e_1 - e_2)$ of F_2. (Note for example that $a_1 a_2 \in \mathrm{Ker}(e_1 - e_2) \setminus \{1\}$.)

Now since the cosets Pa_1^m and Pa_1^n are different when $m \neq n$ it follows that the index $[F_2 \colon P] = \infty$. By Theorem 2.10 on Page 104 of [178] it follows that P cannot be finitely generated. If $H_1 \cap H_2$ had a finite set

$$\{(w_1(a_1, a_2), b^{n_1}), \ldots, (w_k(a_1, a_2), b^{n_k})\}$$

of generators then

$$\{w_1(a_1, a_2), \ldots, w_k(a_1, a_2)\}$$

would generate P, a contradiction. Therefore G does not satisfy the Howson property. □

Theorem 10.4.13. *The Howson property is not first order.*

Proof. Let $G_1 = \langle a_1, a_2; \ \rangle$ be a free group of rank 2 and $G_2 = \langle b \rangle$ infinite cyclic. Since both G_1, G_2 are residually finite then so is $G = G_1 \times G_2$. It follows that the set $S = \{S_n : n \in \mathbb{N}\}$ of symmetric groups discriminates G. Since every homomorphism $\phi : G \to S_n$ may be viewed as a homomorphism into the infinite symmetric S_ω (where only finitely many integers are moved by each permutation) we see that S_ω discriminates G. Note that S_ω is locally finite so, since G is finitely generated, the image of every homomorphism

$$\phi : G \to S_\omega$$

is finite.

Let $\Phi = \operatorname{Hom}(G, S_\omega)$. For each $g \in G$ with $g \neq 1$ let its support be

$$\operatorname{Supp}(g) = \{\phi \in \Phi : \phi(g) \neq 1\}.$$

We let $\operatorname{Supp}(1)$ be the empty set. Since S_ω discriminates G the family

$$\{\operatorname{Supp}(g) : g \in G \setminus \{1\}\}$$

satisfies the finite intersection property and so extends to an ultrafilter D on Φ. Now form the ultraproduct

$$U = \Big(\prod_{\phi \in \Phi} \phi(G)\Big)/D.$$

Each $\phi(G)$ is finite so certainly satisfies the Howson property.

We claim that U violates the Howson property which would imply that the Howson property is not first order.

Since the Howson property is inherited by subgroups it then is sufficient to exhibit a subgroup of U that violates the Howson property to show that U is not Howson. We show that G embeds in U which would finish the argument since from the previous lemma G is not Howson.

Observe that $\psi(g) = [(\phi(g)_{\phi \in \Phi}]_D$ is easily shown to be a homomorphism $\psi : G \to U$. Suppose that $\psi(g) = 1$ in U. Then the complement of the support

$$\operatorname{Supp}(g)' = \{\phi \in \Phi : \phi(g) = 1\} \in D.$$

But, if $g \neq 1$ it follows that $supp(g) \in D$. Thus from $\operatorname{Supp}(g)' \in D$ we infer that $g = 1$ and $\operatorname{Ker}(\psi) = \{1\}$. Hence G embeds in U completing the proof. □

The fact that the group G used in the previous two proofs is not Howson leads to the following very nice result.

Theorem 10.4.14. *A residually free group is fully residually free if and only if it satisfies the Howson property.*

Proof. If Γ is fully residually free then from the result of Kapovich it satisfies the Howson property. Suppose that Γ is residually free and not fully residually free. A result of B. Baumslag [6] says that Γ must contain a copy of the group G of Lemma 10.4.1 and thus contain a subgroup which is not Howson. Since the Howson property is certainly inherited by subgroups this completes the proof. □

Notice that this result fits nicely with our other equivalences on fully residually free and we can add the Howson Property.

Theorem 10.4.15. *Let G be a residually free group. Then the following are equivalent*

(1) *G is fully residually free;*
(2) *G is CT;*
(3) *G is CSA;*
(4) *G is universally free if non-Abelian;*
(5) *G satisfies the Howson property.*

10.5 The Lyndon properties

We now turn to the second type of question, that is what first order free group results are true beyond the class of elementary free groups and survey a large number of recent results. In particular we first consider classes of groups satisfying certain quadratic properties that we call *Lyndon properties*. We then show that the class of groups satisfying these properties is closed under many amalgam constructions. Elementary free groups satisfy these properties and these amalgam results extend the class of groups satisfying Lyndon properties beyond the class of elementary free groups. We first define the *basic Lyndon properties*.

10.5.1 The basic Lyndon properties

Vaught proposed the question whether in a free group a solution of the equation $x^2y^2z^2 = 1$ must generate an Abelian (and hence cyclic) subgroup. This was proved by Lyndon and Schutzenberger and then generalized by Baumslag. Based on Lyndon's result in any group G we define the following *Lyndon Properties*

Definition 10.5.1. The following are called *Lyndon properties*. Let G be a group. Then G satisfies Property

(1) *LZ* if whenever $x^2y^2z^2 = 1$ for $x, y, z \in G$ then $\langle x, y, z\rangle$ is cyclic.
(2) *LA* if whenever $x^2y^2z^2 = 1$ for $x, y, z \in G$ then $\langle x, y, z\rangle$ is Abelian.
(3) *LPZ* if whenever $x^p y^q z^r = 1$ for $x, y, z \in G$ with $2 \leq p, q, r \in \mathbb{N}$ then $\langle x, y, z\rangle$ is cyclic.
(4) *LPA* if whenever $x^p y^q z^r = 1$ for $x, y, z \in G$ with $2 \leq p, q, r \in \mathbb{N}$ then $\langle x, y, z\rangle$ is Abelian.
(5) *LCZ* if whenever $[x^p, y^q]z^r = 1$ for $x, y, z \in G$ with $1 \leq p, q \in \mathbb{N}$, $2 \leq r \in \mathbb{N}$ then $\langle x, y, z\rangle$ is cyclic.
(6) *LCA* if whenever $[x^p, y^q]z^r = 1$ for $x, y, z \in G$ with $1 \leq p, q \in \mathbb{N}$, $2 \leq r \in \mathbb{N}$ then $\langle x, y, z\rangle$ is Abelian.

All of these properties hold in free groups and Properties LA, LPA and LCA are given by universal sentences. Hence these hold in any fully residually free group. Since LA, LPA, LCA are first order they hold in an elementary free group.

We show that the Lyndon properties extend beyond the class of elementary free groups by showing that the property is preserved under some general group amalgams.

10.5.2 Lyndon properties in amalgams

Using Nielsen cancellation methods it can be proved that several of the Lyndon properties are preserved under special free product with amalgamation construtions (see [100]). Since these constructions are not always fully residually free it makes the class of groups satisfying the Lyndon properties wider than the class of limit groups.

Theorem 10.5.1. *Suppose that H_1 and H_2 are groups with no elements of order 2 and that G is the amalgamated product $G = H_1 *_A H_2$ with $H_1 \neq A \neq H_2$ and A is malnormal in both H_1 and H_2. Then*

(1) *if both H_1 and H_2 satisfy Property LZ then G also satisfies Property LZ,*
(2) *if both H_1 and H_2 satisfy Property LA then G also satisfies Property LA.*

Theorem 10.5.2. *Suppose that H_1 and H_2 are torsion-free groups and that G is the amalgamated product $G = H_1 *_A H_2$ with $H_1 \neq A \neq H_2$ and A is malnormal in both H_1 and H_2. Then*

(1) *if both H_1 and H_2 satisfy Property LPZ then G also satisfies Property LPZ,*
(2) *if both H_1 and H_2 satisfy Property LPA then G also satisfies Property LPA,*
(3) *if both H_1 and H_2 satisfy Property LCZ then G also satisfies Property LCZ,*
(4) *if both H_1 and H_2 satisfy Property LCA then G also satisfies Property LCA.*

In particular if $x, y \in G$ with $[x, y] \neq 1$ then if both H_1, H_2 have property LCZ or LCA then $[x, y]$ is never a proper power.

We note that the malnormality condition in both theorems is essential. For example in

$$G = \langle a, b, c; a^2b^2c^2 = 1\rangle$$

we have an equation $x^2y^2z^2 = 1$ such that $\langle x, y, z\rangle$ is not Abelian.

Theorem 10.5.3. *Suppose that G is a cyclically pinched one-relator group*

$$G = \langle a_1, \ldots, a_p, b_1, \ldots, b_q; WV = 1\rangle$$

where $p \geq 2, q \geq 2$, $1 \neq W = W(a_1, \ldots, a_p)$ is not a proper power nor a primitive element in the free group on $a_1, \ldots, a_p$ and $1 \neq V = V(b_1, \ldots, b_q)$ is not a proper power nor a primitive element in the free group on $b_1, \ldots, b_q$. Then G has properties LZ, LPZ, LCZ.

The key idea in the above theorem is that a cyclically pinched one-relator group of the above form is 3-free, that is any subgroup generated by 3 or fewer elements must be free group. The following is then immediate.

Lemma 10.5.1. *Let G be a 3-free group. Then G satisfies properties LZ, LA, LPZ, LPA, LCZ and LCA.*

In [100] the theorem on 3-freeness was further extended.

Theorem 10.5.4. *Suppose that $G = H_1 *_A H_2$ is an amalgamated free product with $H_1 \neq A \neq H_2$. Suppose further that A is malnormal in both H_1 and H_2 and that both H_1 and H_2 are 3-free. Then G is 3-free.*

Combining this result with the lemma we get.

Corollary 10.5.1. *Suppose that $G = H_1 *_A H_2$ is an amalgamated free product with $H_1 \neq A \neq H_2$. Suppose further that A is malnormal in both H_1 and H_2 and that both H_1 and H_2 are 3-free. Then G satisfies LZ, LA, LPZ, LPA, LCZ, LCA.*

10.5.3 The Lyndon properties and HNN constructions

The situation for HNN groups is much more complicated.

Theorem 10.5.5. *Suppose that G is an HNN extension of the base B so that G has the form*

$$G = \langle B, t; \text{rel}(B), t^{-1}K_1t = K_2\rangle.$$

Suppose further that K_1 and K_2 are both malnormal in B and that B does not contain an element of order 2. Suppose further that B satisfies the basic Lyndon Property LZ. Then if $x^2y^2z^2 = 1$ in G and $\mathfrak{U} = \{x, y, z\}$ is regular then $\langle x, y, z\rangle$ is cyclic.

Let $G = \langle B, t; \text{rel}(B), t^{-1}K_1t = K_2\rangle$ be an HNN group with base group B, stable letter t and associated subgroup K_1, K_2. An ordered set $\mathfrak{U} = \{u_1, \ldots, u_n\} \subset G$ is *regular* if there is no Nielsen transformation from $\mathcal{U}$ to a system $\mathcal{U}' = \{u_1', \ldots, u_n'\}$ in which one of the elements is conjugate to an element of K_1 or K_2. However the analogous statement to the above does not hold for the HNN group when $\mathcal{U} = \{x, y, z\}$ is not regular.

Let $G = \langle x, y, z; x^2y^2z^2 = 1\rangle$ be the nonorientable surface group of genus $g = 3$. In G the equation $x^2y^2z^2 = 1$ holds trivially and $\langle x, y, z\rangle$ is non-Abelian and hence noncyclic. G can be written as an amalgamated free product $G = H_1 *_A H_2$ with

$$H_1 = \langle x, y; \ \rangle, H_2 = \langle z; \ \rangle \quad \text{and} \quad A = \langle x^2y^2\rangle = \langle z^{-2}\rangle .$$

However here G does not contradict any of our results since A is not malnormal in H_2. This again shows that malnormality is essential in the amalgamated free product case.

On the other hand using straightforward Tietze transformations ($z = t^{-1}, y = ut, x = tv^{-1}t^{-1}$) the nonorientable surface group

$$G = \langle x, y, z; x^2y^2z^2 = 1\rangle$$

can also be written as an HNN group

$$G = \langle H, t; t^{-1}ut = v^2u^{-1}\rangle$$

with $H = \langle u, v; \ \rangle$. The element u is not conjugate in the base H to v^2u^{-1} and both associated subgroups are malnormal in the base. However the system $\{x, y, z\}$ is not regular (see the definition above) showing that in the HNN case regularity is essential.

10.5.4 The Lyndon properties in certain one-relator groups

A large subclass of the class of one-relator groups satisfies the Lyndon properties since cyclically pinched one-relator groups do. The following due to Fine, Rosenberger and Rosenberger extends the class even further to some one-relator groups with torsion.

Theorem 10.5.6 ([FRR 2]). *Let G be the one-relator group*

$$G = \langle a, b, c, \ldots; R^m = 1\rangle$$

with $m \geq 3$ and m odd and R a cyclically reduced word, not a proper power in the free group on $a, b, c, \ldots$. Let $w(x_1, x_2, x_3)$ be a regular quadratic word in the free group F on x_1, x_2, x_3 and let $\phi\colon F \to G$ be a homomorphism from F into G with $\phi(x_i) = u_i$ for $i = 1, 2, 3$. If $w(u_1, u_2, u_3) = 1$ in G then the subgroup $\langle u_1, u_2, u_3\rangle$ is cyclic.

In particular G satisfies the Lyndon properties LZ and LCZ.

Recall that the quadratic word $W(x_1, x_3, x_3)$ is regular if there is no automorphism $\alpha\colon F \to F$ such that $w' = \alpha(w)$ as a word in x_1, x_2, x_3, contains less of the x_1, x_2, x_3 than w.

10.5.5 The Lyndon properties and tree-free groups

N. Brady, L. Ciobanu, A. Martino and S. O'Rourke considered the Lyndon properties in groups acting freely on Λ-trees. Using the concept of translation length in such groups they were able to prove the following.

Theorem 10.5.7 ([BCMO]). *Let Λ be an ordered Abelian group and let G act freely on a Λ-tree. Then if $x^p y^q = z^r$ with $p, q, r \geq 4$ it follows that x, y and z commute. That is such groups satisfy LPA.*

A study was initiated by Ciabanu, Fine and Rosenberger [54] to consider the smaller cases. It was shown that for small cases you can have tree-free groups that do not satisfy the Lyndon properties.

Theorem 10.5.8. *Let F be a finitely generated non-cyclic free group, and let u and v be elements in F which are not proper powers. Let*

$$G = \langle F, t : tut^{-1} = v \rangle$$

and $r \geq 2$ be a given integer. Then for particular choices of u and v there exist non-commuting elements $a, b, c \in G$ such that

$$a^2 b^2 c^r = 1.$$

Corollary 10.5.2. *There exist Λ-free groups in which*

$$a^2 b^2 c^r = 1$$

holds for non-commuting $a, b, c \in G$, and $r \geq 2$.

Theorem 10.5.9. *Let F be a finitely generated free group, u and v elements in F that are not proper powers and u is not conjugate to v^{-1}, and $G = \langle F, t; tut^{-1} = v \rangle$. Then if for $a, b, c \in G$ and $p \geq 2, q \geq 3, r \geq 3$ the equality*

$$a^p b^q c^r = 1$$

holds, the elements a, b, c must commute.

10.6 The class of B$\mathcal{X}$-groups

In this section we consider classes of groups which generalize the theorems of Benjamin Baumslag [21], Gaglione and Spellman [108] and Remeslennikov [214]. Recall that Remeslennikov [214] and independently Gaglione and Spellman [108] proved the following remarkable theorem which became one of the cornerstones in the proof of the Tarski problems (see [151] and [232].)

Theorem 10.6.1. *Suppose G is non-Abelian and residually free. Then the following are equivalent:*
(1) *G is fully residually free,*
(2) *G is commutative transitive,*
(3) *G is universally free.*

Therefore the class of non-Abelian fully residually free groups coincides with the class of residually free universally free groups. The equivalence of (1) and (2) in the theorem above was proved originally by Benjamin Baumslag ([21]), where he introduced the concept of fully residually free. Any residually free elementary free group being universally free must satisfy this theorem and hence be fully residually free.

Now let $\mathcal{X}$ be a class of groups. Then a group G is *residually* $\mathcal{X}$ if given any nontrivial element $g \in G$ there is a homomorphism $\phi\colon G \to H$ where H is a group in $\mathcal{X}$ such that $\phi(g) \neq 1$. A group G is *fully residually* $\mathcal{X}$ if given finitely many nontrivial elements $g_1, \ldots, g_n$ in G there is a homomorphism $\phi\colon G \to H$, where H is a group in $\mathcal{X}$, such that $\phi(g_i) \neq 1$ for all $i = 1, \ldots, n$. As we have seen, fully residually free groups have played a crucial role in the study of equations and first-order formulas over free groups.

Recall that *universal theory* of a group G consists of all universal sentences true in G. All non-Abelian free groups share the same universal theory and a group G is called *universally free* if it shares the same universal theory as the class of non-Abelian free groups. We recall two additonal concepts that are needed. A group G is *commutative transitive* or *CT* if commutativity is transitive on the set of nontrivial elements of G. That is if $[x, y] = 1$ and $[y, z] = 1$ for nontrivial elements $x, y, z \in G$ then $[x, z] = 1$. A group G is *CSA* if maximal Abelian subgroups are malnormal. CSA implies commutative transitivity but there exist CT groups that are not CSA (see Section 6.2).

We now consider classes of groups $\mathcal{X}$ for which being fully residually $\mathcal{X}$ is equivalent to being residually $\mathcal{X}$ and commutative transitive, thus extending Baumslag's result. The study of this class was initiated in [55].

Definition 10.6.1. A class of groups $\mathcal{X}$ satisfies $B\mathcal{X}$ if a group G is fully residually $\mathcal{X}$ if and only if G is residually $\mathcal{X}$ and CT.

With this definition B. Baumslag's original theorem says that the class of free groups $\mathcal{F}$ satisfies $B\mathcal{F}$. The classes of $B\mathcal{X}$ groups are fairly extensive as the next theorem shows.

Theorem 10.6.2 ([CiFR 2]). *Let $\mathcal{X}$ be a class of groups such that each non-Abelian $H \in \mathcal{X}$ is CSA. Let G be a non-Abelian and residually $\mathcal{X}$ group. Then the following are equivalent*
(1) *G is fully residually $\mathcal{X}$,*
(2) *G is CSA,*
(3) *G is CT.*
Therefore the class $\mathcal{X}$ has the property $B\mathcal{X}$.

Hence a class of groups $\mathcal{X}$ satisfies $B\mathcal{X}$ if each non-Abelian $H \in \mathcal{X}$ is CSA. Examples of $B\mathcal{X}$ classes abound. In particular we list the following.

Theorem 10.6.3 ([CiFR 2]). *Each of the following classes satisfies $B\mathcal{X}$:*
1. *The class of non-Abelian free groups.*
2. *The class of noncyclic torsion-free hyperbolic groups (see [90]).*
3. *The class of noncyclic one-relator groups with only odd torsion (see [90]).*
4. *The class of cocompact Fuchsian groups with only odd torsion.*
5. *The class of noncyclic groups acting freely on Λ-trees where Λ is an ordered Abelian group (see [129]).*
6. *The class of noncylic free products of cyclics with only odd torsion.*
7. *The class of noncyclic torsion-free RG-groups (see [98] and [2]).*
8. *The class of conjugacy pinched one-relator groups of the following form*

$$G = \langle F, t; tut^{-1} = v\rangle$$

where F is a free group of rank $n \geq 1$ and u, v are nontrivial elements of F that are not proper powers in F and for which $\langle u\rangle \cap x\langle v\rangle x^{-1} = \{1\}$ for all $x \in F$.

The theorem follows from the fact that each of these classes has the property that each non-Abelian group in them is CSA.

Since CSA always implies CT we have the following corollary.

Corollary 10.6.1. *Let $\mathcal{X}$ be a class of CSA groups. Then if G is a non-Abelian residually $\mathcal{X}$ group, then CT is equivalent to CSA.*

Commutative transitivity (CT) has been shown to be equivalent to many other properties (see [2]) under the additional condition that Abelian subgroups are locally cyclic (ALC). A group G is *power commutative* if $[x, y^n] = 1$ implies that $[x, y] = 1$ whenever $y^n \neq 1$. Two elements $a, b \in G$ are in *power relation* to each other if there exists an $x \in G \setminus \{1\}$ with $a = x^n, b = x^m$ for some $n, m \in \mathbb{Z}$. G is *power transitive* or *PT* if this relation is transitive on nontrivial elements. Hence we get the corollary.

Corollary 10.6.2. *Let $\mathcal{X}$ be a class of groups such that each non-Abelian $H \in \mathcal{X}$ is CSA. Let $\mathcal{Y}$ be the subclass of $\mathcal{X}$ consisting of those groups in $\mathcal{X}$ which are ALC. Let G be a non-Abelian residually $\mathcal{Y}$ group which is ALC and has trivial center. Then the following are equivalent.*
1. *G is fully residually $\mathcal{Y}$.*
2. *G is CSA.*
3. *G is CT.*
4. *G is Power commutative.*
5. *G is Power transitive.*

This follows directly from the equivalences given in [2].

10.6.1 Big powers groups and univeral freeness

The results of the previous section showed the equivalence of fully residually-$\mathcal{X}$ and commutative transitivity for any class $\mathcal{X}$ of CSA groups. To prove an equivalence with universally-$\mathcal{X}$ groups in [55] the *big powers condition* was used. This was introduced originally by G.Baumslag in [7]. The big powers condition was discussed in Chapter 6. We review some of the ideas here in a slightly different more general manner.

Definition 10.6.2. Let G be a group and $u = (u_1, \ldots, u_k)$ be a sequence of nontrivial elements of G. Then
(1) u is *generic* if neighboring elements in u do not commute, that is $[u_i, u_{i+1}] \neq 1$ for every $i \in \{1, \ldots, k-1\}$.
(2) u is *independent* if there exists an $n = n(u) \in \mathbb{N}$ such that for any $\alpha_1, \ldots, \alpha_k \geq n$ we have $u_1^{\alpha_1} \ldots u_k^{\alpha_k} \neq 1$.
(3) A group satisfies the *big powers condition* or *BP* if every generic sequence in G is independent. We call such groups *BP*-groups.

G. Baumslag proved that free groups are BP-groups [7] while Olshansky [207] showed that torsion-free hyperbolic groups are BP-groups. For BP groups the following results are known.

Lemma 10.6.1 ([158]). *A subgroup of a BP-group is itself a BP-group.*

Lemma 10.6.2 ([207]). *Every torsion-free hyperbolic group is a BP-group.*

A stronger version of this lemma for relatively hyperbolic groups is given in [158].

Lemma 10.6.3. *A free product of CSA BP-groups is also a BP-group.*

Lemma 10.6.4. *Let $G = F_1 \underset{U=V}{*} F_2$ where F_1, F_2 are finitely generated free groups and U, V are nontrivial elements of F_1, F_2 respectively with not both proper powers. Then G is a CSA and BP-group.*

If G and H are groups then we say that G is an H-group or H-domain if G contains an isomorphic copy of H. Being an H-group is crucial for considering the next equivalence. We consider a class of groups $\mathcal{Z}$ in which each finitely generated non-Abelian group H in $\mathcal{Z}$ is CSA and BP. Reinterpreting a result in [16] and [17] (see also [158]) and following the same proof the next theorem was proved in [55].

Theorem 10.6.4 ([CiFR 2]). *Let $\mathcal{Z}$ be a class of finitely presented groups such that each non-Abelian $H \in \mathcal{Z}$ is CSA and BP. Let $H \in \mathcal{Z}$ and G a finitely presented non-Abelian H-group. Then the following are equivalent.*
(1) *G is fully residually H,*
(2) *G is universally equivalent to H.*

Note that being an H-group was not necessary in the case of the class of free groups since a non-Abelian free group and a non-Abelian fully residually free group contain copies of free groups of all countable ranks. It was noticed by D. Spellman that while the BP and CSA conditions were necessary in [16] for embedding a given hyperbolic group into its Lyndon completion and then a modification of this proof with the given conditions was used in the proof of Theorem 8.4 in [55] they were not really necessary for universal equivalence.

In alternative language if a group G is fully residually H then we say that H *discriminates* G. Further if we append to the basic language L_0 appropriate for group theory constants from the group H then we say that G is H-*universally equivalent* to H if G and H have the same universal theory in this extended language. We actually have the following theorem which says that if G is an H-group then H discriminating G is equivalent to H being universally equivalent to G. Further if H if finitely generated then G is H-universally equivalent to H if and only if there is a discriminating family of retractions from G onto H.

Theorem 10.6.5. *Let G be a finitely presented H-group. Then the following are equivalent:*
(1) *G is fully residually H, that is H discriminates G.*
(2) *G is universally equivalent to H.*
Further if H is finitely generated then G is H-universally equivalent to H if and only if there is a discriminating family of retractions $G \to H$.

Proof. Suppose first that G is an H-group and that G is fully residually H. We show that G is universally equivalent to H. To show this we prove that any universal sentence true in H is also true in G. Hence the universal theory of H is contained in the universal theory of G. However H is a subgroup of G so the universal theory of G is contained in the universal theory of H. The equivalence then follows.

To show that every universal sentence true in H is also true in G we show that every existential sentence true in G must also be true in H. Suppose the following existential sentence, which we label $(\star)$, and whose matrix is written in disjunctive normal form, is true in G:

$$\exists x_1,\ldots,x_n\Big(\bigvee_i\Big(\bigwedge_j(u_{ij}(x_1,\ldots,x_n)=1)\Big)\wedge\Big(\bigwedge_k(w_{ik}(x_1,\ldots,x_n)\neq 1)\Big)\Big) \qquad (\star)$$

The sentence $(\star)$ is equivalent to the sentence below which we label $(\star\star)$:

$$\bigvee_i \exists x_1,\ldots,x_n\Big(\Big(\bigwedge_j(u_{ij}(x_1,\ldots,x_n)=1)\Big)\wedge\Big(\bigwedge_k(w_{ik}(x_1,\ldots,x_n)\neq 1)\Big)\Big) \qquad (\star\star)$$

Since $(\star\star)$ holds in G it follows that at least one disjunct must be true in G. Suppose that

$$\exists x_1,\ldots,x_n\Big(\Big(\bigwedge_{j=1}^{r}(u_{i_0j}(x_1,\ldots,x_n)=1)\Big)\wedge\Big(\bigwedge_{k=1}^{q}(w_{i_0k}(x_1,\ldots,x_n)\neq 1)\Big)\Big) \qquad (\star\star\star)$$

holds in G. Let $(g_1, \dots, g_n) \in G^n$ be an n-tuple such that

$$u_{i_0 1}(g_1, \dots, g_n) = \dots = u_{i_0 r}(g_1, \dots, g_n) = 1$$
$$\wedge w_{i_0 1}(g_1, \dots, g_n) \neq 1 \wedge \dots \wedge w_{i_0 q}(g_1, \dots, g_n) \neq 1.$$

Since G is fully residually H there is a map $\phi\colon G \to H$ such that

$$\phi(w_{i_0 k}(g_1, \dots, g_n)) = w_{i_0 k}(\phi(g_1), \dots, \phi(g_n)) \neq 1 \quad \text{for all } k = 1, \dots, q.$$

Further clearly

$$u_{i_0 j}(\phi(g_1), \dots, \phi(g_n)) = \phi(u_{i_0 j}(g_1, \dots, g_n)) = \phi(1) = 1 \quad \text{for all } j = 1, \dots, r.$$

Therefore $(\star\star\star)$ is true in H and working backwards it follows that $(\star)$ holds in H.

Therefore every existential sentence true in G is also true in H and hence they are universally equivalent.

Further if the discrimination is by retractions onto H then the result holds in the extended language where the elements of H are appended as constants.

Now we show that G being fully residually H is necessary for universal equivalence. Let us assume than that G and H are universally equivalent and we wish to show that H discriminates G.

Suppose that the finitely presented H-group G is universally equivalent to H and let $\langle a_1, \dots, a_m; R_1(a_1, \dots, a_m) = \dots = R_n(a_1, \dots, a_m) = 1\rangle$ be a finite presentation for G. Let $g_j = w_j(a_1, \dots, a_m)$, $j = 1, \dots, k$ be nontrivial elements in G. Then the following existential sentence that we denote by $(\star)$ is true in G.

$$\exists x_1, \dots, x_n \Big(\Big(\bigwedge_{i=1}^{n} (R_i(x_1, \dots, x_n) = 1) \Big) \wedge \Big(\bigwedge_{j=1}^{k} (w_j(x_1, \dots, x_n) \neq 1 \Big) \Big) \qquad (\star)$$

Therefore $(\star)$ is also true in H. Let $(h_1, \dots, h_m) \in H^m$ be such that

$$R_1(h_1, \dots, h_m) = \dots = R_n(h_1, \dots, h_m) = 1$$

and

$$w_1(h_1, \dots, h_m) \neq 1 \wedge \dots \wedge w_k(h_1, \dots, h_m) \neq 1.$$

Then since the relations are preserved the map $a_\nu \mapsto h_\nu$, $\nu = 1, \dots, m$ extends to a homomorphism $\phi\colon G \to H$. Then

$$\phi(g_j) = \phi(w_j(a_1, \dots, a_m)) = w_j(\phi(a_1), \dots, \phi(a_m)) = w_j(h_1, \dots, h_m) \neq 1$$

for all $j = 1, \dots, k$.

Therefore G is fully residually H and the first set of equivalences are completed.

Now we consider H to be finitely generated and we want to consider the extended language where we adjoin the elements of H as constants. From the comments after the first part of the proof we know that if there is a family of discriminating retractions then G is H-universally equivalent to G. Now we assume that G is H-universally equivalent to G and we show that there is a discriminating family of retractions from G onto H.

Let $a_1, \ldots, a_p$ be a set of generators for H and these extend to a finite set $a_1, \ldots, a_p, b_1, \ldots, b_q$ of generators of G. Let

$$\langle a_1, \ldots, a_p, b_1, \ldots, b_q; R_1(a_1, \ldots, a_p, b_1, \ldots, b_q) = \cdots = R_n(a_1, \ldots, a_p, b_1, \ldots, b_q) = 1\rangle$$

be a finite presentation for G.

Suppose $g_j = w_j(a_1, \ldots, a_p, b_1, \ldots b_q) \neq 1$ for $j = 1, \ldots, k$ are nontrivial elements of G. Then the following existential sentence is true in G:

$$\exists x_1, \ldots, x_q\Big(\Big(\bigwedge_{i=1}^{n}(R_i(a_1, \ldots, a_p, x_1, \ldots, x_q) = 1)\Big) \wedge \Big(\bigwedge_{j=1}^{k}(w_j(a_1, \ldots, a_p, x_1, \ldots, x_q) \neq 1)\Big)\Big).$$

Since G and H are assumed to be H-universally equivalent this must also hold in H.

Let $(h_1, \ldots, h_q) \in H^q$ be such that

$$R_1(a_1, \ldots, a_p, h_1, \ldots, h_q) = \cdots = R_n(a_1, \ldots, a_p, h_1, \ldots, h_q) = 1$$

and

$$w_1(a_1, \ldots, a_p, h_1, \ldots, h_q) \neq 1 \wedge \cdots \wedge w_k(a_1, \ldots, a_p, h_1, \ldots, h_q) \neq 1.$$

Then since the relations are preserved the maps

$$a_\nu \mapsto a_\nu, \quad \nu = 1, \ldots, p, \qquad b_\mu \mapsto h_\mu, \quad \mu = 1, \ldots, q$$

extend to a retraction $\phi\colon G \to H$.

Furthermore, for all $j = 1, \ldots, k$ we have

$$\begin{aligned}\phi(g_j) &= \phi(w_j(a_1, \ldots, a_p, b_1, \ldots, b_q)) = w_j(\phi(a_1), \ldots, \phi(a_p), \phi(b_1), \ldots, \phi(b_q))\\ &= w_j(a_1, \ldots, a_p, h_1, \ldots, h_q) \neq 1.\end{aligned}$$

Therefore G is discriminated by retractions completing the proof. □

Summarizing our results of this whole section:

Theorem 10.6.6. *Let $\mathcal{Z}$ be a class of finitely presented groups such that each non-Abelian $H \in \mathcal{Z}$ is CSA. Let G be a finitely presented non-Abelian residually $\mathcal{Z}$ group. Then the following are equivalent*
(1) *G is fully residually $\mathcal{Z}$,*
(2) *G is CSA,*
(3) *G is CT.*
If in addition G is an H-group for some $H \in \mathcal{Z}$ then the following are equivalent.
(a) *G is fully residually H,*
(b) *G is universally equivalent to H.*

11 Discriminating and squarelike groups

11.1 Discriminating groups

As we have seen, discrimination and residual properties play a crucial role in the application of algebraic geometry over groups to the solution of the Tarski problems. As a further outgrowth of the study of algebraic geometry over groups, G. Baumslag, Myasnikov and Remslennikov introduced the concept of a *discriminating group* ([18]). Many important universal type groups such as Higman's universal group which embeds in itself every finitely presented group, the commutator subgroup of the Gupta–Sidki–Grigorchuk group and Thompson's group F were shown to be discriminating. In the theory it is important to also know when groups are not discriminating. A systematic study of discriminating groups was initiated in [18] and continued by Baumslag, Fine, Gaglione, Myasnikov, Spellman and others (see [11, 12, 75, 76, 77, 78, 79, 80, 81, 82, 73, 74, 200] and the references there). In an attempt to better capture the axiomatic properties of discriminating groups the class of *squarelike groups* was introduced ([76]). Subsequently it was proved ([79]) that the class of squarelike groups is precisely the axiomatic closure of the class of discriminating groups. In the course of these studies the question was raised as to the relationship of these classes of groups and an older notion of discrimination, which we call *varietal discrimination* that was also introduced by G. Baumslag jointly with Bernard, Hanna and Peter Neumann (see [202]). The properties of this type of discrimination are described in the book of Hanna Neumann [202] and play a role in the structure of product varieties. In this chapter we introduce and discuss the extensive research on this important and relevant class of groups. First we present the formal definition.

Definition 11.1.1. Let G and H be groups. G *separates* H provided that to every nontrivial element h of H there is a homomorphism $\varphi_h : H \to G$ such that $\varphi_h(h) \neq 1$. G *discriminates* H if to every finite nonempty set S of nontrivial elements of H there is a homomorphism $\varphi_S : H \to G$ such that $\varphi_S(s) \neq 1$ for all $s \in S$. The group G is *discriminating* provided that it discriminates every group it separates.

In analogy with classical algebraic geometry we may view the discrimination of H by G as an approximation to H much like the localization of a ring at a prime; think of a set of generators for H as a set of variables. Hence a discriminating group is analogous to the situation when any element can be localized in G then any finite set of nontrivial elements can be localized in G.

The main criterion for determining whether a group is discriminating is provided by the following lemma.

Lemma 11.1.1 ([16, 17, 18, 19]). *A group G is discriminating if and only if G discriminates $G \times G$.*

Proof. Since G clearly separates $G\times G$, if it is discriminating it will discriminate $G\times G$.

Conversely suppose G discriminates $G \times G$. First we show by induction then G will discriminate G^n. We are given that G discriminates $G^2 = G \times G$. Assume that G discriminates G^n and consider $G^{n+1} = G^n\times G$. Let $g_1, \ldots, g_k$ be finitely many nontrivial elements of G^{n+1}. Without loss of generality, we may assume that $g_i = (g_{i1}, g_{i2})$ where $g_{i1} \in G^n$, $g_{i2} \in G$ and $g_{i2} = 1$ for $1 \le i \le m$, but $g_{i2} \ne 1$ for $m+1 \le i \le k$. Then since G discriminates G^n, there is a homomorphism $\theta\colon G^n \to G$ such that $\theta(g_{i1}) \ne 1$ for $1 \le i \le m$. Consider the homomorphism $\phi\colon G^{n+1} \to G^2$ defined by $\phi(g,h) = (\theta(g),h)$ where $g \in G^n$ and $h \in G$. Clearly, $\phi(g_1), \ldots, \phi(g_k)$ are then nontrivial elements of G^2. But G discriminates G^2 by hypothesis. So there exists another homomorphism $\varphi\colon G^2 \to G$ which does not annihilate any of the $\phi(g_i)$ for $1 \le i \le k$. Taking $\delta = \varphi \circ \phi$ completes the induction.

Now suppose that G separates a group H; we want to show that it discriminates H. Let $h_1, \ldots, h_n$ be nontrivial elements in H. Then for each i there is a homomorphism $\phi_i\colon H \to G$ such that $\phi_i(h_i) \ne 1$. Taking $\phi = \phi_1 \times \cdots \times \phi_n$ and using that G discriminates G^n gives us that G discriminates H completing the proof. □

It is clear from the Lemma that if $G \times G$ embeds in G, then G is discriminating. If this is the case we say that G is *trivially discriminating* which we abbreviate as *TD*. In [75] the question was asked if there are finitely generated non-Abelian nontrivially discriminating groups. Subsequently this was answered in the affirmative in [11] and we will present some examples of these later in this chapter.

Before giving both positive and negative examples of discriminating groups we mention some ties to both logic and algebraic geometry. The tie to logic will motivate the introduction of squarelike groups which we introduce shortly and then look at in depth in Sections 11.3 and 11.4.

Recall that if G is a group, then the *universal theory* of G, written $\mathrm{Th}_\forall(G)$, is the set of all universal sentences of L_0 true in G and that two groups G and H are *universally equivalent* if they have the same universal theory. We denote this by $G \equiv_\forall H$, provided $\mathrm{Th}_\forall(G) = \mathrm{Th}_\forall(H)$. Two groups satisfy precisely the same universal sentences if and only if they satisfy precisely the same existential sentences. Discriminating groups must be universally equivalent to their direct squares. We first need the following lemma.

Lemma 11.1.2. *Two groups G and H are universally equivalent ($G \equiv_\forall H$) if and only if every finite coefficientless system*

$$p_i(x_1, \ldots, x_n) = 1\,, \quad 1 \le i \le I$$
$$q_j(x_1, \ldots, x_n) \ne 1\,, \quad 1 \le j \le J$$

of equations and inequations has a solution in G if and only if it has a solution in H.

Proof. As previously mentioned, G and H satisfying the same universal sentences is equivalent to them satisfying the same existential sentences. Consider the existential sentence, $\exists\overline{x}\varphi(\overline{x})$, of L_0 where $\overline{x} = (x_1, \ldots, x_n)$ is a tuple of distinct variables. A *primitive sentence* of L_0 is an existential sentence of the form

$$\exists\overline{x}\Big(\bigwedge_j (p_j(\overline{x}) = 1) \wedge \bigwedge_k (q_k(\overline{x}) \neq 1)\Big).$$

Write the *matrix* $\varphi(\overline{x})$ of the existential sentence $\exists\overline{x}\varphi(\overline{x})$ in *disjunctive normal form*

$$\bigvee_i \Big(\bigwedge_j (p_{i,j}(\overline{x}) = 1) \wedge \bigwedge_k (q_{i,k}(\overline{x}) \neq 1)\Big)$$

It is well known that any formula of L_0 can be so written. We thus see that $\exists\overline{x}\varphi(\overline{x})$ is logically equivalent to the disjunction $\bigvee_i \Psi_i$ where Ψ_i is the primitive sentence

$$\exists\overline{x}\Big(\bigwedge_j (p_{i,j}(\overline{x}) = 1) \wedge \bigwedge_k (q_{i,k}(\overline{x}) \neq 1)\Big).$$

(Note the existential quantifiers can always be pulled through a disjunction to keep the logical equivalence.) But a disjunction will hold if and only if at least one of the disjuncts does. Thus, two groups G and H are universally equivalent if and only if they satisfy precisely the same primitive sentences of L_0. But that is the same as the condition given in this lemma. □

Theorem 11.1.1 ([75]). *If G is discriminating, then $G \times G \equiv_\forall G$.*

Proof. Let G be a discriminating group. From Lemma 11.1.2 groups G and H have the same universal theory if and only if every finite coefficientless system

$$\begin{aligned} p_i(x_1, \ldots, x_n) &= 1, \quad 1 \leq i \leq I \\ q_j(x_1, \ldots, x_n) &\neq 1, \quad 1 \leq j \leq J \end{aligned}$$

of equations and inequations has a solution in G if and only if it has a solution in H.

From this it is clear that if $G \subseteq H$, then a sufficient condition for G and H to have the same universal theory is that G discriminates H. Now $G \subset G \times G$ and since G is discriminating it discriminates its direct square. Therefore $G \times G$ and G have the same universal theory □

In order to better capture the axiomatic properties of discriminating groups we define.

Definition 11.1.2. The group G is *squarelike* provided $G \times G \equiv_\forall G$.

Hence, every discriminating group is squarelike (but, as we shall see, *not* conversely). Squarelike groups will be discussed in detail in Sections 11.3 and 11.4.

Recall from Chapter 4 that a universal sentence of the form

$$\forall\overline{x}\Big(\bigwedge_i (u_i(\overline{x}) = 1) \to (w(\overline{x}) = 1)\Big)$$

is called a *quasilaw* or *quasi-identity*. Note that every identity

$$\forall \overline{x}(w(\overline{x}) = 1)$$

is equivalent to a quasilaw

$$\forall \overline{x}, y\left((y \cdot y^{-1} = 1) \rightarrow (w(\overline{x}) = 1)\right) .$$

A *quasivariety* is the model class of a set of quasilaws. If $\mathcal{X}$ is a nonempty class of groups then the quasivariety qvar($\mathcal{X}$) *generated by* $\mathcal{X}$ is the model class of those quasilaws φ true in every group in $\mathcal{X}$. The *universal closure* ucl($\mathcal{X}$) of $\mathcal{X}$ is the model class of the set of all universal sentences φ true in every group G in $\mathcal{X}$. If $\mathcal{X} = \{G\}$ is a singleton, then we write ucl(G) for the universal closure of G and qvar(G) for the quasivariety generated by G.

Quasivarieties are examples of *axiomatic classes*. That is, they are model classes of sets of sentences of L_0, the language of group theory. Quasivarieties are closed under (unrestricted) direct products. They are also closed under subgroups since they have a set of universal axioms. Put another way, quasivarieties are axiomatic *prevarieties* (see Chapter 4). The class of all groups is a quasivariety of groups and the intersection of any family of quasivarieties is again a quasivariety. For that matter, the intersection of any family of universally axiomatizable model classes (i.e. axiomatic classes having a set of universal axioms) is again a universally axiomatizable model class. It follows that if $\mathcal{X}$ is any class of groups there is a least quasivariety (qvar $\mathcal{X}$) containing $\mathcal{X}$ and a least universally axiomatizable model class ucl($\mathcal{X}$) containing $\mathcal{X}$. (qvar $\mathcal{X}$) is the quasivariety generated by $\mathcal{X}$ and ucl($\mathcal{X}$) is the universal closure of $\mathcal{X}$. If $\mathcal{X} = \{G\}$ is a singleton we write (qvar G) and ucl(G) for (qvar $\mathcal{X}$) and ucl($\mathcal{X}$) respectively. (qvar G) is the model class of the quasi-identities satisfied by G and ucl(G) is the model class of the universal sentences satisfied by G. In general ucl(G) is a proper subclass of (qvar G). The more properties a structure is required to satisfy the fewer structures can satisfy them in general.

We need to recall one final concept from the algebraic geometry over groups. If $S = 1$ is a system of group equations in finitely many variables then $V_G(S)$ denotes the set of solutions in the group G. The group G is *equationally Noetherian* (see Chapter 8) provided for every $n \in \mathbb{N}$ and every system $S = 1$ of equations in n unknowns there is a finite subset $S_0 \subseteq S$ such that $V_G(S) = V_G(S_0)$ where $V_G(S)$ is the variety of solutions in the group G of the equations S.

Examples of equationally Noetherian groups are groups linear over a commutative Noetherian ring with identity. In particular, every group linear over a field is equationally Noetherian (see Chapter 8).

The following results ties all these ideas to discriminating groups.

The following result comes from Baumslag, Myasnikov and Remeslennikov [16].

Theorem 11.1.2. *Let G and H be finitely generated groups and let G be equationally Noetherian. Then G is universally equivalent to H if and only if G discriminates H and H discriminates G.*

This implies the following result.

Lemma 11.1.3. *Let G be a finitely generated equationally Noetherian group. Then G is discriminating if and only if G and $G \times G$ are universally equivalent.*

Next we have

Lemma 11.1.4. *Let G and H be finitely generated groups and suppose that G is equationally Noetherian and discriminating. If H is universally equivalent to G then H is also discriminating.*

Proof. Suppose that $G \equiv_\forall H$. From Theorem 11.1.2 G discriminates H and therefore $G \times G$ discriminates $H \times H$. Since G is discriminating G discriminates $G \times G$ and hence again by Theorem 11.1.2 H discriminates G. Therefore H discriminates $H \times H$ and hence H is discriminating. □

We now can prove a theorem giving a complete characterization of a discriminating group when it is equationally Noetherian. Later in this chapter we will remove the equationally Noetherian criterion when the group is squarelike.

Theorem 11.1.3 ([76]). *Let G be a finitely generated equationally Noetherian group. Then G is discriminating if and only if* $(\mathrm{qvar}\, G) = \mathrm{ucl}(G)$.

Proof. Suppose that G is a finitely generated equationally Noetherian discriminating group. For a class of groups $\mathcal{K}$ we denote by $\mathcal{K}_\omega$ the subclass of finitely generated subgroups in $\mathcal{K}$. To prove that $\mathrm{qvar}(G) = \mathrm{ucl}(G)$ it suffices to show that $\mathrm{qvar}(G)_\omega = \mathrm{ucl}(G)_\omega$. This follows from the fact that every group is embeddable into an ultraproduct of its finitely generated subgroups. Since $\mathrm{ucl}(G) \subseteq \mathrm{qvar}(G)$ the inclusion $\mathrm{ucl}(G)_\omega \subseteq \mathrm{qvar}(G)_\omega$ is obvious. For an equationally Neotherian group one has $\mathrm{qvar}(G)_\omega = pvar(G)_\omega$ (see [192, 193, 194, 195, 196, 197] and Chapter 8) and therefore it suffices to show that $pvar(G)_\omega \subseteq \mathrm{ucl}(G)_\omega$. Let H be a finitely generated group from $pvar(G)_\omega$. Then $H \subset G^\alpha$ for some cardinal α. It is easy to see that if G is discriminating then G is universally equivalent to any Cartesian product of G and hence $G^\alpha \in \mathrm{ucl}(G)$. This implies that $H \in \mathrm{ucl}(G)$ because universal classes are closed under taking subgroups. Therefore $pvar(G)_\omega \subseteq \mathrm{ucl}(G)_\omega$ as desired.

Conversely suppose that G is equationally Noetherain and $\mathrm{qvar}(G) = \mathrm{ucl}(G)$. Then $G \times G \in \mathrm{ucl}(G)$ so $G \times G$ satisfies all the universal sentences true in G. On the other hand G is a subgroup of $G \times G$ so G satisfies all the universal sentences true in $G \times G$. Therefore G is universally equivalent to $G \times G$ and therefore from Lemma 11.1.3 G is discriminating. □

11.2 Examples of discriminating groups

We now present examples of discriminating groups. Clearly if G is a free Abelian group of countably infinite rank then $G \times G$ is isomorphic to G and so G is discriminating. In [16, 17, 18, 19] the question was asked whether the Abelian discriminating groups can be completely characterized. Therefore we look first at the Abelian case.

11.2.1 Abelian discriminating groups

We have immediately that if the torsion subgroup of a group G is finite and nontrivial then the group cannot be squarelike and therefore cannot be discriminating.

Lemma 11.2.1. *Suppose the set of nontrivial elements of finite order in the group G is finite and nonempty. Then G is not squarelike and therefore G is not discriminating.*

Proof. Suppose G contains exactly n elements of finite order where n is an integer greater than 1. Let the least common multiple of the orders of these elements be m. Then G satisfies

$$\forall x_1, \ldots, x_{n+1} \Big(\bigwedge_{1 \le i \le n+1} (x_i^m = 1) \to \bigvee_{1 \le i < j \le n+1} (x_i = x_j) \Big)$$

but $G \times G$ does not. □

We immediately deduce that no nontrivial finitely generated nilpotent group with torsion can be squarelike. In particular, no nontrivial finitely generated Abelian group with torsion and no nontrivial finite group can be squarelike. However torsion free Abelian groups must be discriminating and therefore squarelike.

Lemma 11.2.2. *Every torsion free Abelian group is discriminating. Hence, every torsion free Abelian group is squarelike.*

Proof. Let A be a torsion free Abelian group. We write A additively. Since A separates $A \oplus A$, given $a_1 \in (A \oplus A) \backslash \{0\}$, there is $f_1 \in \mathrm{Hom}(A \oplus A, A)$ such that $f_1(a_1) \neq 0$. Suppose that, given $a_1, \ldots, a_n \in (A \oplus A) \backslash \{0\}$, there is $f_n \in \mathrm{Hom}(A \oplus A, A)$ such that $f_n(a_i) \neq 0, i = 1, \ldots, n$. Now suppose $a_1, \ldots, a_{n+1} \in (A \oplus A) \backslash \{0\}$. Then, by inductive hypothesis, there are $f, g \in \mathrm{Hom}(A \oplus A, A)$ such that $f(a_i) \neq 0, i = 1, \ldots, n$, and $g(a_{n+1}) \neq 0$. We then choose an integer N sufficiently large so that $h = f + Ng$ does not annihilate any of $a_1, \ldots, a_{n+1}$. That completes the induction and proves the lemma. □

So, among finitely generated Abelian groups, the discriminating groups and the squarelike groups are precisely the torsion free ones. A moment's reflection produces discriminating Abelian groups, necessarily not finitely generated, with torsion. For example, every group free of infinite rank in the variety of Abelian groups of exponent dividing n for any fixed integer $n > 1$ is trivially discriminating.

A partial answer to the characterization of Abelian discriminating groups was given in [17]. In order to present it, we must introduce the *Szmielew invariants* of an Abelian group. Given an integer $m > 0$ and a family of elements (a_i) in an additively written Abelian group A, (a_i) is *linearly independent modulo m* provided $\Sigma_i n_i a_i = 0$ implies $n_i \equiv 0\ (mod\ m)$ for all i; (a_i) is *linearly independent modulo m in the stronger sense* provided $\Sigma_i n_i a_i \in mA$ implies the coefficients $n_i \equiv 0\ (mod\ m)$ for all i. For each prime p and positive integer k we define three ranks each of which is a nonnegative integer or the symbol ∞.

(1) $\rho^{(1)}[p,k](A)$ is the maximum number of elements of A of order p^k and linearly independent modulo p^k
(2) $\rho^{(2)}[p,k](A)$ is the maximum number of elements of A linearly independent modulo p^k in the stronger sense.
(3) $\rho^{(3)}[p,k](A)$ is the maximum number of elements of A of order p^k and linearly independent modulo p^k in the stronger sense.

The next result is due to Szmielew [245].

Theorem 11.2.1 ([245]). *Let A and B be Abelian groups. Then A and B are elementarily equivalent if and only if the following two properties are satisfied.*
(1) *A and B either both have finite exponent or both have infinite exponent.*
(2) *For all primes p and positive integers k,* $\rho^{(i)}[p,k](A) = \rho^{(i)}[p,k](B)$ *for* $i = 1, 2, 3$.

For the definition of p-height see the paragraph directly after Theorem 11.2.2 in Section 11.4.

Theorem 11.2.2 ([17]). *Let A be a torsion Abelian group such that, for each prime p, the p-primary component of A modulo its maximal divisible subgroup contains no nontrivial elements of infinite p-height. Then A is discriminating if and only if, for each prime p, the following two properties are satisfied.*
(1) *For all positive integers k,* $\rho^{(1)}[p,k](A)$ *is either 0 or* ∞.
(2) *The rank of the maximal divisible subgroup of the p-primary component of A is either 0 or infinite.*

We do have a complete characterization of Abelian squarelike groups.

Theorem 11.2.3 ([76]). *Let A be an Abelian group. Then A is squarelike if and only if, for each prime p and positive integer k,* $\rho^{(1)}[p,k](A)$ *is either 0 or* ∞.

11.2.2 Trivially discriminating groups and universal type groups

The theory of discriminating groups requires both positive examples, that is groups that are discriminating and negative examples, that is groups that are nondiscriminating. Recall that a group is trivially discriminating if it embeds its direct square. The earliest examples of discriminating groups were either Abelian or trivially discriminating. These included many *universal type* groups. In this section we present some non-Abelian positive examples.

Suppose G is any group whatsoever and I is an infinite index set. Then $G^I \times G^I \cong G^I$ so G^I is surely trivially discriminating. If $G \neq \{1\}$, then G^I is uncountable. Nontrivial countable examples of groups isomorphic to their direct squares are easy enough to construct. Just take the restricted direct power of a nontrivial countable group with respect to a countably infinite index set. Further there exist examples of nontrivial finitely generated groups isomorphic to their direct squares. Such examples were first constructed by Tyler Jones [141] and subsequently by Hirshon and Meier [134]. The question of whether or not there exists a finitely presented group isomorphic to its direct square remains open. None the less, if we relax the condition $G \times G \cong G$ to $G \times G$ embeds in G, finitely presented such groups $G \neq \{1\}$ do exist. G. Baumslag observed the following.

Lemma 11.2.3 ([11, 12]). *A simple discriminating group G is trivially discriminating.*

Proof. Suppose that G is a simple discriminating group. Let g be any nontrivial element of G. Consider the elements $x = (g, 1)$, $y = (1, g) \in G \times G$. Since G is discriminating there exists a homomorphism ϕ from $G \times G$ into G which maps x and y into nontrivial elements of G. Since G is simple it follows that ϕ is monic on $G_1 = G \times \{1\}$ and $G_2 = \{1\} \times G$. If we now set $H_1 = \phi(G_1)$, $H_2 = \phi(G_2)$ and $H = gp(H_1, H_2)$, then H_1 and H_2 commute element-wise. Hence their intersection is central in H_1 and therefore trivial. It follows that $H = H_1 \times H_2 \cong G \times G$. So G is TD. □

Many of the earliest discovered discriminating groups were *universal type* groups. This is evident in our first example.

Example 11.2.1 (Higman's Universal Group). A group G is a *universal finitely presented group* provided it embeds every finitely presented group. Higman proved the existence of such groups in [131]. Since the direct square of a finitely presented group is also finitely presented it is clear that every universal finitely presented group is trivially discriminating. It is shown in [11] that if U is a recursively presentable Abelian group and T is a universal finitely presented group, then the standard wreath product $UwrT$ is trivially discriminating. On the other hand, it is shown in that same paper that if U is a non-Abelian finitely presented group and T is a universal finitely presented group, then $UwrT$ is not discriminating and therefore not trivially discriminating.

Another example, also due to Higman, is the following.

Example 11.2.2 (Higman's Groups $G_{n,r}$). For each integer $n \geq 2$ and each integer $r \geq 1$, Higman in [132] defined a group $G_{n,r}$ in the following manner. Let $\mathcal{V}_n$ be the variety of all algebras with one n-ary operation λ and n unary operations $\alpha_1, \ldots, \alpha_n$ subject to the laws $\lambda(\alpha_1(x), \ldots, \alpha_n(x)) = x$ and $\alpha_i(\lambda(x_1, \ldots, x_n)) = x_i$, $i = 1, \ldots, n$. If $V_{n,r}$ is an algebra free on r generators in $\mathcal{V}_n$, then we let $G_{n,r} = \mathrm{Aut}(V_{n,r})$ be its group of automorphisms. Higman proved that the $G_{n,r}$ are finitely presented. Further, when n is even, $G_{n,r}$ is simple and when n is odd, $G_{n,r}$ contains a simple subgroup $G^+_{n,r}$ of index 2. Setting $G^+_{n,r} = G_{n,r}$ when n is even, he showed that, for fixed r, $G^+_{m,r} \cong G^+_{n,r}$ implies $m = n$. Thus he found an infinite family of finitely presented infinite simple groups. Higman remarks that one of the groups in the family $G^+_{n,r}$ was discovered by Richard Thompson, whose name we shall repeat soon. The $G_{n,r}$ are discriminating; so, when n is even, $G_{n,r}$ is a finitely presented trivially discriminating group by Lemma 11.2.3.

We show directly, regardless of the parity of n, that $G_{n,r}$ is trivially discriminating. To do so we repeat here Higman's observations that if X is a set of free generators for an algebra V in $\mathcal{V}_n$ then, for each $x \in X$, so is $(X \backslash \{x\}) \cup \{\alpha_1(x), \ldots, \alpha_n(x)\}$ and hence $r \equiv s \ (mod\ (n-1))$ implies $V_{n,r} \cong V_{n,s}$. Let us fix $n \geq 2$ and $r \geq 1$. Let $s \geq 2r$ be such that $s \equiv r \ (mod\ (n-1))$. Let $\{x_1, \ldots, x_s\}$ be an s element set of free generators for $V_{n,r}$. For every ordered pair of automorphisms (θ, φ) of $V_{n,r}$ we define an automorphism Ψ of $V_{n,r}$ by

$$\begin{aligned} \Psi(x_i) &= \theta(x_i), \quad i = 1, \ldots, r \\ \Psi(x_i) &= \varphi(x_i), \quad i = r+1, \ldots, 2r \\ \Psi(x_i) &= x_i, \quad i = 2r+1, \ldots, s \end{aligned}$$

where θ acts on the free algebra $\langle x_1, \ldots, x_r \rangle \cong V_{n,r}$ and φ acts on the free algebra $\langle x_{r+1}, \ldots, x_{2r} \rangle \cong V_{n,r}$. Clearly the assignment $(\theta, \varphi) \mapsto \Psi$ is an embedding $G_{n,r} \times G_{n,r} \hookrightarrow G_{n,r}$. Thus the $G_{n,r}$ are all trivially discriminating.

A finitely presented infinite simple group K which embeds its direct square was also given in [129]. The jury is (as of this writing) still out on whether or not $K \cong G_{2,1}$. None the less, many interesting properties of K and characterizations of K are contained in [129].

Example 11.2.3 (Thompson's Group F). Thompson's Group F consists of all orientation preserving piecewise linear homeomorphisms from the unit interval [0,1] onto itself that are differentiable except at finitely many dyadic rational numbers and such that, on intervals of differentiability, the derivatives are powers of 2. This group was introduced by Thompson and has been extensively studied (see [248] and the references there). This group embeds its direct square and hence is trivially discriminating (see [75] for a proof). Further this is finitely presented.

Example 11.2.4 (The Commutator Subgroup of the Gupta–Sidki Group). For each odd prime p, Gupta and Sidki in [108, 109, 110, 111, 112, 113] constructed a group H_p which is a subgroup of the group of automorphisms of a rooted tree. H_p is a 2-generator infinite p-group. Consider the commutator subgroup H_p' of H_p. It can be shown that H_p', while finitely generated, is not finitely presentable. Moreover, it can be shown that H_p' is trivially discriminating.

Example 11.2.5 (Non-Equationally Noetherian Example). The existence of a finitely generated but not finitely presentable trivially discriminating group G_1 (e.g. the commutator subgroup of a Gupta–Sidki group) together with the existence of a universal finitely presented group G_2 allowed us to construct in [78] a group $G = G_1 \times G_2$ which is proven in that paper to be a finitely generated discriminating group which is neither finitely presented nor equationally Noetherian.

Example 11.2.6 (Certain Permutation Groups). The group S_ω of all permutations of $\mathbb{N}$ which move only finitely many integers and the subgroup A_ω of even permutations are each trivially discriminating as we shall presently show; moreover, S_ω embeds in A_ω. Given an integer $n \geq 2$, we define maps $f_i\colon S_n \to S_{2n}$, $i = 1, 2$, and $g\colon S_n \to A_{2n}$ as follows

$$\begin{aligned} f_1(\pi)(2k-1) &= 2\pi(k) - 1, \quad & f_1(\pi)(2k) &= 2k; \;\; k = 1, \dots, n \\ f_2(\pi)(2k-1) &= 2k - 1, \quad & f_2(\pi)(2k) &= 2\pi(k); \;\; k = 1, \dots, n\,. \end{aligned}$$

Then $f\colon S_n \times S_n \to S_{2n}$ defined by $f(\pi_1, \pi_2) = f_1(\pi_1) f_2(\pi_2)$ is easily seen to be an embedding. We define g by $g(\pi) = f(\pi, \pi)$. Observe that $\operatorname{sgn} f(\pi_1, \pi_2) = \operatorname{sgn}(\pi_1)\operatorname{sgn}(\pi_2)$ so that f restricts to an embedding $A_n \times A_n \hookrightarrow A_{2n}$ and, moreover, the image of g is indeed contained in A_{2n}. Since S_ω and A_ω are the direct limits

$$S_\omega = \varinjlim_{\{n \geq 2\}} S_n$$

and

$$A_\omega = \varinjlim_{\{n \geq 2\}} A_n\,,$$

we get embeddings

$$\begin{aligned} S_\omega \times S_\omega &\hookrightarrow S_\omega \\ A_\omega \times A_\omega &\hookrightarrow A_\omega \end{aligned}$$

and

$$S_\omega \hookrightarrow A_\omega\,.$$

Thus we have proven all the claims above.

Let $\mathcal{F}$ be the class of all finite groups. Then, since every finite group embeds in S_ω and in A_ω, every universal sentence true in S_ω or in A_ω must be true in every finite group. On the other hand, universal sentences are preserved in direct unions and each of S_ω and A_ω is a direct union of finite groups. It follows that $\mathrm{Th}_\forall(\mathcal{F}) = \mathrm{Th}_\forall(S_\omega) = \mathrm{Th}_\forall(A_\omega)$ where $\mathrm{Th}_\forall(\mathcal{F})$ is the set of all universal sentences of L_0 true in every finite group.

It is worth mentioning that, if G is any trivially discriminating group, an easy induction shows that G^n embeds in G for every positive integer n. Conversely, if G^n embeds in G for some integer $n \geq 2$, then G is trivially discriminating since clearly, in that event, $G \times G \leq G^n$.

Example 11.2.7 (Groups Isomorphic to their Direct Squares). Examples of nontrivial finitely generated groups isomorphic to their direct squares were first constructed by Tyler Jones [141] and subsequently by Hirshon and Meier [134]. The question of whether or not there exists a finitely presented group isomorphic to its direct square remains open.

Example 11.2.8 (The Grigorchuk Group of Intermediate Growth). In [75], due to conversations with Grigorchuk, it was shown that the Grigorchuk group of intermediate growth is discriminating. Whether it is TD or not is open.

These examples led to the obvious question:

Question. *Does there exist a finitely presented non-Abelian nontrivially discriminating group?*

In the next subsection we give a positive answer to this question.

11.2.3 Nontrivially discriminating groups

In the last section we posed the question of whether there exists a finitely generated non-Abelian nontrivially discriminating group. That is, does there exist a non-Abelian finitely generated discriminating group which does not embed its direct square. One possibility (still open) was the Grigorchuk groups of intermediate growth. We now present two classes of examples of such groups. The groups in the first class are finitely presented. The second is a class of groups studied by B. H. Neumann which are nontrivially discriminating and finitely generated but probably not finitely presented.

To describe the first class of groups, let X be a non-Abelian, finitely generated, torsion free, nilpotent group and let Y be one of the infinite simple groups, $G_{n,r}$, as described by Higman. The following properties of X and Y are needed:

(a) X and Y are finitely presented;
(b) X is residually finite;
(c) every finite group is embeddable in Y;
(d) X is not embeddable in Y; in fact, a finitely generated nilpotent of class 2 subgroup of Y is finite by Abelian;
(e) $Y \times Y \times Y$ is embeddable in Y. (This follows from the fact that $\mathcal{V}_{n,r} \cong \mathcal{V}_{n,s}$ in the notation of [132] if and only if $r \equiv s \pmod{(n-1)}$.)

Our class $\mathcal{G}_1$ consists of the groups $X \times Y$.

Theorem 11.2.4. *Each group $G \in \mathcal{G}_1$ is a non-Abelian, finitely presented nontrivially discriminating group.*

Proof. Clearly each group $G \in \mathcal{G}_1$ is non-Abelian and finitely presented. Then suppose $G \in \mathcal{G}_1$. We must show show that G is discriminating but that G does not embed its direct square.

Since $G = X \times Y$ with X and Y as described above to prove that G is discriminating it suffices to show that Y discriminates any group of the form $W \times Y \times Y$ where W is residually finite. This is sufficient for then $G = X \times Y$ will discriminate its direct square and hence be discriminating. For this purpose, let $h_1, \ldots, h_n$ be finitely many non-identity elements of $W \times Y \times Y$. Write $h_i = (a_i, b_i)$ where $a_i \in W$ and $b_i \in Y \times Y$. Without loss of generality, we may suppose that $a_i \neq 1$ for $1 \leq i \leq m$ and $a_i = 1$ for $m + 1 \leq i \leq n$. Since W is residually finite, there exists a finite group V and a homomorphism $\alpha \colon W \to V$ such that $\alpha(a_i) \neq 1$ for $1 \leq i \leq m$. By (c) above we may embed V into Y and so we get a homomorphism $\beta \colon W \to Y$ such that $\beta(a_i) \neq 1$ for $1 \leq i \leq m$. Let γ be the homomorphism

$$\gamma \colon W \times Y \times Y \to Y \times Y \times Y$$

defined by

$$(w, y, z)) \xrightarrow{\gamma} (\beta(w), y, z).$$

By (e) above let δ be an embedding of $Y \times Y \times Y$ into Y. If we put $\varphi = \delta \circ \gamma$, then clearly φ is a homomorphism such that $\varphi(h_i) \neq 1$ for all $i = 1, \ldots, n$. Thus Y discriminates $W \times Y \times Y$ and therefore it follows that G discriminates $G \times G$. Thus G is discriminating.

To show that is nontrivially discriminating we must show that $G \times G$ is not embeddable into G. Assume to deduce a contradiction that $G \times G \hookrightarrow G$. Restricting the embedding we get an embedding $X \times X \hookrightarrow X \times Y$. Letting A be the projection of $X \times X$ into Y, then $X \times X \leq X \times A$, so $X' \times X' \leq X' \times A'$. But this tells us that if $h =$ the Hirsch length of the commutator subgroup X', then $h \leq 2h \leq h$ since A' is finite. This contradicts the fact that X is non-Abelian and torsion free nilpotent. Therefore $G \times G$ is not embeddable into G and hence is nontrivially discriminating.

Reacll the if a group has a polycyclic series then its *Hirsch length* is a group invariant which gives the number of infinite factors in its subnormal series. ☐

For the next example we need the following Lemma.

Lemma 11.2.4. *Let G be any group.*

(A) *If G contains an element of infinite order but no subgroup free Abelian of rank 2, then $G \times G$ cannot embed in G.*

(B) *If G is torsion by cyclic, then G cannot contain any subgroup free Abelian of rank 2.*

(C) *If G is an extension of a torsion group by an infinite cyclic group, then $G \times G$ cannot embed in G.*

Proof. Part (A) is obvious and (C) follows from (A) and (B). Thus it suffices to show (B). Assume G is torsion by cyclic. Observe that, if the commuting elements b_1 and b_2 of G were linearly independent over $\mathbb{Z}$, then for all $(p,q) \in \mathbb{Z}^2\backslash\{(0,0)\}$ and all $n \in \mathbb{N}$, $(b_1^p b_2^q)^n$ has infinite order so is nontrivial modulo any normal torsion subgroup. Thus, $b_1^p b_2^q$ would be nontrivial in the image for all $(p,q) \in \mathbb{Z}^2\backslash\{(0,0)\}$. But no cyclic group can contain a subgroup free Abelian of rank 2. □

Now we come to the second class of nontrivally discriminating groups, which we will denote $\mathcal{G}_2$. This class of groups was introduced by B.Neumann in a different context. Let $n_1, n_2, \ldots$ be a strictly increasing sequence of odd positive integers with $n_1 \geq 5$. Let $\Lambda_r = \{1, 2, \ldots, n_r\}$ and let G_r be the alternating group on Λ_r. In G_r consider the two elements

$$x_r = (123), \quad y_r = (12\ldots n_r)\,.$$

In the unrestricted direct product $\prod_r G_r$ define

$$x = (x_1, x_2, \ldots), \quad y = (y_1, y_2, \ldots)$$

and let

$$G = \langle x, y\rangle \subseteq \prod_r G_r\,.$$

B. H. Neumann proved that the restricted direct product

$$D = G_1 \times G_2 \times \ldots$$

is a subgroup of G. The class $\mathcal{G}_2$ consists of all the groups G constructed as above.

Theorem 11.2.5. *Each group $G \in \mathcal{G}_2$ is a finitely generated nontrivially discriminating group.*

Proof. Let $G \in \mathcal{G}_2$. We show that G is discriminating but not TD.

Let $g_1, \ldots, g_m$ be nontrivial elements of $G \times G$. Since G is residually finite there is a finite homomorphic image G_0 of $G \times G$ in which none of $g_1, \ldots, g_m$ is annihilated. Then, for sufficiently large r, G_0 embeds in $A_{n(r)}$ and $A_{n(r)} \leq K_0 \leq G$ where K_0 is the restricted direct product of the family $(A_{n(r)})_{r\in\mathbb{N}}$. (For a proof that K_0 is a subgroup of G see [144, Vol. II] or [201].) Hence, there is a homomorphism

$$\varphi\colon G \times G \to G$$

for which

$$\varphi(g_i) \neq 1\,, \quad i = 1, \ldots, m\,.$$

Therefore G is discriminating.

We show that G is not trivially discriminating by proving it is torsion by cyclic and applying Lemma 11.2.4. Let T be the normal closure of $\langle x\rangle$ in G, here denoted by $\langle x\rangle^G$. For each integer $k \geq 0$ a product

$$z = y^{m(1)} x^{\varepsilon(1)} y^{-m(1)} \ldots y^{m(k)} x^{\varepsilon(k)} y^{-m(k)}$$

of k conjugates of $x^{\pm 1}$ is such that, for each $r \in \mathbb{N}$, at most $3k$ integers are moved by the r-th coordinate $z(r)$ of z. Consequently, for all $r \in \mathbb{N}$, we can view $z(r)$ as lying in an isomorphic copy of the symmetric group S_{3k}. Thus, $z^{(3k)!} = 1$ and this is clearly sufficient for T to be a torsion subgroup of G. Since y has infinite order, no positive power of y can lie in the torsion subgroup T. Clearly, G/T is infinite cyclic generated by the image of y modulo T. Hence the group G is not trivially discriminating. □

The methods used in constructing these non-TD examples can be generalized via the use of ascending chains of subgroups.

Theorem 11.2.6. *Let U be the union of a properly ascending chain of subgroups*

$$U_1 \subset U_2 \subset \cdots \subset U_n \subset \ldots$$

and let P be the unrestricted direct product of the U_i:

$$P = \prod_{i=1}^{\infty} U_i$$

Furthermore let Q be the restricted direct product

$$Q = \overline{\prod_{i=1}^{\infty}} U_i$$

Then every subgroup G of P containing Q is discriminating (where here we view Q as a subgroup of P).

Proof. The proof of this result is straightforward. To this end, let $D = G \times G$ and let $d_1 = (a_1, b_1), \ldots, d_k = (a_k, b_k)$ be finitely many nontrivial elements of D. Each of the elements of G can be viewed as a sequence of elements whose n^{th} term is contained in U_n. Since G contains Q and the series of subgroups U_j is increasing, there exists an integer α such that all of $a_1, b_1, \ldots, a_k, b_k \in U_\alpha$. We define now a homomorphism of D into G by projecting the first coordinates of the elements of D to U_α and the second coordinates to $U_{\alpha+1}$. The upshot of this is that we have defined a homomorphism θ of $G \times G$ into $U_\alpha \times U_{\alpha+1}$. If we view $U_\alpha \times U_{\alpha+1}$ as a subgroup of G, it follows that θ is a homomorphism of $G \times G$ into G mapping the given elements $d_1, \ldots, d_k$ nontrivially. This completes the proof. □

This final theorem can be used to recover the proofs of the preceding examples and to obtain the fact that there exists continuously many 2-generator non-TD discriminating groups. In particular let A_i denote the alternating group of degree i and let G_r be the subgroup defined before the statement of Theorem 11.2.5. Then G_s contains the restricted direct product of the A_{n_i} and that

$$G_s \cong G_t \quad \text{if and only if } s = t$$

This implies that there are continuously many non–isomorphic 2-generator groups. Theorem 11.2.6 then applies to these groups. Further B. H. Neumann proved that the only finite normal subgroups of G_r are direct products of finitely many of the A_{n_i}. This implies, remembering that the alternating groups involved here are all simple, that none of the G_r are TD. We remark here that it has recently been proved that none of the groups G_s is finitely presented (G. Baumslag and C. F. Miller).

Corollary 11.2.1. *The groups G_r are discriminating and further are not TD. Thus there exist continuously many 2-generator nontrivially discriminating groups.*

11.3 Negative examples: nondiscriminating groups

In the application of algebraic geometry it is as important to know both whether a group is discriminating and whether it is not discriminating. Therefore knowledge of the nondiscrimination of various classes of groups, such as nilpotent groups or solvable groups, is very important.

There are two general ideas needed to prove nondiscrimination. The first is to show that some universal property which is true in G cannot be true in $G \times G$. The second is to find a number, such as dimension, that is additive so that this number cannot hold in both $G \times G$ and in G.

Recall that the group G is commutative transitive or CT if the relation "commutes with" is transitive on the nonidentity elements of G. This is equivalent to the fact that G satisfies the universal sentence

$$\forall x, y, z(((y \neq 1) \wedge (xy = yx) \wedge (yz = zy)) \rightarrow (xz = zx)) .$$

Lemma 11.3.1. *A non-Abelian CT group is nondiscriminating.*

Proof. Recall that if G is discriminating then G has the same universal theory as its direct square. However if G is non-Abelian then $G \times G$ cannot satisfy the above universal sentence. If it did then $G \times G$ would be Abelian and so would G. To see this suppose that $x, y \in G$. Then $(x, 1)$ commutes with $(1, y)$ which commutes with $(y, 1)$ and hence $(x, 1)$ commutes with $(y, 1)$ and x commute with y. Therefore G would have to be Abelian. Hence G cannot discriminate $G \times G$. □

This provides a wide array of examples of nondiscriminating groups. In the following lemma we describe the nondiscrimination of certain CT groups.

Corollary 11.3.1. *The following groups are all nondiscriminating.*
(1) *Any torsion-free hyperbolic group and in particular any non-Abelian free group is nondiscriminating.*
(2) *Non-Abelian free solvable groups and their non-Abelian subgroups are nondiscriminating.*

Lemma 11.3.2. *The free product of two nondiscriminating groups is nondiscriminating.*

Proof. Suppose that G and H are two nondiscriminating groups. Let $K = G * H$. If K were discriminating then $K \times K$ would be discriminated by K. This would imply that the direct square of each factor is discriminated by that factor. In particular $G \times G$ would be discriminated by G a contradiction. □

The following is just a restatement of Lemma 11.2.1.

Lemma 11.3.3. *If the torsion elements of a group G form a finite nontrivial subgroup then G is nondiscriminating.*

Corollary 11.3.2. *Any nontrivial finite group is nondiscriminating.*

In a finitely generated nilpotent group the torsion elements form a finite subgroup. Hence any finitely generated nilpotent group with nontrivial torsion is nondiscriminating

Corollary 11.3.3. *Any finitely generated nilpotent group with nontrivial torsion is nondiscriminating*

Theorem 11.3.1. *A finitely generated nilpotent group is discriminating if and only if it is torsion–free Abelian.*

This originally was proved in several stages. A proof using an extension of commutative transitivity was given in [75] for free nilpotent groups. Here we first give a direct version due to Myasnikov and Shumyatsky [200] using what is termed *centralizer dimension*. This method also yields that linear groups are nondiscriminating. We then give an alternative proof using Malcev completions due to Baumslag, Fine, Gaglione and Spellman [11]. For the first proof we initally define centralizer dimension.

Definition 11.3.1. For a group G, its *centralizer dimension*, or *c-dimension*, which we denote $\dim_c(G)$, is the length of the longest strictly descending chain of centralizers in G. G has *finite c-dimension* if $\dim_c(G)$ is finite. We also denote $\dim_C(G)$ by $cd(G)$.

In commutative transitive groups proper centralizers are maximal Abelian and hence CT groups have finite centralizer dimension. In particular free groups and torsion-free hyperbolic groups have finite c-dimension.

Lemma 11.3.4. *General linear groups $GL(m, K)$ over a field K have finite c-dimension.*

Proof. Let $A_i \in GL(m, K), i \in I$, be a finite set of matrices. Then the system $T = 1$ of matrix equations $[X, A_i] = 1, i \in I$, where X is an indeterminate matrix is equivalent to a system $S_T = 0$ of linear equations over K with m^2 variables (the entries of X). The system $S_T = 0$ has at most m^2 independent equations and hence the sytem $T = 1$ is equivalent to its own subsystem of at most m^2 equations. This implies that the length of any strictly descending chain of centralizers in $GL(m, K)$ is at most $m^2 + 1$ and thus $GL(m, K)$ has $\dim_c(GL(m, K)) \leq m^2 + 1$. □

Lemma 11.3.5. *If $H \subset G$ then $cd(H) \leq cd(G)$.*

Proof. Notice that if

$$C_H(A_1) \supset C_H(A_2) \supset \cdots \supset C_H(A_\alpha)\ldots$$

is a strictly decreasing chain of centralizers in H then

$$C_G(A_1) \supset C_G(A_2) \supset \cdots \supset C_G(A_\alpha)\ldots$$

is a strictly decreasing chain of centralizers in G. Hence the longest possible chain in H is no greater than the longest possible chain in G. □

Lemma 11.3.6. *If $\dim_c(G) < \infty$ and $\dim_c(H) < \infty$ then*

$$\dim_c(H \times G) = \dim_c(G) + \dim_c(H) - 1.$$

Proof. Let

$$C_G(A_1) \supset C_G(A_2) \supset \cdots \supset C_G(A_d)\ldots$$

$$C_H(B_1) \supset C_H(B_2) \supset \cdots \supset C_H(B_c)\ldots$$

be strictly decreasing chains of centralizers in G and H respectively. Then

$$C(A_1 \times B_1) \supset C(A_2 \times B_1) \supset \cdots \supset C(A_d \times B_1) \supset C(A_d \times B_2) \supset$$

$$C(A_d \times B_2) \supset \cdots \supset C(A_d \times B_c)$$

is a strictly decreasing chains of centralizers in $G \times H$ which has length $d + c - 1$. Hence

$$\dim_c(G \times H) \geq \dim_c(G) + \dim_c(H) - 1.$$

To prove the reverse inequality we use induction on $\dim_c(G) + \dim_c(H)$.

If $\dim_c(G) = \dim_c(H) = 1$ then G and H are Abelian and hence $G \times H$ is also Abelian and

$$\dim_c(G \times H) = 1 = \dim_c(G) + \dim_c(H) - 1.$$

Now let

$$C_{G\times H}(z_1) \supset C_{G\times H}(z_2) \supset \cdots \supset C_{G\times H}(z_k)$$

be a strictly decreasing chains of centralizers in $G \times H$ of length k. Then

$$C_{G\times H}(z_2) = C_G(B) \times C_G(D)$$

for suitable subsets $B \subset G, D \subset H$. Strict inequality

$$C_{G\times H}(z_1) \supset C_{G\times H}(z_2)$$

implies that

$$G \geq C_G(B), H \geq C_H(D)$$

where at least one of these inclusions is proper. By the inductive hypothesis

$$\begin{aligned}\dim_c(C_{G\times H}(z_2)) &= \dim_c(C_G(B) \times C_H(D)) = \dim_c(C_G(B)) + \dim_c(C_H(D)) - 1\\ &\leq \dim_c(G) + \dim_c(H) - 2\,.\end{aligned}$$

Clearly

$$k \leq \dim_c(C_{G\times H}(z_2)) + 1 \leq \dim_c(G) + \dim_c(H) - 1$$

as required proving the lemma. □

Corollary 11.3.4. *Any linear group or finite direct product of linear groups has finite c-dimension.*

It is known that finitely generated nilpotent groups are linear and therefore we get the following.

Lemma 11.3.7. *Any finitely generated nilpotent group has finite c-dimension.*

The following is a generalization of Lemma 11.3.6 and can be proved in the same manner.

Lemma 11.3.8. *Let $R = K_1 \times \cdots \times K_n$ be a finite direct product of fields. Then the general linear group $GL(m, R)$ has finite c-dimension.*

Wehrfritz [254] proved that any finitely generated metAbelian group can be embedded in a linear group of the form of Lemma 11.3.8. Therefore:

Lemma 11.3.9. *Any finitely generated metabelian group has finite c-dimension.*

Lemma 11.3.10. *Let G be a group with $\dim_c(G) < \infty$. Then*
(1) *If a group H discriminates G then $\dim_c(H) \geq \dim_c(G)$.*
(2) *If a group H is universally equivalent to G then $\dim_c(H) = \dim_c(G)$.*

Proof. Let

$$C(A_1) \supset C(A_2) \supset \cdots \supset C(A_d)$$

be a strictly decreasing chains of centralizers in G. There are elements $g_i \in C(A_i)$ amd $a_{i+1} \in A_{i+1}$ such that $[g_i, a_{i+1}] \neq 1$ for $i = 1, \ldots, d-1$. Since H discriminates G there exists a homomorphism $\phi\colon G \to H$ such that

$$\phi([g_i, a_{i+1}]) \neq 1.$$

This shows that the chain of centralizers

$$C(\phi(A_1)) \supset C(\phi(A_2)) \supset \cdots \supset C(\phi(A_d))$$

is strictly decreasing in H proving assertion (1).

The argument given in the proof of (1) can be described by an existential formula. Since universal equivalence is the same as existential equivalence it follows that (2) holds. □

What ties this concept to discrimination is the following theorem which also holds for squarelike groups.

Theorem 11.3.2. *Let G be a group with finite centralizer dimension. If G is discriminating or squarelike then G is Abelian.*

Proof. Let G be a group of finite c-dimension. If G is discriminating then G discriminates $G \times G$. Hence from Lemma 11.3.6 and 11.3.10

$$\dim_c(G) \geq \dim_c(G \times G) = 2\dim_c(G) - 1.$$

This implies that $\dim_c(G) = 1$ and hence G is Abelian.

Recall that G is squarelike if G is universally equivalent to $G \times G$. In this case

$$\dim_c(G) = \dim_c(G \times G) = 2\dim_c(G) - 1.$$

Again this implies that $\dim_c(G) = 1$ and so G is Abelian. □

We can now put these together to obtain the proof of Theorem 11.3.1

Proof. (Theorem 11.3.1) Suppose that G is a finitely generated nilpotent group. Then from Lemma 11.3.7 G has finite centralizer dimension and so from Theorem 11.3.3 if G is discriminating it is Abelian. □

Using the same technique of centralizer dimension the following are clear.

Theorem 11.3.3.
(1) *A finitely generated linear group is discriminating only if it is torsion–free Abelian.*
(2) *A finitely generated metAbelian group is discriminating only if its Abelian.*

An independent proof using vector space dimension was given by Baumslag, Fine, Gaglione and Spellman [12]. Again the general idea is that we cannot map $G \times G$ into G for a finitely generated non-Abelian nilpotent group.

We denote by $\gamma_n(G)$ the nth term of the lower central series of the group G for $n = 1, 2, 3, \ldots$, where $\gamma_1(G) = G$ and $\gamma_{i+1}(G) = [\gamma_i(G), G]$. Recall that a group G is nilpotent of class c if c is the least integer such that $\gamma_{c+1}(G) = 1$. If G is any nilpotent group and x is any nontrivial element of G, then the *weight* of x, denoted by $wt(x)$, is the unique integer n such that $x \in \gamma_n(G)$ but $x \notin \gamma_{n+1}(G)$. Letting $F = F(r)$ be the absolutely free group of rank $r > 0$, we denote by $F_r(\mathcal{N}_c) = F/\gamma_{c+1}(F)$ the free group of rank r in the variety of groups nilpotent of class c.

In order to continue, we briefly discuss Mal'cev completions. Let G be a finitely generated torsion free nilpotent group of class $c \geq 1$. Any such group embeds into a finitely generated nilpotent group which admits extraction of roots. This is its Mal'cev completion, denoted here by $G^{\mathbb{Q}}$ (see Theorem 2.4 [8]). Suppose F is a free group of rank r. Let $F^{\mathbb{Q}}$ be the tensor $\mathbb{Q}$-completion of F. It is an open problem as to whether or not the Magnus formal power series representation M is a faithful representation of $F^{\mathbb{Q}}$ (See [178] Chapter 5 and also see [115]). None the less if we take the Magnus power series representation and truncate by deleting all the terms of degree at least $c+1$, i.e., factor out by the lower central ideal $d_{c+1}(M)$ which is the least normal $\mathbb{Q}$-subgroup containing the weight $(c+1)$-commutators, we do get a faithful representation of the tensor $\mathbb{Q}$-completion (= Mal'cev completion), relative to $\mathcal{N}_c$, $F_r(\mathcal{N}_c)^{\mathbb{Q}}$ (see Theorem 2.3 [115]). It follows that every element is uniquely a product of rational powers of basic commutators (see Section 3.3 of [7] for the definition of basic commutators; also see Proposition 2.2 in [115]). The Mal'cev completion, $G^{\mathbb{Q}}$, of G is a $\mathbb{Q}$-homomorphic image of $F_r(\mathcal{N}_c)^{\mathbb{Q}}$ for suitable r and c. Thus every element of $G^{\mathbb{Q}}$ is (not necessarily uniquely) a product of rational powers of basic commutators. Also note the class of $G^{\mathbb{Q}}$ is also c. We note that the Mal'cev completion of $F_r(\mathcal{N}_c)$ is not the tensor $\mathbb{Q}$-completion of $F_r(\mathcal{N}_c)$ relative to the category of all groups.

We first get the following result which is of independent interest.

Theorem 11.3.4. *Let G be a finitely generated torsion-free nilpotent group and let $G^{\mathbb{Q}}$ be its Mal'cev completion. Then G is discriminating implies that $G^{\mathbb{Q}}$ is.*

Proof. Write an $x \in G^{\mathbb{Q}}$ as a product of rational powers of basic commutators in a finite set of generators for G. Say $x = c_1^{r(1)} \dots c_k^{r(k)}$ where $r(i) = m_i/n_i$ with $(m_i, n_i) \in \mathbb{Z} \times \mathbb{N}$ and $\gcd(m_i, n_i) = 1$ for all $i = 1, \dots, k$. Let $s(i) = 1/n_i, i = 1, \dots, k$. Now let $G_0 = \langle c_1^{s(1)}, \dots, c_k^{s(k)} \rangle^{\mathbb{Q}}$ and $H_0 = \langle c_1, \dots, c_k \rangle$ – of course, we only need the weight 1 basic commutators to generate G but throwing in all those of weight $\leq c$ causes no harm. Now applying Lemma 2.8 of [7] with G and H there replaced by our G_0 and H_0, respectively, we see that since $c_i^{s(i)}$ raised to the power n_i equals $c_i \in H_0, i = 1, \dots, k$, we can conclude there is an $n \in \mathbb{N}$ such that $x^n \in H_0 = G$. Now x was an arbitrary element of $G^{\mathbb{Q}}$; so given $x \in G^{\mathbb{Q}}$ there is an $n \in \mathbb{N}$ such that $x^n \in G$. Thus powers of nontrivial elements of $G^{\mathbb{Q}} \times G^{\mathbb{Q}}$ lie in $G \times G$ and a homomorphism $G \times G \to G$ which does not annihilate those powers will induce a $\mathbb{Q}$-homomorphism $G^{\mathbb{Q}} \times G^{\mathbb{Q}} \to G^{\mathbb{Q}}$ which does not annihilate the original elements. This completes the proof of Theorem 11.3.4. □

Using this result we can give the alternative proof of Theorem 11.3.1 on torsion-free nilpotent groups.

Theorem 11.3.5. *A finitely generated nilpotent group is discriminating if and only if it is torsion free Abelian.*

Proof. Since torsion free Abelian groups are discriminating [16, 17, 18, 19], it remains to prove the converse. We will show that any finitely generated nilpotent group of class $c > 1$ is nondiscriminating. By using Theorem 3.1 in [75] we may restrict ourselves to the torsion free case. Let us denote the commutator ideal $d_2(G^{\mathbb{Q}})$ of the Mal'cev completion by $[G^{\mathbb{Q}}, G^{\mathbb{Q}}]$. Clearly $G^{\mathbb{Q}}/[G^{\mathbb{Q}}, G^{\mathbb{Q}}]$ can be viewed as a vector space V over $\mathbb{Q}$ whose dimension is the rank of G. Thus it makes sense to say that a subset $X \subseteq G^{\mathbb{Q}}$ is (linearly) independent modulo $[G^{\mathbb{Q}}, G^{\mathbb{Q}}]$. Next we claim that if $X \subseteq G^{\mathbb{Q}}$

projects modulo the commutator ideal $[G^{\mathbb{Q}}, G^{\mathbb{Q}}]$ to a basis for V, then X is a set of $\mathbb{Q}$-generators for $G^{\mathbb{Q}}$. To see this we use induction on the nilpotency class c of G. If $c = 1$, then $[G^{\mathbb{Q}}, G^{\mathbb{Q}}] = 1$ and the claim is trivial. Suppose that the claim is true for all n, $1 \leq n < c$. Supoose also that B is a finite set of $\mathbb{Q}$-generators for $G^{\mathbb{Q}}$. Let $H = G^{\mathbb{Q}}/d_c(G^{\mathbb{Q}})$. Clearly $H/d_2(H)$ is also (up to an obvious $\mathbb{Q}$-isomorphism) V. So by inductive hypothesis, X projects modulo $d_2(G^{\mathbb{Q}})$ to a set of $\mathbb{Q}$-generators for $G^{\mathbb{Q}}$. So to finish up, it suffices to show that every basic commutator of weight c in B lies in the smallest $\mathbb{Q}$-subgroup of $G^{\mathbb{Q}}$ containing X. Suppose for this purpose that $[b_1, \ldots, b_c]$ is such a commutator where for notational convenience we write the commutator as though it were left-normed. Then $b_i = w_i(\overline{x})\delta_i$ where $\overline{x}$ is a tuple of elements of X, w_i is a $\mathbb{Q}$- word and δ_i lies in the commutator ideal $d_2(G^{\mathbb{Q}}) = [G^{\mathbb{Q}}, G^{\mathbb{Q}}]$, $i = 1, \ldots, c$. Then $[b_1, \ldots, b_c] = [w_1(\overline{x})\delta_1, \ldots, w_c(\overline{x})\delta_c] = [w_1(\overline{x}), \ldots, w_c(\overline{x})] \in \langle X\rangle_{\mathbb{Q}}$. That completes the induction and proves the claim.

We now apply Theorem 11.3.4 in order to complete the proof.

Suppose that G is nilpotent of class $q+2$ where $q \geq 0$. Let us suppose to deduce a contradiction that G is discriminating. Thus so is $G^{\mathbb{Q}}$ by Theorem 11.3.4. As a matter of fact $(G^{\mathbb{Q}})^n$ is then discriminated by $G^{\mathbb{Q}}$ for every integer $n > 0$. Choose $a, b \in G^{\mathbb{Q}}$ such that there exist $x(1), \ldots, x(q)$ in $G^{\mathbb{Q}}$ with $c = [a, b, x(1), \ldots, x(q)] \neq 1$. Then neither a nor b lie in the commutator ideal $[G^{\mathbb{Q}}, G^{\mathbb{Q}}]$ since otherwise $wt(c) \geq q+3$ and so $c = 1$ – a contradiction. So if V is the vector space $G^{\mathbb{Q}}/[G^{\mathbb{Q}}, G^{\mathbb{Q}}]$, then $\dim_{\mathbb{Q}}(V) \geq 2$. Let n be a positive integer and let H_i be a $\mathbb{Q}$-isomorphic copy of $G^{\mathbb{Q}}$ for $i = 1, \ldots, n$. Let φ be a $\mathbb{Q}$-homomorphism

$$\varphi\colon H_1 \times \cdots \times H_n \to G^{\mathbb{Q}}$$

which maps each of the elements $c_i = [a_i, b_i, x_i(1), \ldots, x_i(q)]$ into nontrivial elements of $G^{\mathbb{Q}}$. (Such a φ exists since $(G^{\mathbb{Q}})^n$ is then discriminated by $G^{\mathbb{Q}}$.) Since $\varphi(c_1) \neq 1$ it follows that $\varphi(a_1)$ and $\varphi(b_1)$ are independent modulo $[G^{\mathbb{Q}}, G^{\mathbb{Q}}]$. This is so, since if $\varphi(a_1)$ and $\varphi(a_2)$ lay in the same 1-dimensional subspace modulo $[G^{\mathbb{Q}}, G^{\mathbb{Q}}]$, then $\varphi(a_1) = u^{\alpha}\delta_a$ and $\varphi(b_1) = u^{\beta}\delta_b$ where α and β are rational and each of δ_a and δ_b lies in $[G^{\mathbb{Q}}, G^{\mathbb{Q}}]$. In that event, we would have $\varphi(c_1) = [u^{\alpha}\delta_a, u^{\beta}\delta_b, \ldots] = [u^{\alpha}, u^{\beta}, \ldots] = 1$ which contradicts the assumption that $\varphi(c_1) \neq 1$.

The same argument shows that $\varphi(a_2)$ and $\varphi(b_2)$ are independent modulo $[G^{\mathbb{Q}}, G^{\mathbb{Q}}]$ since $\varphi(c_2) \neq 1$. Next we claim that $\{\varphi(a_1), \varphi(b_1), \varphi(a_2)\}$ is linearly independent modulo $[G^{\mathbb{Q}}, G^{\mathbb{Q}}]$. To see this, we first note that if $\varphi(a_1)$ and $\varphi(a_2)$ were dependent modulo $[G^{\mathbb{Q}}, G^{\mathbb{Q}}]$, then there would be a nonzero rational number ρ and an element $\delta \in [G^{\mathbb{Q}}, G^{\mathbb{Q}}]$ such that $\varphi(a_2) = \varphi(a_1)^{\rho}\delta$. (This is so since $\varphi(c_1) \neq 1$ and $\varphi(c_2) \neq 1$ imply that neither $\varphi(a_1)$ nor $\varphi(a_2)$ is trivial modulo $[G^{\mathbb{Q}}, G^{\mathbb{Q}}]$.) Then

$$\begin{aligned}&[\varphi(a_2), \varphi(b_1), \varphi(x_1(1)), \ldots, \varphi(x_1(q))]\\ =&[\varphi(a_1)^{\rho}\delta, \varphi(b_1), \ldots] = [\varphi(a_1)^{\rho}, \varphi(b_1), \ldots]\\ =&[\varphi(a_1), \varphi(b_1), \ldots]^{\rho} \neq 1\,.\end{aligned}$$

But a_2 and b_1 lie in different factors of $H_1 \times \cdots \times H_n$ so we cannot have

$$[\varphi(a_2), \varphi(b_1), \ldots] \neq 1 .$$

This contradiction shows that $\varphi(a_1)$ and $\varphi(a_2)$ are independent modulo $[G^{\mathbb{Q}}, G^{\mathbb{Q}}]$. Thus the only way that $\{\varphi(a_1), \varphi(b_1), \varphi(a_2)\}$ could be linearly dependent modulo $[G^{\mathbb{Q}}, G^{\mathbb{Q}}]$ is if there are nonzero rational numbers α and β and an element $\delta \in [G^{\mathbb{Q}}, G^{\mathbb{Q}}]$ such that $\varphi(a_2) = \varphi(a_1)^{\alpha}\varphi(b_1)^{\beta}\delta$. Then

$$\begin{aligned}[\varphi(a_2), \varphi(b_1), \varphi(x_1(1)), \ldots, \varphi(x_1(q))] &= [\varphi(a_1)^{\alpha}\varphi(b_1)^{\beta}\delta, \varphi(b_1), \ldots] \\ &= [\varphi(a_1), \varphi(b_1), \ldots]^{\alpha} \neq 1 .\end{aligned}$$

But again a_2 and b_1 lying in different direct factors entails that

$$[\varphi(a_2), \varphi(b_1), \ldots] \neq 1$$

is impossible. This contradiction shows that $\{\varphi(a_1), \varphi(b_1), \varphi(a_2)\}$ is independent modulo $[G^{\mathbb{Q}}, G^{\mathbb{Q}}]$. A similar argument shows that $\{\varphi(a_1), \varphi(b_1), \varphi(b_2)\}$ is independent modulo $[G^{\mathbb{Q}}, G^{\mathbb{Q}}]$.

Next we claim that $\{\varphi(a_1), \varphi(b_1), \varphi(a_2), \varphi(b_2)\}$ is linearly independent modulo $[G^{\mathbb{Q}}, G^{\mathbb{Q}}]$. If not there would exist nonzero rational numbers $\alpha(1), \alpha(2)$ and β and an element $\delta \in [G^{\mathbb{Q}}, G^{\mathbb{Q}}]$ such that $\varphi(b_2) = \varphi(a_1)^{\alpha(1)}\varphi(b_1)^{\beta}\varphi(a_2)^{\alpha(2)}\delta$ and so

$$\begin{aligned}[\varphi(a_2), \varphi(b_2), \varphi(x_2(1)), \ldots, \varphi(x_2(q))] &= [\varphi(a_2), \varphi(a_1)^{\alpha(1)}\varphi(b_1)^{\beta}\varphi(a_2)^{\alpha(2)}\delta, \ldots] \\ &= [\varphi(a_2), \varphi(a_1)\varphi(b_1), \ldots]^{\alpha(1)\beta} = 1\end{aligned}$$

since $\varphi(a_2)$ commutes with each $\varphi(a_1)$ and $\varphi(b_1)$ since $\langle a_2\rangle$ and $\langle a_1, b_1\rangle$ lie in different direct factors. But this contradicts the fact that

$$[\varphi(a_2), \varphi(b_2), \varphi(x_2(1)), \ldots, \varphi(x_2(q))] = \varphi(c_2) \neq 1 .$$

This contradiction shows that the set $\{\varphi(a_1), \varphi(b_1), \varphi(a_2), \varphi(b_2)\}$ is linearly independent modulo $[G^{\mathbb{Q}}, G^{\mathbb{Q}}]$ as claimed. Clearly, we can iterate this argument to show that $\{\varphi(a_1), \varphi(b_1), \ldots, \varphi(a_n), \varphi(b_n)\}$ is linearly independent modulo $[G^{\mathbb{Q}}, G^{\mathbb{Q}}]$. But this is impossible since n may be taken arbitrarily large and $V = G^{\mathbb{Q}}/[G^{\mathbb{Q}}, G^{\mathbb{Q}}]$ is finite dimensional. The contradiction shows that $G^{\mathbb{Q}}$ cannot be discriminating and so by Theorem 11.3.4 G cannot be discriminating. □

Kassabov [144] proved the corresponding result for finitely generated solvable groups. We note that Myasnikov and Shumyatsky [200] gave an independent proof of this using centralizer dimension.

Theorem 11.3.6 (Kassabov). *A finitely generated solvable group is discriminating only if it is torsion–free Abelian.*

11.3.1 Further negative examples in varieties

Recall that a U-group is a group which has unique roots. Further if $\mathcal{V}$ is a variety given by a set of laws V then in a free group $V(F)$ denotes the verbal subgroup generated by the laws V. In [11] Baumslag, Fine, Gaglione and Spellman proved some more general results on nondiscrimination of relatively free groups. In particular we have the following theorems.

Theorem 11.3.7. *Let $\mathcal{V}$ be a nontrivial variety of groups having infinite exponent. Suppose that F is a non-Abelian free group and R is a nontrivial normal subroup of F such that*
(1) *F/R is a U-group and contains a rank 2 free Abelian subgroup;*
(2) *$R/V(R)$ is a U-group.*
Then F/R is not discriminating.

Theorem 11.3.8 ([11]). *Let $\mathcal{V}$ and $\mathcal{U}$ be two varieties. If the variety $\mathcal{V}$ contains the infinite cyclic group, then the non-Abelian free groups in the product variety $\mathcal{U}\mathcal{V}$ are not discriminating.*

Theorem 11.3.9 ([11]). *For all sufficiently large primes p, the non–Abelian free groups G in the variety of all groups of exponent p are not discriminating.*

The rest of this section will be devoted to proving these results. Throughout this section we let $\mathcal{V}$ be a variety. We begin with the proof of the almost obvious

Lemma 11.3.11. *Let G be a group and suppose that the centralizer of every nontrivial element of G belongs to $\mathcal{V}$. If there exits elements $a, b \in G$ such that $H = gp(a, b) \notin \mathcal{V}$, then G is not discriminating.*

Proof. Let G_1 and G_2 be isomorphic copies of G. If $g \in G$, we denote the image of g in G_i by g_i for $i = 1, 2$. Since $H \notin \mathcal{V}$, there exists a law $v(x, y)$ in $\mathcal{V}$ such that $v(a, b) \neq 1$. Suppose that G discriminates $G_1 \times G_2$. There exists a homomorphism $\phi\colon G_1 \times G_2 \longrightarrow G$ such that $\phi(a_1) \neq 1$ and also $\phi(v(a_2, b_2) \neq 1$. Let us put $\phi(\alpha) = a_1$. Then the centralizer of α contains the image J of the copy H_2 of H in G_2, which does not satisfy the laws of $\mathcal{V}$. This completes the proof. □

Now we prove the following result which is of independent interest. Recall that a U-group is a group which has unique roots. Further if $\mathcal{V}$ is a variety given by a set of laws V then in a free group $V(F)$ denotes the verbal subgroup generated by the laws V.

Theorem 11.3.10. *Let F be a non-Abelian free group and let R be a nontrivial, normal subgroup of F and let $\mathcal{V}$ be a nontrivial variety of infinite exponent. Suppose that F/R and $R/V(R)$ are U-groups. Then the centralizer of an element of $F/V(R)$ which is not in $R/V(R)$ is infinite cyclic.*

The proof will be accomplished by means of four lemmas.

First we need a "positioning" lemma. We note that as a consequence of the work of Baumslag [7, 8, 9, 10] the hypotheses imply that $F/V(R)$ is a U-group.

Lemma 11.3.12. *Let F be a free group and let R be a normal subgroup of F. Suppose that F/R is a finitely generated free Abelian group with free basis $\{u_1R, \dots, u_mR\}$. Then there exists a free basis $X = W \cup Y$ of F where*

$$W = \{u_1d_1, \dots, u_md_m\},$$

$d_1, \dots, d_m \in F'$ and Y is contained in R.

Proof. To prove Lemma 11.3.12, suppose that F is free on a set $Z = \{z_j;\ j \in J\}$. Then there exist a finite subset $z_1, \dots, z_n$ of Z such that

$$u_1, \dots, u_m \in H = gp(z_1, \dots, z_n).$$

Since $u_1, \dots, u_m$ freely generate a free Abelian group modulo R, it follows that they generate, modulo F', a direct factor of F. Therefore they generate a direct factor of H modulo H'. Consequently there is an automorphism α of H/H' which maps $z_1H', \dots, z_mH'$ onto $u_1H', \dots, u_mH'$. Now every automorphism of the finitely generated free Abelian group H/H' is induced by an automorphism of H. This implies that there is a second set $w_1, \dots, w_n$ of free generators of H such that w_i is congruent to u_i modulo H' for $i = 1, \dots, m$. In other words,

$$w_1 = u_1d_1, \dots, w_m = u_md_m$$

where each $d_i \in H'$. Now observe that

$$W \cup (Z - \{z_1, \dots, z_n\}) \cup \{w_{m+1} = z_{m+1}\alpha, \dots, w_n = z_n\alpha\}$$

freely generates F. Let us set

$$V = (Z - \{z_1, \dots, z_n\}) \cup (\{w_{m+1}, \dots, w_n\}) = \{v_\ell : \ell \in L\}.$$

Since W generates F modulo R, each v_ℓ can be expressed in terms of the elements of W modulo R, i.e., using functional notation

$$v_\ell = f_\ell(w_1, \dots, w_m)r_\ell, r_\ell \in R.$$

Now put

$$Y = \{v_\ell^{-1}f_\ell : \ell \in L\}.$$

Then Y is a subset of R and $X = W \cup Y$ freely generates F, as required. □

We notice, adopting the notation and hypothesis above, that R is the normal closure in F of Y together with the derived group of F and that W freely generates, modulo R, the free Abelian group F/R.

We continue now to adopt the notation and hypothesis of Theorem 11.3.10.

Lemma 11.3.13. *Suppose that a and b generate, modulo R, a free Abelian subgroup of F. Then $aV(R)$ and $bV(R)$ do not commute.*

Proof. We can assume that $F = gp(a, b, R)$. Then, F/R is free Abelian on aR and bR. So, by Lemma 11.3.12, we can find a free basis $X = W \cup Y$ for F where $W = \{aa', bb'\}(a', b' \in F')$. Now let E be the subgroup of F generated by W. Then E is free on W. Furthermore, let ϕ be the homomorphism of F onto E defined by mapping all of Y to the identity and W identically onto itself. If we put $S = \phi(R)$, then $S = E'$ and therefore $V(S) = V(E')$. It follows that $D = E/V(S)$ is free of rank two in the product variety $\mathcal{V}\mathcal{A}$, where here $\mathcal{A}$ consists of the variety of all Abelian groups. It follows that if D'' denotes the second derived group of D, then D/D'' is free in the product variety $\mathcal{V}_e\mathcal{A}$, where $\mathcal{V}_e$ is the variety of Abelian groups in $\mathcal{V}$, which is therefore the variety of Abelian groups of exponent e, where we allow $e = \infty$. Consequently, if $e = 2$, then the dihedral group of order 8 is a quotient group of D and if e is not even, then there is an odd prime p such that the non-Abelian group of order p^3 is a quotient group of D.

Now ϕ induces a homomorphism ϕ_* of $F/V(R)$ onto $E/V(S)$. It follows therefore that there exists a prime p and a homomorphism θ of $F/V(R)$ onto a non-Abelian group P of order p^3. Since aa' and bb' generate E, $\theta(aa')$ and $\theta(bb')$ generate P. Since the Frattini subgroup of a nilpotent group contains the derived group, $a\theta$ and $b\theta$ generate P and therefore they do not commute. Consequently, neither do a and b. This completes the proof of Lemma 11.3.13. □

Next we prove

Lemma 11.3.14. *If an element $tV(R) \in F/V(R)$ commutes with a nontrivial element of $R/V(R)$, then $t \in R$.*

Proof. Suppose the contrary. We may assume that $F = gp(t, R)$, where t is of infinite order modulo R. It follows from a theorem of Wagner [253] (see also Federer and Johnson [71]) that there is a free basis for F which takes the form

$$\{t\} \cup Y = \{y_i : i \in I\},$$

with the property that R is the normal closure of Y. Put

$$y_{i,j} = t^{-j} y_i t^j (i \in I, j = 0, \pm 1, \pm 2, \dots).$$

Then $R/V(R)$ is a free group in the variety $\mathcal{V}$, freely generated, modulo $V(R)$ by the $y_{i,j}$. Suppose that $w \in R$ and that $t^{-1}wt = wv$ where $v \in V(R)$. Express $wV(R)$ as a word in the free generators $y_{i,j}V(R)$ of $R/V(R)$. Choose n sufficiently large to ensure that the second subscripts that appear on conjugating $wV(R)$ by $t^n V(R)$ do not appear in $wV(R)$. Now we have $t^{-n}wt^nV(R) = wV(R)$. We now define a homomorphism σ of $R/V(R)$ into $R/V(R)$ by mapping all of the generators that arise in the word $t^{-n}wt^nV(R)$ to the identity and all of the other generators to themselves. Then $\sigma(wV(R))$ is simultaneously $wV(R)$ and the identity. This is possible only if $w \in V(R)$, which completes the proof of Lemma 11.3.14. □

Lemma 11.3.15. *The centralizer $C/V(R)$ of an element $aV(R)$ in $F/V(R)$, where $a \notin R$, is cyclic.*

Proof. We can assume at the outset that $F = CR$. By Lemma 11.3.14, $C \cap R \leq V(R)$, i.e., $C \cap R = V(R)$. Therefore the canonical homomorphism of $C/V(R)$ onto $CR/R = F/R$ is a monomorphism. By Lemma 11.3.13, $C/V(R)$ is locally, infinite cyclic. It follows that $R/V(R)$ is not a torsion group and hence that $V(R) \leq R' \leq F'$. Now

$$F/R = CR/R \cong C/C \cap R = C/V(R)\,.$$

This implies that there is a canonical homomorphism of $C/V(R)$ into F/F' obtained by mapping C identically into F and then taking its image modulo F'. Since $C/V(R)$ is locally cyclic this means that either $C/V(R)$ is infinite cyclic or else that $C \leq F'$. However $C \leq F'$ implies that $C \leq R$, which in turn implies that $F = R$, a contradiction. This completes the proof of Lemma 11.3.15 and therefore also that of Theorem 11.3.10. □

Next we prove

Theorem 11.3.11. *Let F be a non-Abelian free group, let R be a normal subgroup of F and let $\mathcal{V}$ be a variety of groups in which the free groups are U-groups. If F/R is torsion-free, then $F/V(R)$ is a U-group.*

Proof. Suppose that n is a positive integer and that

$$(xV(R))^n = (yV(R))^n, x, y \in F\,.$$

If $x \notin R$, then $y \notin R$ because F/R is torsion-free. By Theorem 11.3.10, the centralizer C of $x^nV(R)$ is infinite cyclic. Since both $xV(F)$ and $yV(F)$ belong to C and C is a U-group, $xV(F) = yV(F)$ as required. On the other hand, if $x \in R$, then $y \in R$ and the required result follows immediately since $R/V(R)$ is a U -group. □

We come now to the proof of Theorem 11.3.7.

Theorem 11.3.12 (Theorem 11.3.7). *Let $\mathcal{V}$ be a nontrivial variety of groups having infinite exponent. Suppose that F is a non-Abelian free group and R is a nontrivial normal subgroup of F such that*
(1) *F/R is a U-group and contains a rank 2 free Abelian subgroup;*
(2) *$R/V(R)$ is a U-group.*
Then F/R is not discriminating.

Proof. If $\mathcal{V}$ is the variety of all groups, then $V(R) = 1$ and $F/V(R) \cong F$ is a non-Abelian CT-group and so is nondiscriminating by Lemma 11.3.1. Assume for the rest of the proof that $\mathcal{V}$ is not the variety of all groups. As in the proof of Lemma 11.3.13, there is a free subgroup E of F of rank 2. So as in the proof of Lemma 11.3.13, E is free on the subset $W = \{aa', bb'\}$ of F. Moreover if

$$\varphi\colon F \to E$$

is defined on the free basis X, which is the disjoint union of W and Y, for F by $\varphi(x) = x$ if $x \in W$ and $\varphi(x) = 1$ if $x \in Y$, then putting $S = \varphi(R)$, we get $S = E'$.

To see that, we first refer to Lemma 11.3.12 to see that Y is contained in R. So F/R is generated by $aa'R$ and $bb'R$. As in the proof of Lemma 11.3.13, F/R is free Abelian of rank 2. Thus $\{aa'R, bb'R\}$ is a basis. It follows that R is normal closure in F of the commutator $[aa', bb']$ and Y. Indeed, if N is that normal closure, then $N \leq R$. Thus we have the short exact sequence of Abelian groups (written additively)

$$0 \to R/N \to F/N \to F/R \to 0\,.$$

But F/N is free Abelian of rank 2. Finitely generated Abelian groups are Hopfian; so the kernel R/N must be trivial and consequently $R = N$.

Now φ induces a homomorphism φ_* of $F/V(R)$ onto $E/V(S) = E/V(E')$. $E/V(E')$ is generated by $aa'V(E)$ and $bb'V(E)$ and so it is the image under φ_* of the subgroup H of $F/V(R)$ generated by $aa'V(R)$ and $bb'V(R)$.

Now $E/V(E')$ does not belong to the variety $\mathcal{V}$ (Note this is false if $\mathcal{V}$ is the variety of all groups!) To see this, suppose the contrary. Then $V(E) \subseteq V(E')$ and thus $V(E) = V(E')$. Now $F_\omega(\mathcal{V}) = E'/V(E') = E'/V(E) \subset E/V(E) = F_2(\mathcal{V})$. From the facts that every variety is generated by its free group of infinite rank and that $F_\omega(\mathcal{V}) \subset F_2(\mathcal{V})$, we conclude that $\mathcal{V}$ is generated by $F_2(\mathcal{V})$. Now $E/V(E) = E/V(E')$. Thus if $\mathcal{A}$ is the Abelian variety, then $F_2(\mathcal{V}\mathcal{A}) \in \mathcal{V}$. This follows from the fact that the verbal subgroup of a group G corresponding to a product variety $\mathcal{V}\mathcal{W}$ is $V(W(G))$. (See 21.12, p. 38 of [202]) Thus $F_2(\mathcal{V}\mathcal{A}) = E/V(E') = E/V(E) = F_2(\mathcal{V}) \in \mathcal{V}$. Then since $\mathcal{V}$ is not the variety of all groups, Corollary 25.33 on p. 67 of [202] applies so that $\mathcal{V}\mathcal{A}$ is generated by $F_2(\mathcal{V}\mathcal{A})$. But clearly $\mathcal{V} \subseteq \mathcal{V}\mathcal{A}$ and since $F_2(\mathcal{V}\mathcal{A}) \in \mathcal{V}$, we must also have $\mathcal{V}\mathcal{A} \subseteq \mathcal{V}$. Therefore $\mathcal{V}\mathcal{A} = \mathcal{V}$. However, we also have $\mathcal{V}\mathcal{E} = \mathcal{V}$. This contradicts unique factorization in product varieties which follows from a theorem of Bernard, Hanna, and Peter Neumann (See Theorem 23.23. on p. 57 of [202]). They showed that the class of all proper varieties of groups is a free monoid where the binary operation is the product of varieties. This contradiction proves our claim that $E/V(E')$ does not belong to the variety $\mathcal{V}$.

Given the above, the subgroup H of $F/V(R)$ generated by $aa'V(R)$ and $bb'V(R)$ also cannot lie in $\mathcal{V}$. If we can show that the centralizer of every nontrivial element of $F/V(R)$ lies in $\mathcal{V}$, then we would be finished by Lemma 11.3.11. Since $\mathcal{V}$ has infinite exponent, it contains the infinite cyclic groups. Thus, by Theorem 11.3.10, the centralizer of every element of $F/V(R)$ not in $R/V(R)$ lies in $\mathcal{V}$. Now suppose $\rho = rV(R)$ lies in $R/V(R)$ and is nontrivial (r lies in R but not in $V(R)$). Let $C(\rho) = \{\tau\colon \tau\rho = \rho\tau\}$ be its centralizer in $F/V(R)$. Suppose $\tau = tV(R)$ lies in $C(\rho)$. Then, by Lemma 11.3.14, t lies in R so $\tau = tV(R)$ lies in $R/V(R)$. Since τ was an arbitrary element of $C(\rho)$, we conclude that $C(\rho)$ is contained in $R/V(R)$. But, of course, $R/V(R) \in \mathcal{V}$. Hence its subgroup $C(\rho)$ must also lie in $\mathcal{V}$ and we are finished. □

The next result is an easy consequence of the above results.

Theorem 11.3.13. *Let F be a non-Abelian free group and let R be a nontrivial, normal subgroup of F such that F/R is infinite and torsion-free. Then $F/\gamma_c(R)$ is not discriminating, for every $c > 1$, where here $\gamma_c(R)$ denotes the cth term of the lower central series of R.*

By Theorem 11.3.10 there exist elements in $F/\gamma_c(R))$ whose centralizers are infinite cyclic. Moreover, the centralizer in $F/\gamma_c(R)$ of a nontrivial element in $R/\gamma_c(R)$ is contained in $R/\gamma_c(R)$ and hence is nilpotent of class at most c. The fact that centralizers in $F/\gamma_c(R))$ can be cyclic implies that $F/\gamma_c(R))$ is not nilpotent. These remarks show that Lemma 11.3.11 applies and the desired result follows.

From the ideas already discussed we can deduce in a straightforward manner Theorem 11.3.8.

Theorem 11.3.14. *Let $\mathcal{V}$ and $\mathcal{U}$ be two varieties. If the variety $\mathcal{V}$ contains the infinite cyclic group, then the non-Abelian free groups in the product variety $\mathcal{U}\mathcal{V}$ are not discriminating.*

Finally we come to Theorem 11.3.9.

Theorem 11.3.15. *For all sufficiently large primes p, the non–Abelian free groups G in the variety of all groups of exponent p are not discriminating.*

Proof. As S. Adian [3] has proved that the centralizers of the nontrivial elements in the groups G involved here are cyclic, Lemma 11.3.11 applies here. □

These results raise the following question.

Question. *Does there exists a non-Abelian, discriminating, relatively free group?*

11.4 Squarelike groups and axiomatic properties

As mentioned in Section 11.1, in order to better capture the axiomatic properties of discriminating groups the squarelike groups were introduced. We repeat the definition.

Definition 11.4.1 ([76]). The group G is *squarelike* provided $G \times G$ has the same universal theory as G, that is $G \times G \equiv_\forall G$.

From Theorem 11.1.1 it follows that if G is discriminating then it has the same universal theory as its direct square. Hence, every discriminating group is squarelike. In this section we will prove that the converse is not true, that is the class of discriminating groups is a proper subclass of the class of squarelike groups. This will be proved in two different ways. First we explicitly construct a nondiscriminating group which is provably squarelike. Second we will prove that the class of squarelike groups is axiomatic but the class of discriminating groups is not axiomatic. Although the class of discriminating groups is a proper subclass of the class of squarelike groups we will prove that they actually correspond in the case of finite presentation, That is, although a general squarelike group need not be discriminating a finitely presented squarelike group is discriminating.

Recall that a class of groups is *axiomatic* if it is the model class of a set of first order sentences (axioms). It can be shown that a class of groups is axiomatic if it is closed under ultraproducts and elementary equivalence (see Chapter 4). We first show that the class of squarelike groups is axiomatic.

Theorem 11.4.1 ([78, 67]). *The class of squarelike groups is axiomatic.*

Proof. In [78] the fact that the squarelike groups are axiomatic was proved by using the closure properties. That is it was proved directly that the class of squarelike groups is closed under elementary equivalence and ultraproducts. However subsequently V. H. Dyson [67] discovered an explicit axiom schema for the class of squarelike groups. To each ordered pair (w, u) of finite tuples of words on a fixed but arbitrary finite set $\overline{x} = \{x_1, \dots, x_n\}$ of distinct variables and their formal inverses we assign the following sentence $\sigma(\overline{w}, \overline{u})$ of L_0.

$$\forall \overline{\mathbf{x}} \bigwedge_i \bigwedge_i (w_i(\overline{\mathbf{x}}) = 1) \to \bigvee_j \bigvee_j \left(u_j(\overline{\mathbf{x}}) = 1\right)$$

$$\to \bigvee_j \bigvee_j \exists \overline{\mathbf{x}} \bigwedge_i \bigwedge_i (w_i(\overline{\mathbf{x}}) = 1) \to \left(u_j(\overline{\mathbf{x}}) = 1\right) .$$

The contrapositive of $\sigma(\overline{x}, \overline{u})$ is (up to logical equivalence) the sentence $\tau(w,u)$ asserting

$$\bigwedge_j \bigwedge_j \exists \overline{\mathbf{x}} \bigwedge_i \bigwedge_i (w_i(\overline{\mathbf{x}}) = 1) \wedge \left(u_j(\overline{\mathbf{x}}) \neq 1\right)$$

$$\to \exists \overline{\mathbf{x}} \bigwedge_i \bigwedge_i (w_i(\overline{\mathbf{x}}) \neq 1) \wedge \bigwedge_i \bigwedge_i \left(u_j(\overline{\mathbf{x}}) \neq 1\right) .$$

The class of squarelike groups is precisely the model class of the group axioms and the sentences $\sigma(\overline{w}, \overline{u})$ given above. □

Next we recall the following result given in Section 11.1 classifying the finitely generated equationally Noetherian discriminating groups.

Theorem 11.4.2 (Theorem 11.1.3 [FGMS 1]). *Let G be a finitely generated equationally Noetherian group. Then G is discriminating if and only if* $\mathrm{qvar}(G) = \mathrm{ucl}(G)$.

For squarelike groups we now show that the equationally noetherian condition is unnecessary. An argument similar to the proof of Theorem 11.4.2 shows that the class of all groups H for which there exists a discriminating group G_H elementarily equivalent to H is axiomatic. Clearly, this class is the least axiomatic class containing the discriminating groups. We now show that being squarelike is equivalent to being universally equivalent to a discriminating group. We need the following three results from Chapter 4.

Lemma 11.4.1. *Let G and H be groups. Then every universal sentence of L true in G is also true in H if and only if H is embeddable in an elementary extension $^*G$ of G.*

Lemma 11.4.2. *Let $\mathcal{X}$ be an axiomatic class of groups. If $\mathcal{X}$ is closed under direct products of two factors, then $\mathcal{X}$ is closed under direct products of an arbitrary number (finite or infinite) of factors*

Lemma 11.4.3. *Let $\mathcal{X}$ be a universally axiomatizable class of groups. Then $\mathcal{X}$ is a quasivariety if and only if the trivial group 1 lies in $\mathcal{X}$, (equivalently: $\mathcal{X}$ is nonempty), and $\mathcal{X}$ is closed under direct products.*

Now we can state and prove the result on squarelike groups.

Theorem 11.4.3. *Let G be a group. Then the following three conditions are equivalent.*
(1) *G is squarelike;*
(2) $\mathrm{ucl}(G) = \mathrm{qvar}(G)$;
(3) *G is universally equivalent to a discriminating group.*

Proof. (1) $\Longrightarrow$ (2). Assume that G is squarelike. Since $\{1\} \subseteq G$ we must have every universal sentence true in G true also in the trivial group $\{1\}$. Hence $1 \in \mathrm{ucl}(G)$. Now let $H, K \in \mathrm{ucl}(G)$. Then by Lemma 4.3.1 there are elementary extensions G_H and G_K of G such that H is embeddable in G_H and K is embeddable in G_K. Then $H \times K$ is embeddable in $G_H \times G_K$. Since elementary equivalence is preserved in direct products, $G_H \times G_K$ is elementarily equivalent to $G \times G$. Hence $Th_\forall(G_H \times G_K) = Th_\forall(G \times G)$. But $Th_\forall(G \times G) = Th_\forall(G)$ since G is squarelike. Hence every universal sentence true in G must also be true in $H \times K$. It follows that $\mathrm{ucl}(G)$ is closed under direct products of two factors. By Lemma 4.3.3, $\mathrm{ucl}(G)$ is closed under arbitrary direct products. Then, by Lemma 4.4.1, $\mathrm{ucl}(G)$ is a quasivariety, which must clearly coincide with the least quasivariety $\mathrm{qvar}(G)$ containing G.

(2) $\Longrightarrow$ (3) Assume $\mathrm{ucl}(G) = \mathrm{qvar}(G)$. Let I be an infinite index set. Let $G \in \mathrm{qvar}(G)$, then we must have $G^I \in \mathrm{qvar}(G) = \mathrm{ucl}(G)$. Hence every universal sentence true in G must also be true in G^I. But every universal sentence true in G^I must also be true in G since G embeds in G^I, It follows that $Th_\forall(G) = Th_\forall(G^I)$. Finally note that G^I is discriminating since $G^I \times G^I$ is isomorphic to G^I.

(3) $\Longrightarrow$ (1) Suppose that H is a discriminating group and $Th_\forall(G) = Th_\forall(H)$. Since H is discriminating it is squarelike and thus $\mathrm{ucl}(H) = \mathrm{qvar}(H)$ by the first implication of this proof. Then $G \in \mathrm{ucl}(H) = \mathrm{qvar}(H)$ implies that $G \times G \in \mathrm{qvar}(H) = \mathrm{ucl}(H) = \mathrm{ucl}(G)$. Hence every universal sentence true in G must also be true in $G \times G$ and so G is squarelike. □

Corollary 11.4.1. *Let G be a finitely generated, equationally noetherian group. Then G is squarelike if and only if it is discriminating.*

Therefore the squarelike groups are the universal closure of the discriminating groups. In fact as we prove in the next section they are the axiomatic closure.

We are now in a position to prove that a general squarelike group need not be discriminating.

Theorem 11.4.4. *The class of discriminating groups is a proper subclass of the class of squarelike groups.*

Proof. As remarked we will prove this result in two different ways. First we show this by explicitly constructing a nondiscriminating group which is provably squarelike. After presenting this example we will give an alternate proof using an axiomatic argument. □

We first present the example. Let H be the subgroup of the group of all permutations of the set $\mathbb{Z}$ of integers generated by the 3-cycle $\xi = (012)$ and the translation $\eta(n) = n+1$ for all $n \in \mathbb{Z}$. The group $H = \langle \xi, \eta \rangle$ can also be described as the semidirect product of the group M, of all even parity permutations within the group N of all permutations of the set $\mathbb{Z}$ of integers which move only finitely many integers, by an infinite cyclic group $C = \langle c; \ \rangle$ where the automorphism $\alpha(c) \colon M \to M$ acts by $\alpha(c)(\pi)(n) = \pi(n-1)+1$. (We say that $\alpha(c)$ acts by translation by 1.) Note that any bijection between $\mathbb{N}$ and $\mathbb{Z}$ induces an isomorphism between the infinite alternating group A_ω and M. H first appeared in print in the same paper of B. H. Neumann in which the uncountably many nontrivially discriminating groups, G_n, exhibited before, were introduced. B. H. Neumann observed that (independent of n) if K_0 is the restricted direct product of the family $(A_{n(r)})_{r \in \mathbb{N}}$ of alternating groups, then the quotient of $G = G_n$ modulo K_0 is isomorphic to H.

Theorem 11.4.5. *The group H is squarelike but nondiscriminating.*

We prove this theorem with a series of lemmas.

Lemma 11.4.4. *The group H is centerless.*

Proof. Since an infinite cyclic group is free, it is projective and the sequence $\{1\} \to M \to H \to H/M \to \{1\}$ must split. Then H is a semidirect product and there is a retraction $H \to \langle \eta \rangle$. Every element of H is uniquely of the form $a\eta^m$ where $a \in M$ and $m \in \mathbb{Z}$. Given a nontrivial element $a \in M \cong A_\omega$ there is clearly an element $b \in M$ which does not commute with it since A_ω is a non-Abelian simple group; hence, it is centerless. So a nontrivial central element in H must be of the form $a\eta^m$ with $m \neq 0$. We cannot have $a = 1$ since $\eta^m \xi \eta^{-m} = (m\; m+1\; m+2) \neq (012) = \xi$. Thus, a nontrivial central element in H must be of the form $a\eta^m$ with $a \in M \backslash \{1\}$ and $m \neq 0$. But, if any $a\eta^m$ were central in H, then both $a\eta^m$ and η would lie in $C_H(\eta^m)$ so $a \in C_H(\eta^m)$. Thus, a commutes with η^m and η^{-m}. Then $\eta^{|m|} a \eta^{-|m|} = a$. But that is absurd since conjugation by $\eta^{|m|}$ moves every integer in the disjoint cycle decomposition of a by $|m|$ units. So if n is the largest integer moved by a it is replaced by $n + |m|$ – a contradiction. □

We say that a group G is *subdirectly irreducible* or *monolithic* provided it has a unique minimal normal subgroup $M \neq \{1\}$. Following Hanna Neumann [202] we shall call the unique minimal normal subgroup of a monolithic group its *monolith*.

Lemma 11.4.5. *H is monolithic with monolith M.*

Proof. We must show every $\{1\} \subsetneqq N \triangleleft H$ contains M. In the following we use $\langle a\rangle^M$ to denote the normal closure of a in$M$.

Case I: $M \cap N \neq \{1\}$. Let $a \in M \cap N$, $a \neq 1$. Then $\langle a\rangle^M \leq \langle a\rangle^H \leq N$. But $\langle a\rangle^M = M$ since $M \cong A_\omega$ is simple.

Case II: $M \cap N = \{1\}$. We shall prove this case cannot occur. If $M \cap N = \{1\}$, then $MN \subset H$ and $MN \cong M \times N$. We have $N \cong MN/M \leq H/M$ is infinite cyclic since $N \neq \{1\}$. Suppose $a\eta^m$ generates N with $a \in M$. We first claim $a = 1$. For suppose $a \neq 1$. Since $a\eta^m \in C_H(M)$ we would have $a\eta^m \in C_H(a)$ and $\eta^m \in C_H(a)$ would follow. But, as we saw before, for $a \neq 1$, this is absurd. Thus, $N = \langle \eta^m\rangle$ with $m \neq 0$. But then η^m would be a nontrivial central element in $H = \langle M, \eta\rangle$. The contradiction shows that Case II cannot occur so M is the monolith of H. □

Lemma 11.4.6. *Every nontrivial normal subgroup*

$$\{1\} \neq N \trianglelefteq H \times H$$

must intersect $M \times M$ nontrivially.

Proof. Assume $\{1\} \neq N \trianglelefteq H \times H$. Suppose not to deduce a contradiction. Let $p_i\colon H \times H \to H$ be the projection onto the i-th coordinate, $i = 1, 2$. We claim that $M \cap p_i(N) = \{1\}$ for $i = 1, 2$. To see that suppose that

$$(a_1, b_2\eta^{m(2)}) \in N$$

where $a_1 \in M\backslash\{1\}$ and $m(2) \neq 0$. Then $\langle b_2\eta^{m(2)}\rangle^H \supset M$ so $a_1 \in \langle b_2\eta^{m(2)}\rangle^H$ and there are $h_1, \ldots, h_k$; $\varepsilon(1), \ldots, \varepsilon(k)$ such that

$$a_1 = h_1^{-1}(b_2\eta^{m(2)})^{\varepsilon(1)}h_1 \ldots h_k^{-1}(b_2\eta^{m(2)})^{\varepsilon(k)}h_k.$$

So that

$$\begin{aligned}(1, h_1)^{-1}(a_1, b_2\eta^{m(2)})^{\varepsilon(1)}(1, h_1)\ldots(1, h_k)^{-1}(a_1, b_2\eta^{m(2)})^{\varepsilon(k)}(1, h_k)\\ = (a_1^{\varepsilon(1)+\cdots+\varepsilon(k)}, a_1)\end{aligned}$$

lies in N and also $M \times M$. Thus, $M \cap p_1(N) \neq \{1\}$ contradicts our assumption that $N \cap (M \times M)$ is trivial. Similarly, if $(b_1\eta^{m(1)}, a_2) \in N$ with $a_2 \in M\backslash\{1\}$ we get a contradiction to the triviality of $N \cap (M \times M)$. Thus, $M \cap p_i(N) = \{1\}$ for $i = 1, 2$ as claimed. From this we see that $p_i(N) = \{1\}$ for $i = 1, 2$ since M is the monolith of H. Then $N = \{1\}$ contrary to hypothesis. This shows that N must intersect $M \times M$ nontrivially. □

Lemma 11.4.7. *Every nontrivial normal subgroup* $\{1\} \neq N \trianglelefteq H \times H$ *must contain either* $M \times \{1\}$ *or* $\{1\} \times M$.

Proof. Suppose (a_1, a_2) is a nontrivial element of $N \cap (M \times M)$. If $a_2 = 1$, then

$$M \times \{1\} = \langle (a_1, 1) \rangle^{M \times \{1\}} \subseteq \langle (a_1, 1) \rangle^{H \times H} \subseteq N.$$

Similarly, if $a_1 = 1$, then $\{1\} \times M \subseteq N$. So we may assume that both a_1 and a_2 are nontrivial. Suppose that the order of a_2 is m. Since $\langle a_1 \rangle^H = M$ we can find $b \in M \cong A_\omega$ of order $n > 1$ prime to m. Now, proceeding in a similar manner as before,

$$b = h_1^{-1} a_1^{\varepsilon(1)} h_1 \dots h_k^{-1} a_1^{\varepsilon(k)} h_k.$$

Then

$$(b, a_2^{\varepsilon(1)+\dots+\varepsilon(k)}) = (h_1, 1)^{-1} (a_1, a_2)^{\varepsilon(1)} (h_1, 1) \dots (h_k, 1)^{-1} (a_1, a_2)^{\varepsilon(k)} (h_k, 1)$$

lies in N. Hence, so does

$$(b, a_2^{\varepsilon(1)+\dots+\varepsilon(k)})^m = (b^m, 1)$$

and $b^m \neq 1$. Then, arguing as before, $M \times \{1\} \subset N$. □

Lemma 11.4.8. *H is nondiscriminating.*

Proof. If H were discriminating, then there would be a homomorphism $\varphi\colon H \times H \to H$ which did not annihilate either of the elements $(\xi, 1)$ or $(1, \xi)$. Then neither $M \times \{1\}$ nor $\{1\} \times M$ can be contained in the kernel of φ. It follows that the kernel of φ is trivial and φ is an embedding of $H \times H \hookrightarrow H$. That is impossible by Lemma 11.4.6. The contradiction shows that H is not discriminating. □

We now complete the proof of Theorem 11.5.1 by showing that H is squarelike.

Lemma 11.4.9. *H is squarelike.*

Proof. Let us consider a group $G = G_n$ in the family of nontrivially discriminating groups of B. H. Neumann discussed in Section 11.2.3. $G \subseteq \prod_{r \in \mathbb{N}} A_{n(r)} \subseteq A_\omega^{\mathbb{N}}$. Let K_0 be, as before, the restricted direct product of the family $(A_{n(r)})_{r \in \mathbb{N}}$ and let K be the restricted direct power of the family $(A_\omega)_{r \in \mathbb{N}}$. Under the epimorphism

$$A_\omega^{\mathbb{N}} \to A_\omega^{\mathbb{N}}/K$$

the group G maps to $GK/K \cong G/(G \cap K) = G/K_0 \cong H$. Moreover, ${}^*A_\omega = A_\omega^{\mathbb{N}}/K$ is the reduced power of A_ω modulo the filter D of cofinite subsets of $\mathbb{N}$. Let $d\colon A_\omega \to {}^*A_\omega$ be the canonical embedding. We claim that A_ω and ${}^*A_\omega$ are universally equivalent with respect to L_0. Since A_ω embeds in ${}^*A_\omega$, every universal sentence true in ${}^*A_\omega$ must also be true in A_ω. Suppose that that u is a universal sentence true in A_ω. Since A_ω is squarelike, it follows from Lemma 11.4.6 that u is a consequence of a set Q of quasi-identities true in A_ω. But Q is preserved in the reduced product ${}^*A_\omega$ so its logical consequence u must also hold in ${}^*A_\omega$. Thus, A_ω and ${}^*A_\omega$ satisfy precisely the same universal sentences of L_0 and are indeed universally equivalent. Since $A_\omega \cong M \subseteq H$, every universal sentence true in H must also be true in A_ω. Suppose u is a universal sentence true in A_ω. Then u is also true in ${}^*A_\omega$. But, since H is embedded in ${}^*A_\omega$, u must also be true in H. Thus, A_ω and H satisfy precisely the same universal sentences and are universally equivalent. Recalling that A_ω is discriminating, it follows, again by Lemma 11.4.6, that H is squarelike. □

We now prove Theorem 11.4,3 proved in an alternative manner using the fact that the squarelike groups are axiomatic.

Theorem 11.4.6. *The class of discriminating groups is not axiomatic. Since the class of squarelike groups is axiomatic it follows that the class of discriminating groups is a proper subclass of the class of squarelike groups.*

Proof. To prove that the discriminating groups are nonaxiomatic we present the following specific example of a nondiscriminating group (however squarelike) which is elementarily equivalent to a discriminating group. Hence the discriminating groups are not closed under elementary equivalence and therefore cannot be axiomatic. □

To construct the example we need the following lemma. Reduced products were defined in Chapter 4.

Lemma 11.4.10. *The class of discriminating groups is closed under forming reduced products but not under direct unions.*

Proof. Let I be a nonempty set and let $(G_i)_{i\in I}$ be a family of discriminating groups indexed by I. Suppose D is a proper filter on I and let $\prod_{i\in I} G_i/D$ be a reduced product (see Chapter 4). We must show that $\prod_{i\in I} G_i/D$ discriminates $(\prod_{i\in I} G_i/D) \times (\prod_{i\in I} G_i/D)$. It suffices to show that $\prod_{i\in I} G_i/D$ discriminates $\prod_{i\in I}(G_i \times G_i)/D$. Let $[(f_1, g_1)]_D, \ldots, [(f_n, g_n)]_D$ be finitely many nontrivial elements of $\prod_{i\in I}(G_i \times G_i)/D$. Then for all j with $1 \le j \le n$,

$$\{i \in I : (f_j(i), g_j(i)) = 1\} \notin D.$$

For each $i \in I$, let $J(i) = \{j : (f_j(i), g_{j'}(i)) \neq 1\}$. If $J(i)$ is empty, let $\varphi_i : G_i \times G_i \to G_i$ be projection onto the first coordinate. Otherwise, choose $\varphi_i : G_i \times G_i \to G_i$ such that $\varphi_i(f_j(i), g_j(i)) \neq 1$ for all $j \in J(i)$. This is possible since each G_i is discriminating. We then get an induced map

$${}^*\varphi : \prod_{i\in I}(G_i \times G_i)/D \to \prod_{i\in I} G_i/D.$$

We claim that ${}^*\varphi$ does not annihilate any of $[(f_j, g_j)]_D; j = 1, 2, \ldots, n$. Suppose this is false. Fix j and write $[(f, g)]_D$ for $[(f_j, g_j)]_D$. If ${}^*\varphi$ annihilated $[(f, g)]_D$, then

$$\{i \in I : \varphi_i(f(i), g(i)) = 1\} \in D.$$

But we claim that

$$\{i \in I\colon \varphi_i(f(i), g(i)) = 1\}$$

coincides with

$$\{i \in I\colon (f(i), g(i)) = 1\}.$$

For if $(f(i), g(i)) = 1$, then certainly $\varphi_i(f(i), g(i)) = 1$, while, on the other hand, if $(f(i), g(i)) \neq 1$ then $j \in J(i)$ and $\varphi_i(f(i), g(i)) \neq 1$ by the very choice of φ_i. But then

$$\{i \in I\colon \varphi_i(f(i), g(i)) = 1\} = \{i \in I\colon (f(i), g(i)) = 1\}$$

does not lie in D; so, $\varphi([(f, g)]_D) \neq 1$, contrary to our assumption. Thus, $\prod_{i\in I} G_i/D$ discriminates $\prod_{i\in I}(G_i \times G_i)/D$ and $\prod_{i\in I} G_i/D$ is discriminating whenever G_i is discriminating for all $i \in I$. In other words, the class of discriminating groups is, as claimed, closed under reduced products. (In particular, it is closed under direct products.).

We must now show that it is not closed under direct unions. Recall Theorem 11.2.2:

Theorem (Theorem 11.2.2). *Let A be a torsion Abelian group. Suppose that for each prime p, the p-primary component of A modulo its maximal divisible subgroup contains no nontrivial element of infinite p-height. Then A is discriminating if and only if the following two conditions are satisfied for each prime p.*

(1) *For every positive integer k, $\rho^{(1)}[p, k](A)$ is either 0 or ∞.*
(2) *The rank of the maximal divisible subgroup of the p-primary component of A is either zero or infinite.*

Here the *rank* of a divisible Abelian p-group is the maximal number of direct summands of the quasicyclic group $\mathbb{Z}_{p^\infty}$; moreover, the p-height of an element a of an Abelian p-group A is (with respect to A) the maximal positive integer n, if it exists, such that the equation $p^n x = a$ has a solution in A. Now for each positive integer k let M_k be a free module of countably infinite rank over the ring $\mathbb{Z}/(2^k\mathbb{Z})$. Let M be the direct sum of the Abelian groups M_k as k varies over the positive integers. Let $D = \mathbb{Z}_{2^\infty}$ be a rank 1 divisible 2-group. Let A be the direct sum $M \oplus D$. Then $A/D \cong M$ is a torsion Abelian 2-group containing no nontrivial elements of infinite 2-height. By Theorem 11.2.2, A is not discriminating since its maximal divisible subgroup D has rank 1. (It clearly suffices to restrict ourselves to the prime 2 since, if p is an odd prime, the p-primary component of A is 0.) But A is the direct union of the family $A_k = M \oplus (\mathbb{Z}/2^k\mathbb{Z})$ of subgroups as k varies over the positive integers. Each A_k is discriminating since, clearly, $A_k \cong M$ for all k, and $M \cong M \times M$ is discriminating. Thus, the class of discriminating groups is not closed under direct unions □

From this last proof we can complete the proof that the class of discriminating groups is nonaxiomatic. By Szmielew's Theorem (Theorem 11.2.1), A and M as given above are elementarily equivalent since they both have infinite exponent and, if $p = 2$, then $\rho^{(i)}[p, k](A) = \rho^{(i)}[p, k](M) = \infty, i = 1, 2, 3$, for all k and, if $p \neq 2$, then

$\rho^{(i)}[p,k](A) = \rho^{(i)},[p,k](M) = 0, i = 1,2,3,$ for all k. Thus, the discriminating group M is elementarily equivalent to the nondiscriminating group A. It follows that the class of discriminating groups is not closed under elementary equivalence and therefore is not axiomatic.

Although the squarelike groups properly contain the discriminating groups they are very close and in fact correspond in the presence of finite presentation. In particular:

Theorem 11.4.7. *Let G be a finitely presented group. Then G is discriminating if and only if G is squarelike.*

Proof. Suppose G is a finitely presented group. If it is discriminating, it is squarelike. Now we suppose that G is squarelike and we must show that it is discriminating.

Let

$$G = \langle x_1,\ldots,x_n; R_1 = \cdots = R_m = 1\rangle$$

be a finite presentation for G where $R_i = R_i(x_1,\ldots,x_n)$ are words in $x_1,\ldots,x_n$. To show that G is discriminating we must show that G discriminates $G \times G$.

A finite presentation for $G \times G$ is then given by

$$\begin{aligned}G\times G = \langle x_1,\ldots,x_n,y_1,\ldots,y_n; R_1(x_1,\ldots x_n) = \cdots = R_m(x_1,\ldots,x_n) = 1,\\ R_1(y_1,\ldots y_n) = \cdots = R_m(y_1,\ldots,y_n) = 1, [x_i,y_j] = 1, i,j = 1,\ldots,n\rangle\end{aligned}$$

Now suppose $W_1,\ldots,W_k$ are nontrivial elements of $G \times G$. Then each $W_i, i = 1,\ldots k$, is given by $W_i = W_i(x_1,\ldots,x_n,y_1,\ldots,y_n)$, a word in the given generators of $G \times G$. Consider now the existential sentence

$$\begin{aligned}\exists x_1,\ldots,x_n,y_1,\ldots,y_n\Big(\Big(\bigwedge_{i=1}^{m} R_i(x_1,\ldots,x_n) = 1\Big) \wedge \Big(\bigwedge_{i=1}^{m} R_i(y_1,\ldots,y_n) = 1\Big)\\ \wedge \Big(\bigwedge_{i,j}[x_i,y_j] = 1\Big) \wedge \Big(\bigwedge_{i=1}^{k} W_i(x_1,\ldots,x_n,y_1,\ldots,y_n) \neq 1\Big)\Big).\end{aligned}$$

This existential sentence is clearly true in $G \times G$. Since G is squarelike, G and $G \times G$ have the same universal theory. Hence they have the same existential theory and therefore the above existential sentence is true in G. Therefore there exists elements $a_1,\ldots,a_n,b_1,\ldots,b_n$ in G such that $R_i(a_1,\ldots,a_n) = 1$ for $i = 1,\ldots,m$; $R_i(b_1,\ldots,b_n) = 1$ for $i = 1,\ldots,m$; $[a_i,b_j] = 1$ for $i,j = 1,\ldots,n$ and $W_i(a_1,\ldots,a_n,b_1,\ldots,b_n) \neq 1$ for $i = 1,\ldots,k$. The map from $G\times G$ to G given by mapping x_i to a_i and y_i to b_i for $i = 1,\ldots,n$ defines a homomorphism for which the images of $W_1,\ldots,W_k$ are nontrivial. Hence G discriminates $G \times G$ and therefore G is discriminating. □

11.5 The axiomatic closure property

The least axiomatic class containing the discriminating groups is the class of those groups G for which there exists a discriminating group H with $G \equiv H$. Here $\equiv$ denotes elementary equivalence. We call this class the *axiomatic closure* of the class of discriminating groups.

The next theorem shows that the squarelike groups are precisely the axiomatic closure of the discriminating groups.

Theorem 11.5.1 ([78]). *Let G be a group. Then G is squarelike if and only if there is a discriminating group H with $G \equiv H$. Hence the class of squarelike groups is the axiomatic closure of the class of discriminating groups.*

Proof. We outline the proof which uses ultrapowers and ultralimits.

If $G \equiv H$, then $G \equiv_{\forall} H$; so, G elementarily equivalent to a discriminating group implies G is squarelike.

Assume now that G_0 is a squarelike group. Then $G_0 \equiv_{\forall} (G_0 \times G_0)$ so there is an ultrapower $G_1 = G_0^I/D$ admitting an embedding $\varphi_1 : G_0 \times G_0 \to G_1$. Let $d_{0,1} : G_0 \to G_1$ be the canonical embedding and let $G_2 = G_1^I/D$. Then φ_1 induces $\varphi_2 : G_1 \times G_1 \to G_2$ and a diagram chase convinces one that the square

$$\begin{array}{ccc} G_0 \times G_0 & \xrightarrow{\varphi_1} & G_1 \\ {\scriptstyle d_{0,1} \times d_{0,1}} \downarrow & & \downarrow {\scriptstyle d_{1,2}} \\ G_1 \times G_1 & \xrightarrow[\varphi_2]{} & G_2 \end{array}$$

is commutative where $d_{1,2} : G_1 \to G_2$ is the canonical embedding.

We may iterate a countable infinity of times. Taking G_ω to be the ultralimit of G_0 with respect to the constant sequence of ultrafilters $(D, D, \ldots, D, \ldots)$, we get an embedding $G_\omega \times G_\omega \to G_\omega$ so that G_ω is discriminating. But the limit map $d_{0,\omega} : G_0 \to G_\omega$ is elementary. Hence, $G_0 \equiv G_\omega$. □

11.6 Further axiomatic information about discriminating and squarelike groups

Recall that a class of groups $\mathcal{G}$ is *axiomatic* if it is the model class of a set of first-order sentences. These sentences are the *axioms* for the class. The class is *finitely axiomatizable* if there is a finite set of axioms. In Section 11.4 we saw that the class of squarelike groups is axiomatic while in the last section we saw that the class of squarelike groups is the least axiomatic class containing the class of discriminating groups. In formal language the class of squarelike groups is the axiomatic closure of the discriminating groups. In this section we consider some further axiomatic aspects of the class of discriminating groups and squarelike groups. In partiulcar we prove that the theory of squarelike groups, while axiomatic, is not finitely axiomatizable and further the theory of squarelike groups is undecidable. Recall that a theory is decidable if there is an algorithm to determine if a sentence in that theory is true or false.

Theorem 11.6.1. *The theory of squarelike groups is undecidable.*

Proof. Given a theory T, we let $T_\forall$ be the set of all universal sentences which are consequences of T. From [129] it follows that the model class of $T_\forall$ consists of substructures of the models of T. If we let T be the theory of squarelike groups, then $T_\forall$ is the set of all universal sentences lying in T and the models of $T_\forall$ are precisely the subgroups of squarelike groups. If T were decidable, then so would be $T_\forall$.

We claim first that any group whatever is embeddable in a squarelike group.

To see this claim let G_0 be any group and let $G = G_0^\omega$ be the unrestricted direct power of a countable infinity of copies of G_0. Then $G \times G \cong G$ so that G is discriminating and therefore certainly squarelike. Since G_0 embeds in G, the claim is established.

But then $T_\forall$ coincides with the universal theory of the class of all groups. The undecidability of the universal theory of groups follows easily from the unsolvability of the word problem for finitely presented groups. Indeed, suppose that $\langle a_1, \ldots, a_m; R_1 = \cdots = R_n = 1\rangle$ is a finite presentation with unsolvable word problem. Let $\overline{x} = (x_1, \ldots, x_m)$ be an m-tuple of distinct variables and let $w(\overline{x})$ denote a word on those variables and their inverses. If the universal theory of groups were decidable, then we would have an algorithm to decide the truth of each of the sentences

$$\sigma_w : \forall \overline{x} \left(\bigwedge_{i=1}^{n} (R_i(\overline{x}) = 1) \rightarrow (w(\overline{x}) = 1 \right)$$

as $w(\overline{x})$ varies over all words on $x_1, \ldots, x_m$ and their inverses; hence, we would have an algorithm to solve the word problem for the given presentation. The contradiction shows that the universal theory of groups is undecidable and so $T_\forall$ is undecidable. Hence, T must be undecidable too. □

The next theorem answers a question posed by Paul Schupp; whether the theory of the squarelike groups is finitely axiomatizable.

Theorem 11.6.2. *The theory of squarelike groups is not finitely axiomatizable.*

Proof. Suppose to deduce a contradiction that $\{\sigma_1, \ldots, \sigma_n\}$ is a finite axiomatization of this theory. Then it would be axiomatized by the single sentence $\sigma : \sigma_1 \wedge \cdots \wedge \sigma_n$. Thus, the class of those L_0-structures which are not squarelike groups would also be axiomatic, being the model class of the single axiom σ. The class of those L_0-structures which are not squarelike groups would then be closed under taking ultraproducts. Let $\mathbb{N}$ be the set of positive integers. For each $n \in \mathbb{N}$ let G_n be an elementary Abelian 2-

group of order 2^n. Since $G_n$is finite and nontrivial, it is an L_0-structure which is not a squarelike group. Now let D be a nonprincipal ultrafilter on $\mathbb{N}$. Form the ultraproduct $G = \prod_{n\in\mathbb{N}} G_n/D$ of the family $(G_n)_{n\in\mathbb{N}}$ with respect to the ultrafilter D. Then G is an infinite Abelian 2-group. (Indeed, G has the power of the continuum.) In particular, $G \times G \cong G$, so that G is discriminating and hence squarelike. Thus, the class of L_0-structures which are not squarelike groups is not closed under ultraproducts and we have arrived at a contradiction. The contradiction shows that the theory of squarelike groups is not finitely axiomatixable. □

Observe that the proof of Theorem 11.6.2 actually shows that the class of Abelian squarelike groups is not finitely axiomatizable. In [79] an alternative proof of this was given which avoids the ultraproduct construction.

11.7 Varietal discrimination

There is an older notion of discrimination, which we call *varietal discrimination* that was also introduced by G. Baumslag jointly with Bernhard, Hanna and Peter Neumann (see [202]). The properties of this type of discrimination are described in the book of Hanna Neumann [202] and play a role in the structure of product varieties.

Definition 11.7.1. Let $\mathcal{V}$ be a variety of groups and $G \in \mathcal{V}$. G *discriminates* $\mathcal{V}$ provided to every finite set $w_i(\overline{x})$ of words in the variables $X \cup X^{-1}$ with none of the equations $w_i(\overline{x}) = 1$ a law in $\mathcal{V}$ there is a tuple $\overline{g}$ of elements of G such that simultaneously $w_i(\overline{g}) \neq 1$ in G for all i. Further a group G is *varietally discriminating* if it discriminates the variety it generates.

In [80] the relationship between these various notions of discrimination was studied. If we let $\mathcal{D}$ denote the class of discriminating groups, $\mathcal{S}$ the class of squarelike groups and $\mathcal{V}$ the class of varietally discriminating groups our main result is that

$$\mathcal{D} \subset \mathcal{S} \subset \mathcal{V}$$

and all inclusions are proper.

In order to prove that $\mathcal{S} \subset \mathcal{V}$ we introduce a further notion of discrimination called *q-discriminating* and then prove that a group is squarelike if and only if it is q-discriminating.

Theorem 11.7.1. *$\mathcal{D} \subset \mathcal{S} \subset \mathcal{V}$ and all inclusions are proper.*

Proof. $\mathcal{D} \subset \mathcal{S}$ is Theorem 11.1.1. In order to prove $\mathcal{S} \subset \mathcal{V}$ we reinterpret squarelike in terms of a discrimination property similar to varietal discrimination but involving quasivarieties rather than varieties. First though we omit the middle inequality and prove the following directly. □

Lemma 11.7.1. *$\mathcal{D} \subset \mathcal{V}$ and the inclusion is proper.*

Proof. Suppose $G \in \mathcal{D}$ so that G is a discriminating group. We show that it is varietally discriminating. Let $V(G)$ denote the group variety generated by G and let $F_\omega(V(G))$ be the free group of countably infinite rank in $V(G)$. Now

$$F_\omega(V(G)) \hookrightarrow G^I$$

for some index I (see [202]). G separates G^I and since G is discriminating it must also discriminate G^I. Hence given a finite set of words $w_i \in F_\omega(V(G) \subset G^I, i = 1, \ldots, n$, there exists a homomorphism $\phi \colon G^I \to G$ such that $\phi(w_i) \neq 1$ for $i = 1, \ldots, n$. Suppose that $\{a_1, a_2, \ldots., a_N, \ldots\}$ is a set of free generators for $F_\omega(V(G)$. Since the set of w_i is finite there is an N such that all w_i lie in $\langle a_0, \ldots, a_N \rangle$. Let $\overline{g} = \langle \phi(a_0), \ldots., \phi(a_N) \rangle$. Then $w_i(\overline{g}) \neq 1$ in G for all i. It follows that G is varietally discriminating. Therefore $G \in \mathcal{V}$ and hence $\mathcal{D} \subseteq \mathcal{V}$.

In [202] it was shown that if G is a residually finite p-group for infinitely many primes p then G is varietally discriminating. It follows that if G is a non-Abelian free nilpotent group then G is varietally discriminating and hence $G \in \mathcal{V}$. However from Theorem 11.3.1 we have that a torsion free nilpotent group is discriminating only if it is Abelian. Therefore a non-Abelian free nilpotent group is not discriminating so $G \notin \mathcal{D}$. It follows that the inclusion $\mathcal{D} \subset \mathcal{V}$ is proper. □

To handle the middle inclusion we reinterpret the property of being squarelike in terms of discrimination of quasivarieties. Recall that a universal sentence of the form

$$\forall \overline{x}\Big(\bigwedge_i (u_i(\overline{x}) = 1) \to (w(\overline{x}) = 1)\Big)$$

is called a *quasilaw* or *quasi-identity*. Note that every identity $\forall \overline{x}(w(\overline{x}) = 1)$ is equivalent to a quasilaw $\forall \overline{x}, y((y \cdot y^{-1} = 1) \to (w(\overline{x}) = 1))$. A *quasivariety* is the model class of a set of quasilaws. A group variety is hence a quasivariety but not conversely. For example the class of torsion-free groups is a quasivariety. In general quasivarieties are closed under subgroups and (unrestricted) direct products but not necessarily homomorphic images. If G is a group, then the quasivariety $\operatorname{qvar}(G)$ is the model class of those quasilaws true in G. The *universal closure* of G denoted $\operatorname{ucl}(G)$ is the model class of the set of all universal sentences true in G. Recall from Section 11.4 the following.

Theorem 11.7.2. *A group G is squarelike if and only if* $\operatorname{qvar}(G) = \operatorname{ucl}(G)$.

We now introduce a quasivarietal version of varietal discrimination.

Definition 11.7.2. Let $\mathcal{Q}$ be a quasivariety and let $G \in \mathcal{Q}$. Then G *q-discriminates* $\mathcal{Q}$ provided that given finitely many quasilaws

$$\forall \overline{x}\left(\bigwedge_i u_i(\overline{x} = 1) \to (w_j(\overline{x}) = 1)\right)$$

with the same antecedents and none of which holds in $\mathcal{Q}$ there exists a tuple $\overline{g}$ from G such that simultaneously $u_i(\overline{g}) = 1$ and $w_j(\overline{g}) \neq 1$ for all i, j. A group G is *q-discriminating* if G q-discriminates $\operatorname{qvar}(G)$.

This almost, except for having the same antecedents, appears to be the translation of varietal discrimination in terms of quasivarieties. The condition of having the same antecedents is necessary for the next theorem. Without this restriction the only q-discriminating group would be the trivial group. The above notion of q-discrimination characterizes squarelike groups. We introduce one more idea.

Definition 11.7.3. A group G is *q-algebraically closed* if and only if whenever a finite system

$$u_i(x_1,\dots,x_n) = 1$$
$$w_j(x_1,\dots,x_n) \neq 1$$

of equations and inequations has a solution in some group $H \in \operatorname{qvar}(G)$ it also has a solution in G.

We can now characterize squarelike groups in terms of q-discrimination.

Theorem 11.7.3. *Let G be a group. The following are pairwise equivalent.*
(1) *G is q-discriminating,*
(2) *G is q-algebraically closed,*
(3) *G is squarelike.*

Proof. (1) $\Longrightarrow$ (2) Suppose that G is q-discriminating and suppose that the finite system

$$u_i(x_1,\dots,x_n) = 1$$
$$w_j(x_1,\dots,x_n) \neq 1$$

of equations and inequations has a solution in some group $H \in \operatorname{qvar}(G)$. It follows that none of the quasilaws

$$\forall \overline{x}\Big(\bigwedge_i u_i(\overline{x} = 1) \to (w_j(\overline{x}) = 1)\Big)$$

can hold in $\operatorname{qvar}(G)$ since they fail in H. Since G is q-discriminating there is a tuple $\overline{g}$ from G such that $u_i(\overline{g}) = 1$ and $w_j(\overline{g}) \neq 1$ for all i, j. Then this tuple is a solution to the finite system above in G and therefore G is q-algebraically closed.

(2) $\Longrightarrow$ (3) Suppose that G is q-algebraically closed and suppose that the finite system

$$u_i(x_1,\dots,x_n) = 1$$
$$w_j(x_1,\dots,x_n) \neq 1$$

of equations and inequations has a solution in $G \times G$. Since quasivarieties are closed under direct products $G \times G \in \operatorname{qvar}(G)$ and hence since G is q-algebraically closed therefore the above system must have a solution in G. But this implies that $G \times G$ has the same universal theory as G and therefore G is squarelike.

(3) $\Longrightarrow$ (1) Suppose that G is squarelike. Then from Theorem 11.4.3 we have that $\mathrm{ucl}(G) = \mathrm{qvar}(G)$. Given finitely many quasilaws

$$\forall \overline{x}\Big(\Big(\bigwedge_i u_i(\overline{x}) = 1\Big) \to (w_j(\overline{x}) = 1)\Big)$$

with the same antecedents and none of which holds in $\mathrm{qvar}(G)$ they don't hold in $\mathrm{ucl}(G)$. Hence none is part of the universal theory of G and therefore there exists a tuple $\overline{g}$ from G such that simultaneously $u_i(\overline{g}) = 1$ and $w_j(\overline{g}) \neq 1$ for all i, j. It follows that G discriminates $\mathrm{qvar}(G)$ and hence G is q-discriminating.

Alternatively, recall that V.Dyson [67] gave, together with the group axioms, the following explicit set of axioms for the class of squarelike groups. To each ordered pair $\overline{w}, \overline{u}$ of finite tuples of words on a fixed but arbitrary finite set $\overline{x} = \{x_1, \dots, x_n\}$ of distinct variables and their formal inverses we assign the following sentence $\sigma(\overline{w}, \overline{u})$ of L_0.

$$\forall \overline{x}\Big(\bigwedge_i (w_i(\overline{x}) = 1) \to \bigvee_j (u_j(\overline{x}) = 1)\Big) \to \bigvee_j \forall \overline{x}\Big(\Big(\bigwedge_i w_i(\overline{x}) = 1\Big) \to (u_j(\mathbf{x}) = 1)\Big).$$

However a group satisfying these axioms directly is q-discriminating. □

With this characterization we can complete the proof of Theorem 11.7.1.

Proof. Since a variety is a quasivariety it follows that any q-discriminating group must be varietally discriminating and hence from Theorem 11.7.2 be squarelike. Therefore $\mathcal{S} \subseteq \mathcal{V}$.

A non-Abelian free nilpotent group again provides an example to show that the inclusion is proper. If G is a non-Abelian free nilpotent group then again from [202] G is varietally discriminating. In Section 11.3 it was shown that a non-Abelian nilpotent group is nondiscriminating and cannot discriminate its direct square. The same proof shows that it cannot have the same universal theory as its direct square and so is not squarelike. This completes the proof of Theorem 11.7.1. □

11.8 Co-discriminating groups and domains

There is a dual concept to that of a discriminating group called a *co-discriminating group* that also arises from algebraic geometry over groups. This concept was introduced in the same paper as that of a discriminating group by Baumslag, Myasnikov and Remslennikov [16, 17, 18, 19]. In distinction to discriminating groups where nontrivial examples are hard to find, co-discriminating groups abound. In fact any nontrivial free product except the infinite dihedral group $\mathbb{Z}_2 * \mathbb{Z}_2$ is a co-discriminating group. Further co-discriminating groups are closely tied to *domains* or *groups with no zero divisors* which we will define below. First we give the definition of a co-discriminating group.

Definition 11.8.1. A group G is a *co-discriminating group* if every group H separating G also discriminates G.

The concept arises from the case of non-Abelian CSA groups. Recall that if a non-Abelian CSA group (see Chapter 6) is residually free it is fully residually free. Far from being unique for CSA groups it was observed that this behavior did not depend on the separating object, that is if a non-Abelian CSA group is residually H for some group G then it is fully residually H. The proof depended upon the following fact. If a, b are any two nontrivial elements of a CSA group G (where a could equal b) then there exists an $x \in G$ such that $[a, b^x] \neq 1$ where $b^x = xbx^{-1}$. Groups satisfying this property are called *domains*. We looked at these earlier in Section 7.3.

Definition 11.8.2. A group G is a *domain* if for any two nontrivial elements $a, b \in G$ (where a could equal b) there exists an $x \in G$ such that $[a, b^x] \neq 1$. Domains are also called *groups without zero divisors*. Elements a, b in a group G where a commutes with every conjugate of b are called *zero divisors*.

The comment above about CSA groups can now be phrased as:

Lemma 11.8.1. *Every non-Abelian CSA group is a domain.*

Theorem 11.8.1. *If G is a domain then G is a co-discriminating group.*

Proof. Let G be a domain and let $g_1, \dots, g_n$ be finitely many nontrivial elements in G. There exists an element $x_1 \in G$ such that $[g_1, g_2^{x_1}] \neq 1$ Hence there exists an element $x_2 \in G$ such that $[[g_1, g_2^{x_1}], g_3^{x_2}] \neq 1$ and so on. It follows that we can find elements $x_1, \dots, x_{n-1} \in G$ such that

$$c = [\dots [[g_1, g_2^{x_1}], g_3^{x_2}] \dots g_n^{x_{n-1}}] \neq 1$$

Now suppose that G is separated by H. Let ϕ be a homomorphism $\phi\colon G \to H$ with $\phi(c) \neq 1$. Then ϕ does not annihilate any of $g_1, \dots, g_n$ and therefore H discriminates G. □

Corollary 11.8.1. *Every non-Abelian CSA group is a co-discriminating group.*

As mentioned in the introductory comments above, domains and co-discriminating groups abound. Every nontrivial free product except $\mathbb{Z}_2 * \mathbb{Z}_2$ is co-discriminating.

Theorem 11.8.2. *Suppose A, B are nontrivial groups. Then the free product $A * B$ is a domain and hence a co-discriminating group unless both A and B are of order 2.*

Proof. Suppose that the order of B is at least 3 and let x, y be nontrivial elements of $A * B$. We claim there exist elements $g, h \in A * B$ such that $[x^g, y^h] \neq 1$. We may assume that both x, y are cyclically reduced. Suppose first that the syllable length of both x, y is one so that both lie in a factor. Then $[x, y] \neq 1$ if x, y lie in different factors. If they both lie in the same factor, say A, then $[x, y^b] \neq 1$ where b is any nontrivial element of B.

Suppose now that the syllable length of x is greater than or equal to 2. Then we may assume (conjugating if necessary) that $x = a_1b_1 \dots a_nb_n$ with $1 \neq a_i \in A$ and $1 \neq b_i \in B$. Again conjugating if necessary we may assume that y is a strictly alternating product with the first and last syllable coming from B. This can be done since the order of B is at least 3. In this event the products xy and yx have different syllable lengths so conjugates of x, y do not commute as desired.

The infinite dihedral group $\mathbb{Z}_2 * \mathbb{Z}_2$ has an nontrivial normal Abelian subgroup and hence is not a domain. □

In the above theorem A and B need not be domains themselves. The case for more complicated amalgams becomes more difficult. The following can be proved.

Theorem 11.8.3. *Suppose that A and B are domains. Suppose that C is a subgroup of both A and B satisfying the following condition*

$$\text{if } c \in C, \ c \neq 1 \text{ then either}$$

$$[c, A] \text{ is not a subset of } C$$

or

$$[c, B] \text{ is not a subset of } C$$

Then the amalgamated free product $G = A *_C B$ *is a domain.*

The following is a generalization of a result of Baumslag [16, 17, 18, 19].

Theorem 11.8.4. *Let A and B be nontrivial groups not both of order 2. Let G be a group which separates its own free square $G * G$ (in particular a torsion-free hyperbolic group). Then the following are equivalent*

(1) *G separates* $A * B$.
(2) *G discriminates* $A * B$.
(3) *G discriminates each of A and B.*

Proof. From Theorem 11.8.2, $A * B$ is a co-discriminating group, so (1) and (2) are equivalent. (3) follows immediately from (2) since A and B are subgroups of $A * B$.

Suppose now that both groups A and B are discriminated by G. Then in the natural way $A * B$ is discriminated by $G * G$. It is then easy to see that if G separates $G * G$ then by transitivity of separation G separates $A * B$. □

Besides all nontrivial free products being co-discriminating all one-relator groups with more than 2 generators are co-discriminating.

Theorem 11.8.5. *Let* $G = \langle X; R \rangle$ *be a one-relator group with* $|X| > 2$. *Then G is a domain and hence a co-discriminating group.*

The proof of Theorem 11.8.5 uses a variation of the Magnus method. First we need the following preliminary Lemma.

Lemma 11.8.2. *Let N be a domain and G a semidirect product of N and an Abelian group. Then G is a domain if and only if G has trivial center.*

Proof. Clearly if G is a domain then the center of G is trivial.

Suppose now that the center of G is trivial. Observe that if x, y are a pair of zero divisors in G then their normal closures $N_G(x)$ and $N_G(y)$ commute

$$[N_G(x), N_G(y)] = 1.$$

If $u, v \in G$ then

$$[x, u] \in N_G(x), \quad [y, v] \in N_G(y)$$

and hence

$$[N_G([x, u]), N_G([y, v])] = 1 .$$

Now the center of G is trivial so therefore there exist elements $u, v \in G$ such that $[x, u], [y, v]$ are nontrivial. But this implies that both $[x, u]$ and $[y, v]$ are zero divisors in G. Since G/N is Abelian both $[x, u]$ and $[y, v]$ are contained in N. It follows that, in particular, $[x, u]$ is a zero divisor in N contradicting the assumption that N is a domain. □

We now give the proof of Theorem 11.8.5.

Proof. The proof uses a variation of Magnus' method for the Freiheitssatz. Let

$$G = \langle X, R\rangle$$

where $X = \{t, b, c, \ldots.\}$ and R is a cyclically reduced word in X. Let $\sigma_x(R)$ denote the exponent sum of x in R.

Suppose first that $\sigma_t(R) = 0$. Denote by N the normal closure in G of $X \setminus t$. Put

$$b_i = t^i b t^{-i}, \quad c_i = t^i c t^{-i}, \ldots .$$

Then N is generated by $Y = \{b_i, c_i, \ldots\}$. Since $\sigma_t(R) = 0$, R can be rewritten as a word $S = S(\ldots, b_j, \ldots, c_k, \ldots)$ in the generators Y where each of the generators x occurring in R is replaced by x_i, where the subscript i is the sum of the t-exponents of the subword of R preceding x. It follows then that

$$t^i R t^{-i} = t^i S(\ldots, b_j, \ldots, c_k, \ldots) t^{-i} = S(, \ldots b_{j+i}, \ldots, c_{k+i}, \ldots), \quad i \in \mathbb{Z}$$

Put

$$S_i = S(\ldots, b_{j+1}, \ldots, c_{k+1}, \ldots), \quad i \in \mathbb{Z} .$$

Using the Reidemeister–Schreier method it follows that N has the presentation

$$N = \langle b_i, c_i, \ldots . (i \in \mathbb{Z}); S_i = 1, i \in \mathbb{Z}\rangle$$

Let α, β be, respectively, the minimal and maximal index of b that occur in the word S. Now for arbitrary nonnegative integers i, j put

$$N_{-i,j} = \langle c_k, \dots (k \in \mathbb{Z}), b_{\alpha-i}, b_{\alpha-i+1}, \dots, b_{\beta+j}; S_{-i} = \dots = S_j = 1 \rangle.$$

It follows that N is the union of the following ascending chain of subgroups

$$N_{0,0} \subseteqq N_{0,1} \subseteq N_{-1,1} \subseteq N_{-1,2} \cdots \subseteq N_{-i+1,i} \subseteq N_{-i,i} \subseteq \dots .$$

Notice that in the presentation for $N_{-i,j}$ there are infinitely many generators and finitely many relators so each $N_{-i,j}$ is a free product of two infinite groups. Then by Theorem 11.8.2 for each $i, j \geq 0$ the group $N_{-i,j}$ is a domain. Since the union of an ascending chain of domains is a again a domain we conclude that N is a domain. Observe further that the group G is a semidirect product of N and the infinite cyclic group generated by t. A result of Murasugi [190] shows that a one-relator presentation with more than two generators defines a group with a trivial center. Then by Lemma 11.8.2 G is a domain.

Now suppose that $\sigma_x(R) \neq 0$ for every $x \in X$. Denote by $F(X)$ the free group on X. By the exponent sum assumption it follows that $R \notin [F, F]$. Consequently there exists a free basis $Y = \{y_1, y_2, \dots .\}$ such that

$$R = y_1^e y'$$

where e is an integer and y' is an element in the derived group of F, expressed in terms of the new basis Y of F. Let S be the word R rewritten in terms of the new basis. Clearly $\sigma_y(S) = 0$ for every $y \in Y \setminus y_1$. Hence if we present G on the generators Y we find that

$$G = \langle Y; S \rangle$$

and we are back to the situation considered previously. Therefore G is a domain as before. □

The condition that there must be more than 2 generators is essential. For example the following two-generator one-relator groups are not domains since they contain nontrivial Abelian normal subgroups;

$$G_1 = \langle x, y; x^k = y^m \rangle, \quad G_2 = \langle x, y; y^{-1}xy = x^k \rangle \text{ with } k, m \geq 2 .$$

Finally, recall from Chapter 6 that a 2-generator group is fully residually free if and only if it is residually free and further in this case it must be free or free Abelian. Hence, if G is a 2-generator one-relator group it is residually free if and only if it is fully residually free. From Theorem 11.8.5 a one-relator with more than 2 generators is co-discriminating and hence if it is residually free it must be fully residually free. Putting these comments together we obtain the following.

Theorem 11.8.6. *Every residually free one-relator group is fully residually free.*

References

[1] P. Ackermann, B. Fine and G. Rosenberger, *On Surface Groups: Motivating Examples in Combinatorial Group Theory*, **Groups St. Andrews 2005**. Cambridge University Press, 2007, 96–129.

[2] P. Ackermann, V. gr. Rebel and G. Rosenberger, *On Power and Commutation Transitive, Power Commutative and Restricted Gromov Groups*, **Cont. Math**, 360, 2004, 1–4.

[3] S. I. Adian, *Periodic Groups of Odd Exponent*, **Proc. Second Int. Conference on Groups Canberra 1973**, Springer Lecture Notes 372, 1974, 8–12.

[4] R. C. Alperin and H. Bass, *Length functions of group actions on Λ-trees*, in **Combinatorial Group Theory and Topology**, S. M. Gersten, J. Stallings editors, 1987, Princeton University Press, 265–378.

[5] K. I. Appel, *One variable equations in free groups*, **Proc. Amer. Math. Soc.**, 19, 1968, 912–918.

[6] H. Bass, *Groups acting on non–archimedian trees.* **Arboreal group theory**, 1991, 69–130.

[7] G. Baumslag, *On generalised free products*, **Math. Z.**, 78, 1962, 423–438.

[8] G. Baumslag, *Groups with the same lower central sequence as a relatively free group. I. The groups.*, **Trans. Amer. Math. Soc.**, 129, 1967, 308–321.

[9] G. Baumslag, *Residually finite one-relator groups*, **Bull.Amer. Math. Soc.**, 73, 1967, 618–620.

[10] G. Baumslag, **Lecture Notes on Nilpotent Groups**, Amer. Math. Soc. Providence, RI., 1969.

[11] G. Baumslag, B. Fine, A. M. Gaglione and D. Spellman, *Reflections on discriminating groups*, **J. Group Theory**, 10, 2007, 87–99.

[12] G. Baumslag, B. Fine, A. M. Gaglione and D. Spellman, *A note on nondiscrimination of nilpotent groups and Malcev completions*, **Cont. Math.**, 421, 2006, 29–34.

[13] G. Baumslag, S. Gersten, M. Shapiro and H. Short, *Automatic Groups and Amalgams* in **Algorithms and Classification in Combinatorial Group Theory** eds. G. Baumslag, C. F. Miller III, Springer-Verlag MSRI Publications, 23, 1992, 179–195.

[14] G. Baumslag and D. Troeger, *Virtually Free-by-Cyclic Groups I*, in **Aspects of Infinite Groups** World Scientific Press, 2009.

[15] G. Baumslag, B. Fine, C. Miller and D. Troeger, *Virtual Properties of Cyclically Pinched One-Relator Groups*, **Int, J. of Alg. and Comp.**, 19, 2009, 1–15.

[16] G. Baumslag, A. G. Myasnikov and V. N. Remeslennikov, *Algebraic Geometry over Groups I*, **J. of Algebra**, 219, 1999, 16–79.

[17] G. Baumslag, A. G. Myasnikov and V. N. Remeslennikov, *Discriminating and co-discriminating groups*, **J. of Group Theory**, 3, 2000, 467–479.

[18] G. Baumslag, A. Myasnikov and V. Remeslennikov, *Discriminating completions of hyperbolic groups*, **Goem. Ded.**, 92, 2002, 115–143.

[19] G. Baumslag, A. Myasnikov and V. Remeslennikov, *Residually hyperbolic groups and approximation theorems for extensions of centralizers*, 1996.

[20] G. Baumslag. A. Myasnikov and V. Roman'kov, *Two theorems about equationally Noetherian groups*, **J. Algebra**, 194, 1997, 654–664.

[21] B. Baumslag, *Residually free groups*, **Proc. London Math. Soc.** 17, 3, 1967, 402–418.

[22] B. Baumslag, *Free groups and free products*, **Comm. Pure and Appl. Math.**, 20, 1967, 635–645.

[23] O. Belegradek, *Discriminating and square-like groups*, **J. Group Theory**, 7, 2004, 521–532.

[24] O. Belegradek, *Theory of abelian square-like groups is decidable*, **Festschrift for U. Felgner**, Algebra, Logic, and Set Theory, College Park, London, 4, 2007, 33–46.

[25] J. L. Bell, A. B. Slomson, **Models and Ultraproducts: An Introduction**, North-Holland, Amsterdam 1972 (Second revised printing).

[26] A. Berzins, B. Plotkin and E. Plotkin, *Algebraic geometry in varieities of algebras with the given algebra of constants*, **J. of Math. Sci.**, 102, 2000, 4039–4070.

[27] M. Bestvina, $\mathbb{R}$-*trees in topology, geometry and group theory*, in **Handbook of geometric topology**, North-Holland, 2002, 55–91.

[28] M. Bestvina and M. Feighn, *A combination theorem for negatively curved groups*, **J. Diff. Geom.**, 35, 1992, 85–101.

[29] M. Bestvina and M. Feighn, *Notes on Sela's Work: Limit Groups and Makanin–Razborov Diagrams*, preprint.

[30] M. Bestvina and M. Feighn, *Stable actions of groups on real trees*, **Inventiones Mathematicae**, 121, 1995, 287–321.

[31] M. Bestvina and M. Feighn, *Bounding the complexity of simplicial group actions on trees*, **Inventiones Mathematicae**, 103, 1991, 449–460.

[32] O. Bogopolski, *A surface analogue of a theorem of Magnus*, **Cont. Math.**, 352, 2005, 55–89.

[33] O. Bogopolski and K. Sviridov, *A Magnus theorem for some one-relator groups*, in **The Zieschang Gedenkschrift** v.14, 2008, 63–73.

[34] W. W. Boone, *The Word Problem*, **Ann. of Math**, 70, 1959, 207–265.

[35] N. Brady, L. Ciobanu, A. Martino and S. O'Rourke, *The Equation* $x^p y^q = z^r$ *and Groups that act freely on* Λ*-trees*, **Trans. Amer. Math. Soc.** 361, 2009, 223–236.

[36] M. Bridson and H. Wilton, *Subgroup separability in residually free groups*, **Math. Z.**, 260, 2008, 25–30.

[37] J. L. Britton, *Solution of the word problem for certain types of groups I, II*, **Proc. Glasgow Math Soc.**, 3, 1956, 45–54, 3, 1957, 68–90.

[38] R. Bryant, *The verbal topology of a group*, **J. of Alg.**, 48, 1977, 340–346.

[39] I. Bumagin, O. Kharlampovich and A. Myasnikov, *Isomorphism Problem for Finitely Generated Fully Residually Free groups*, **J. Pure and Applied Algebra**, 208, 2007, 961–977.

[40] C. C. Chang, H. J. Keisler, **Model Theory**, North-Holland, Amsterdam, 1973.

[41] C. Carstensen, B. Fine, G. Rosenberger, **Abstract Alegbra**, DeGruyter, 2012.

[42] T. Camps, V. grosse Rebel, G. Rosenberger, **Einführung in die kombinatorische und die geometrische Gruppentheorie**, Heldermann-Verlag, 2008.

[43] T. Camps, S. Kühling and G. Rosenberger, **Einführung in die mengentheoretische und die algebraische Topologie**, Heldermann-Verlag 2006.

[44] J. W. Cannon, *The combinatorial structure of cocompact discrete hyperbolic groups*, **Goem. Dedicata**, 16, 1984, 123–148.

[45] S. Chagas and P. Zalesskii, *Limit Groups are Conjugacy Separable*, **IJAC**, 17, 2007, 851–857.

[46] C. Champetier and V. Guirardel, *Limit Groups as Limits of Free Groups*, **Israel J. of Math.** 146, 2005, 1–75.

[47] I. Chiswell, **Introduction to** Λ**-trees**, World Scientific 2001.

[48] I. Chiswell, *Abstract length functions in groups*, **Math. Proc. Cambridge Philos. Society**, 80, 1976, 451–463.

[49] I. Chiswell, *Harrison's Theorem for* Λ*-trees*, **Quarterly J. Math Oxford**, 2, 45, 1994, 1–12.

[50] I. Chiswell, *Non-trivial group actions on* Λ*-trees*, **Bull. London Math. Soc.**, 24, 1992.

[51] I. Chiswell, *Introduction to* Λ*-trees*, in **Semigroups, formal languages and groups** (J. Fountain ed.), Kluwer, 1995.

[52] I. Chiswell, *Length Functions and* Λ*-trees*, **Cont. Math.**

[53] I. Chiswell and T. Muller, *A Universal Construction for Groups Acting Freely on Real Trees*, Cambridge University Press, 2012.

[54] L. Ciobanu, B. Fine and G. Rosenberger, *On Lyndon's equation in some Lambda-free groups and HNN Extensions*, **J. Group Theory**, 1, 2010, 1–7.

[55] L. Ciobanu, B. Fine and G. Rosenberger, *Groups Generalizing a Theorem of Benjamin Baumslag*, to appear.

[56] D. Cohen, **Combinatorial Group Theory: a topological approach**, Cambridge University Press, 1989. 1965.

[57] P. M. Cohen, **Universal Algebra**, Harper and Row, New York, 1965.

[58] M. Cohen and M. Lustig, *Very Small Actions on* $\mathbb{R}$*-trees and Dehn Twist Automorphisms*, **Topology**, 34, 1985, 575–617.

[59] D. Collins and H. Zieschang, *On the Nielsen Method in Free products with Amalgamation*, **Math. Z**, 197, 1987, 97–118.

[60] L. Comerford and C. Edmonds, *Solutions of Equations in Free Groups*, **Proc. of Conf. in Group Theory Singapore 1987**, Springer-Verlag, 347–355, 1989.

[61] P. Csorgo, B. Fine and G. Rosenberger, *On Certain Equations in Free Groups*, **Acta Sci. Math.**, 68, 2002, 895–905.

[62] F. Dahmani and V. Guirardel, *The isomorphism problem for all hyperbolic groups*, **Geom. and Func. Anal**, 21, 2011, 223–300.

[63] T. Delzant and L.Potyagailo, *Accesibilite hierarique des groups de presentation finis*, **Topology**, 40, 2001, 617–629.

[64] M. Dehn, *Über unendliche diskontinuerliche Gruppen*, **Math. Ann.**, 69, 1911, 116–144.

[65] M. Dehn, *Transformation der Kurven auf zweiseitigen Flächen*, **Math. Ann.**, 72, 1912, 413–420.

[66] W. Dicks, **Groups, trees and projective modules**, Lecture Notes in Mathematics 790, Springer, Berlin, 1980.

[67] V. H. Dyson, Private Communication.

[68] D. B. A. Epstein, J. W. Cannon, D. F. Holt, M. S. Paterson, W. P. Thurston, **Word processing and group theory**, Jones and Bartlett, 1997.

[69] Y. L. Ershov, *Elementary group theories*, **Soviet Math. Dokl.**, 13, 1972, 528–532.

[70] A. Ehrenfeucht, *An application of games to the completeness problem for formalized theories*, **Fund. Math.**, 49, 129–141.

[71] H. Federer and B. Jonsson, *Some properties of free groups*, **Trans. Amer. Math. Soc.**, 68, 1950, 1–27.

[72] B. Fine, **Algebraic Theory of the Biachi Groups**, Marcel Dekker, 1989.

[73] B. Fine, A. Gaglione, A. Myasnikov, G. Rosenberger and D. Spellman, *A Classification of Fully Residually Free Groups of Rank Three or Less*, **J. of Algebra**, 200, 1998, 571–605.

[74] B. Fine, A. Gaglione, A. Myasnikov, G. Rosenberger and D. Spellman, *A On the Elementary Theory of Groups*, **Groups St Andrews 2001**, London Mat. Soc. Lecture Series Notes, 304, Cambridge University press, 203, 197–232.

[75] B. Fine, A. M. Gaglione, A. G. Myasnikov and D. Spellman, *Discriminating groups*, **J. of Group Theory**, 4, 2001, 463–474.

[76] B. Fine, A. M. Gaglione, A. G. Myasnikov and D. Spellman, *Groups Whose Universal Theory is Axiomatizable by Quasi-Identities*, **J. of Group Theory**, 5, 2002, 365–381.

[77] B. Fine, A. M. Gaglione and D. Spellman, *Every abelian group universally equivalent to a discriminating group is elementarily equivalent to a discriminating group*, **Cont. Math.**, 296, 2002, 129–137.

[78] B. Fine, A. M. Gaglione and D. Spellman, *The axiomatic closure of the class of discriminating groups*, **Archiv. der Math.**, 83, 2004, 106–112.

[79] B. Fine, A. M. Gaglione and D. Spellman, *Discriminating and squarelike groups I: axiomatics*, **Cont. Math.**, 360, 2004, 35–46.

[80] B. Fine, A. M. Gaglione and D. Spellman, *Discriminating and squarelike groups II: examples*, **Houston J. of Math.**, 31, 2005, 649–673.

[81] B. Fine, A. M. Gaglione and D. Spellman, *Notions of discrimination*, **Comm. in Alg**, 2006, 2175–2182.

[82] B. Fine, A. M. Gaglione and D. Spellman, *Discriminating Groups: A Comprehensive Overview*, **Proc. Groups St Andrews**, 2010.

[83] B. Fine, A. M. Gaglione and D. Spellman, *Discrimination in a General Alegbraic Setting*, In preparation.

[84] B. Fine, A. M. Gaglione, S. Lipschutz and D. Spellman, *Turner Groups and First Order Theory*, preprint.

[85] B. Fine, A. M. Gaglione, G. Rosenberger and D. Spellman, *n-free Groups and Questions about Universally-Free Groups* **Proc. of Groups St Andrews 1993**, London Math. Soc. Lecture Notes Series 211, 191–204.

[86] B. Fine, A. M. Gaglione, G. Rosenberger and D. Spellman, *The Commutative Transitive Kernel*, **Algebra Colloquium**, 2, 1997, 141–152.

[87] B. Fine, A. M. Gaglione, G. Rosenberger and D. Spellman, *Elementary Free Groups*, **Cont. Math.** to appear.

[88] B. Fine, A. M. Gaglione, G. Rosenberger and D. Spellman, *Something for Nothing: Some Consequences of the Solutions of the Tarski Problems*, **Proc. of Groups St. Andrews 2013** to appear.

[89] B. Fine, M. Kreuzer and G. Rosenberger, *Faithful Real Representations of Cyclically Pinched One-Relator Groups*, **International Journal of Group Theory**, 3, 2014, 1–8.

[90] B. Fine, G. Rosenberger, **Algebraic Generalizations of Discrete Groups**, Marcel Dekker, 1999.

[91] B. Fine, G. Rosenberger, On Restricted Gromov Groups, **Communications in Algebra**, 20, Number 8, 1992, 2171–2182.

[92] B. Fine and G. Rosenberger, *Surface Groups within Baumslag Doubles*, **Proceedings of the Edinburgh Math. Soc.**, 54, 2011, 91–97.

[93] B. Fine and G. Rosenberger, *A Note on Faithful Representations of Limit Groups*, **Groups, Complexity and Cryptology**, 3, 2011, 349–355.

[94] B. Fine and G. Rosenberger, *Faithful Representations of Limit Groups 2*, **Groups, Complexity and Cryptology**, 5, 2013, 91–96.

[95] B. Fine and G. Rosenberger, *The Classification of One-Relator Limit Groups and the Surface Group Conjecture*, **Cont. Math.**, 582, 2012, 107–129.

[96] B. Fine and G. Rosenberger, *Reflections on Commutative Transitivity* in **Aspects of Infinite Groups**, World Scientific Press, 2009, 112–130.

[97] B. Fine and G. Rosenberger, *Generalising Algebraic Properties of Fuchsian Groups*, **Groups St Andrews 1989, Vol 1**, (1991), 124–148.

[98] B. Fine, A. Myasnikov, V. gr. Rebel and G. Rosenberger, *A Classification of Conjugately Separated Abelian, Commutative Transitive and Restricted Gromov One-Relator Groups*, **Result. Math.**, 50, 2007, 183–193.

[99] B. Fine, O. Kharlampovich, A. Myasnikov, V. Remeslennikov, G. Rosenberger, On the Surface Group Conjecture, **Scienta: Math Series A**, 1, 2008, 1–15.

[100] B. Fine, A. Rosenberger and G. Rosenberger, *Quadratic Properties in Group Amalgams*, **J. Group Theory**, 2011, 1–15.

[101] B. Fine, A. Rosenberger and G. Rosenberger, *A Note on Lyndon Properties in One Relator Groups*, **Results in Math.**, 2012.

[102] B. Fine, F. Röhl and G. Rosenberger, *Two Generator Subgroups of Certain HNN Groups*, **Contemporary Math.**, 109, 1990, 19–25.

[103] B. Fine, F. Röhl and G. Rosenberger, *A Three-Free Theorem for Certain HNN Groups*, in **Infinite Groups and Group Rings** edited by J. Corson, M. Dixon, M. Evans, F. Röhl, World Scientific Press, 1993, 13–37.

[104] B. Fine, G. Rosenberger and M. Stille, *Conjugacy Pinched and Cyclically Pinched One-Relator Groups*, **Revista Math. Madrid**, 10, 1997, 207–227.

[105] B. Fine, G. Rosenberger and M. Stille, *Nielsen Transformations and Applications: A Survey* **Groups Korea 94**, DeGruyter, 1995, 69–105.

[106] B. Fine, G. Rosenberger, D. Spellman and M. Stille, *Test Elements, Generic Elements and Almost Primitivity in Free groups*, **Pacific J. of Math.**, 190, 1999, 277–297.

[107] D. Gaboriau, G. Levitt and F. Paulin, *Pseudogroups of isometries of R and Rips' theorem on free actions on R-trees*, **Israel Journal of Mathematics**, 87, 1994, 403–428.

[108] A. Gaglione and D. Spellman, *Generalizations of free groups: some questions*, **Comm. in Alg.**, 22, 8, 1993, 3159–3169.

[109] A. Gaglione and D. Spellman, *Parametric words and models of the elementary theory of non-abelian free groups*, **Proceedings Groups St Andrews/Galway 1993**. London Math Soc. Lecture Notes Series 211, 1995, 233–248.

[110] A. Gaglione and D. Spellman, *Some Model Theory of Free Groups and Free Algebras*, **Houston J. Math**, 19, 1993, 327–356.

[111] A. Gaglione and D. Spellman, *More Model Theory of Free Groups*, **Houston J. Math**, 21, 1995, 225–245.

[112] A. Gaglione and D. Spellman, *Every Universally Free Group is Tree Free*, **Proc. Ohio State Conference for H. Zassenhaus**, World Scientific, 1993, 149–154.

[113] A. Gaglione and D. Spellman, *Even More Model Theory of Free Groups*, in **Infinite Groups and Group Rings**, World Scientific, 1993, 37–40.

[114] A. Gaglione, S. Lipschutz and D. Spellman, *Almost Locally Free Groups and a Theorem of Magnus*, **J. of Groups, Complexity and Cryptology**, 1, 2009, 181–198.

[115] A. Gaglione, A. Myasnikov, V. Remeslennikov and D. Spellman, *Formal PowerSeries Representations of Free Exponential Groups*, **Comm. in Algebra**, 25, 1997, 631–648.

[116] D. Gildenhuys, O. Kharlampovich and A. Myasnikov, *CSA Groups and Separated Free Constructions*, **Bull. Austral. Math. Soc.**, 52, 1995, 63–84.

[117] R. Gilman, *On the definition of word hyperbolic groups*, **Math. Z.**, 242, 2002, 529–541.

[118] G. Gratzer, **Universal Algebra**, Van-Nostrand, Princeton, 1968.

[119] M. Greendlinger, *Dehn's Algorithm for the Word Problem*, **Comm. Pure and Applied Math.**, 13, 1960, 67–83.

[120] R. I. Grigorchuk and P. F. Kurchanov, *On quadratic equations in free groups*, **Cont. Math.**, 131, 1992, 159–171.

[121] R. Grigorchuk and P. Kurchanov, *Some questions of group theory related to geometry.* in **Itogi Nauki i Techniki, Sovremennye problemy matematiki. Fundamental'nye napravlenia, VINITI, 58. Encyclopedia of math. sciences**, 1990, English translation in 1993.

[122] M. Gromov, *Hyperbolic Groups* in **Essays in Group Theory** S. Gersten ed. MSRI Publication 8, Springer-Verlag, 1987.

[123] V. Guba, *Equivalence of infinite systems of equations in free groups and semigroups to finite subsystems*, **Mat. Zametki**, 40, 1986, 321–324.

[124] V. Guirardel, *Limit groups and groups acting freely on $\mathbb{R}^n$-trees*, **Geom. Topol**, 8, 2004, 1427–1470.

[125] C. K. Gupta and V. Shpilrain, *Lifting automorphisms: a survey*, **Proc. Groups '93 Galway/St. Andrews, Vol 1 (Galway, 1993)**, London Math. Soc. Lecture Note Ser., 211, Cambridge Univ. Press, Cambridge, 1995, 249–263.

[126] N. Gupta and S. Sidki, *On the Burnside Problem for Periodic Groups*, **Math. Z.**, 182, 1983, 385–388.

[127] M. Hall, **Theory of Groups**, Macmillan, 1959.

[128] J. D. Halpern, *The independence of the axiom of choice from the Boolean prime ideal theorem*, **Fund. Mat.**, 55(1964), 57–66.

[129] N. Harrison, *Real length functions in groups*, **Trans. Amer. Math. Soc.**, 174, 1972, 77–106.

[130] R. Hartshorne, **Algebraic geometry**. Springer-Verlag, New York, 1977.

[131] G. Higman, *Subgroups of finitely presented groups*, **Proc. Royal Soc. London** Ser A, 264, 1961, 455–475.

[132] G. Higman, *Finitely Presented Infinite Simple Groups*, **Notes on Pure Math.**, 8, I. A.S., Austral. Nat Univ., Canberra, 1974.

[133] G. Higman and E. Scott, **Existentially Closed Groups**, Clarendon Press, Oxford, 1988.

[134] R. Hirshon and D. Meier, *Groups with a quotient that contains the original group as a direct factor*, **Bull. Austral. Math Soc.** 45, 1992, 513–520.

[135] J. Howie, *Some Results on One-Relator Surface Groups*, **Boletin de la Sociedad Matematica Mexicana**, 10, 2004, 255–262.

[136] W. Jaco and P. Shalen, *Seifert fibered spaces in 3-manifolds*, **Mem. Amer. Math. Soc.**, 21, 1979.

[137] W. Jaco and P. Shalen, *Seifert fibered spaces in 3-manifolds*, **Geometric topology: Proc. Georgia Topology Conf.**, Academic Press, New York-London, 1979, 91–99.

[138] W. Jaco and P. Shalen, *A new decomposition theorem for irreducible sufficiently-large 3-manifolds*, . **Algebraic and geometric topology**, (Proc. Sympos. Pure Math., Stanford Univ., Stanford, Calif., 1976), Part 2, pp. 71–84, Proc. Sympos. Pure Math., XXXII, Amer. Math. Soc., Providence, R. I., 1978.

[139] K. Johannson, *Homotopy equivalences of 3-manifolds with boundaries*, **Lecture Notes in Mathematics**, 761, Springer, 1979.

[140] J. M. Tyrer Jones, *Direct products and the Hopf property*, **J. Austral. Math. Soc.**, 17, 1974, 174–196.

[141] D. L. Johnson, **Presentations of Groups**, Cambridge University Press, 1990.

[142] A. Juhasz and G. Rosenberger, *On the Combinatorial Curvature of Groups of F-type and Other One-Relator Products of Cyclics*, **Cont. Math.**, 169, 1994, 373–384.

[143] R. N. Kalia and G. Rosenberger, *Über Untergruppen ebener diskontinuierlicher Gruppen*, **Cont. Math.**, 33, 1984, 308–327.

[144] I. Kapovich, *Subgroup Properties of Fully Residually Free Groups*, **Trans. Amer. Math, Soc.**, 354, 2002, 335–362.

[145] I. Kapovich and A. Myasnikov, *Stallings Foldings and Subgroups of Free Groups*, **J. of Algebra**, 248, 2002, 608–668.

[146] A. Karrass and D. Solitar, *The Subgroups of a Free Product of Two Groups with an Amalgamated Subgroup*, **Trans. Amer. Math. Soc.**, 150, 1970, 227–255.

[147] A. Karrass and D. Solitar, *Subgroups of HNN Groups and Groups with One Defining Relation*, **Can. J. Math. Soc.**, 23, 1971, 627–643.

[148] A. Karrass and D. Solitar, *The Subgroups of a Free Product of Two Groups with a Malnormal Amalgamated Subgroup*, **Can. J. Math.**, 23, 1971, 933–959.

[149] A. Karrass. J. Pietrowski and D. Solitar, *An Improved Subgroup Theorem for HNN Groups with some Applications*, **Can. J. Math. Soc.**, 26, 1974, 214–224.

[150] M. Kasabov, *Weak Identities in Finitely Generated Groups*, arXiv:math/0311494v1 [math. GR] 27 Nov 2003.

[151] O. Kharlamapovich and A. Myasnikov, *Equations over Free Groups and Fully Reisdually Free groups*, to appear.

[152] O. Kharlamapovich and A. Myasnikov, *Irreducible affine varieties over a free group: I. Irreducibility of quadratic equations and Nullstellensatz*, **J. of Algebra**, 200, 1998, 472–516.

[153] O. Kharlamapovich and A. Myasnikov, *affine varieties over a free group: II. Systems in triangular quasi-quadratic form and a description of residually free groups*, **J. of Algebra**, 200, 1998, 517–569.

[154] O. Kharlamapovich and A. Myasnikov, *The Implicit Function Theorem over Free groups*, **J. Alg.**, 290, 2005, 1–203.

[155] O. Kharlamapovich and A. Myasnikov, *Effective JSJ Decompositions*, **Cont. Math.**, 378, 2005, 87–211.

[156] O. Kharlamapovich and A. Myasnikov, *Elementary Theory of Free Nonabelian Groups*, **J. Alg.**, 302, 2006, 451–552.

[157] O. Kharlamapovich and A. Myasnikov, *Hyperbolic Groups and Free Constructions*, **Trans. Amer. Math. Soc.**, 350, 2, 1998, 571–613.

[158] O. Kharlamapovich, A. Myasnikov, V. Remeslennikov and D. Serbin, *Subgroups of Fully Residually Free Groups*, **Cont. Math.**, 360, 2004, 63–101.

[159] O. Kharlamapovich, A. Myasnikov and D. Serbin, *Actions, Length Functions and Non-Archimedean Words*, **Int. J. Algebra Comput.**, 23, 2013, 325–350.

[160] N. Koblitz, **Algerbaic Aspects of Cryptography**, Springer-Verlag, 1998.

[161] A. G. Kurosh, **The Theory of Groups**, Vol. 's I, II, Chelsea, New York, 1960.

[162] A. V. Kvaschuk, A. G. Myasnikov and D. E. Serbin, Pregroups with the Big Powers Condition **Algebra and Logic**, 48. 2009, 193–213.

[163] S. Lang, **Abstract Algebra**, Addison-Wessley, 1965.

[164] F. Levin and G. Rosenberger, *On Power Commutative and Commutation Transitive Groups*, **Proc. Groups St Andrews 1985**, Cambridge University Press, 1986, 249–253.

[165] G. Levitt, *La dynamique des pseudogroupes de rotations*, **Invent. Math.**, 113, 1993, 633–670.

[166] A. A. Lorenc, *The solution of systems of equations in one unknown in free groups*, **Dokl. Akad. Nauk SSSR**, 148, 1963, 262–266.

[167] A. Lubotzky, private communication.

[168] R. C. Lyndon, *Length Functions in Groups*, **Math. Scand.**, 12, 1963, 209–234.

[169] R. C. Lyndon, *On Dehn's Algorithm*, **Math. Ann.**, 166, 1966, 208–228.

[170] R. C. Lyndon, *A maximum principle for graphs*, **J. Combinatorial Theory**, 3, 1967, 34–37.

[171] R. C. Lyndon, *On the Freiheitssatz*, **J. London Math. Soc.**, 5, 1972, 95–101.

[172] R. C. Lyndon, *The equation* $a^2b^2 = c^2$ *in free groups*, **Michigan Math J.**, 6, 1959, 155–164.

[173] R. C. Lyndon, *Groups with parametric exponents*, **Trans. Amer. Math. Soc.**, 96, 1960, 518–533.

[174] R. C. Lyndon, *Equations in free groups*, **Trans. Amer. Math. Soc.**, 96, 1960, 445–457.

[175] R. C. Lyndon, *Problems in combinatorial group theory*, **Annals of Mathemical Studies**, 111, Princeton U. Press, 1987, 3–33.

[176] R. C. Lyndon and P. E. Schupp, **Combinatorial Group Theory**, Springer-Verlag 1977.

[177] J. Los and R Suszko, *On the infinite sums of models* **Bull. Acad. Polon. Sci. Sèr. Sci. Math. Asctroum. Phys.** 3, 1955, 201–202.

[178] W. Magnus, A. Karrass and D. Solitar, **Combinatorial Group Theory**, Wiley 1966, Second Edition, Dover Publications, New York, 1976.

[179] G. S. Makanin, *Equations in a free group (Russian)*, **Izv. Akad. Nauk SSSR, Ser. Mat.**, 46, 1982, 1199–1273 transl. in Math. USSR Izv., V. 21, 1983; MR 84m:20040.

[180] G. S. Makanin, *Decidability of the universal and positive theories of a free group*, **Math. USSR Izvestiya**, 25, 1985, 75–88.

[181] A. I. Mal'cev, *The elementary properties of linear groups*, **Certain problems in Math. and Mechanics, Izdat. Sibirsk. Otdel. Acad.**, Nauk SSSR, Novosibirsk 1961, 110–132, English translation in The metamathematics of algebraic systems. Collected papers: 1936–1967, North-Holland Amsterdam, 1971.

[182] A. I. Mal'cev, *On free soluble groups*, **Dokl. Akad. Nauk SSR**, 130, 1960, 495–498 (English transl. (Soviet Math. Dokl., 1960, 65–68.)).

[183] L. Markus–Epstein, *Stallings' Foldings and Subgroups of Amalgams of Finite Groups*, **Int. J. Algebra Comput.**, 17, 2007, 1493–1535.
[184] A. Martino and S. O'Rourke, *Free actions on $\mathbb{Z}^n$-trees: a survey*, **Cont. Math.**, 372, 2005, 11–25.
[185] Y. I. Merzlyakov, *Positive Fromulas on Free Groups*, **Algebra i Logika**, 5, 1966, 25–42.
[186] C. F. Miller III, **On Group Theoretic Decision Problems and their Classification**, Princeton, 1971.
[187] J. W. Morgan, *Λ-trees and their applications*, **Bull. Amer. Math. Soc.**, 26, 1992, 87–112.
[188] J. W. Morgan and P. Shalen, *Valuations, trees and degenerations of hyperbolic structures: I*, **Annals of Math.**, 122, 1985, 398–476.
[189] D. E. Muller and P. E. Schupp, *Groups, the theory of ends and context-free languages*, **J. Comput. and Sys. Sci.**, 26, 1983, 295–310.
[190] K. Murasugi, *The center of a group with a single defining relation*, **Math. Ann.**, 155, 1964, 246–251.
[191] A. Myasnikov, *Structure of models and decidability criterion for complete theories of finite-dimensional algebras*, **Izvest. Akademii Nayk USSR, ser. math.**, 53, 1989, 379–397.
[192] A. Myasnikov and V. N. Remeslennikov, *Elementary properties of powered nilpotent groups*, **Doklady Akademii Nayuk**, 258, 5, 1981, 1056–1059.
[193] A. G. Myasnikov and V. N. Remeslennikov, *Definability of the set of Mal'cev bases and elementary theories of finite dimensional algebras I*, **Siberian Math. Jour.** 23, 1983, 711–724.
[194] A. G. Myasnikov and V. N. Remeslennikov, *Definability of the set of Mal'cev bases and elementary theories of finite dimensional algebras II*, **Siberian Math. Journ.**, 24, 1983, 231–246.
[195] A. Myasnikov and V. N. Remeslennikov, *Algebraic geometry 2: logical foundations*, **J. of Algebra**, 234, 2000, 225–276.
[196] A. G. Myasnikov and V. N. Remeslennikov, *Exponential groups 1: Foundations of the theory and tensor completion*, **Siberian Mat. J.**, 5, 1994, 1106–1118.
[197] A. G. Myasnikov and V. N. Remeslennikov, *Exponential groups 2: Extensions of centralizers and tensor completion of CSA groups*, **Interntional Journal of Algebra and Computation**, 6, 1996, 687–711.
[198] A. Myasnikov, V. N. Remeslennikov and D. Serbin, *Regular Free Length Functions on Lyndon's free $\mathbb{Z}[t]$-group*, **Cont. Math.**, 2005, 378, 37–77.
[199] A. Myasnikov, V. N. Remeslennikov and D. Serbin, *Fully Residually Free Groups and Graphs Labeled by Infinite Words*, **IJAC**, 16, 4, 2006, 689–737.
[200] A. Myasnikov, P. Shumyatsky, **Discriminating groups and c-dimension**, **J. Group Theory**, 7, 1, 2004, 135–142.
[201] B. H. Neumann, *Some remarks on infinite groups*, **J. London Math. Soc.**, 12, 1937, 120–127.
[202] H. Neumann, **Varieties of Groups**, Springer-Verlag, New York, 1967.
[203] J. Nielsen, *Die Isomorphismengruppe der freien Gruppen*, **Math. Ann.**, 91, 1924, 169–209.
[204] G. A. Noskov, *The elementary theory of a finitely generated almost solvable group*, **Math. USSR-Izv.**, 22, 1984, 465,-482.
[205] P. S. Novikov, *On the algorithmic undecidability of the word problem in group theory*, **T. Math. Inst. Steklov**, 44, 1955.
[206] F. Oger, *Elementary equivalence for abelian-by-finite and nilpotent groups*, **J. of Symbolic Logic**, 66, 2001, 1471–1480.
[207] A. Ol'Shansky, *On Relatively Hyperbolic and G-subgroups of Hyperbolic Groups*, **Int. J. Alg. and Compt.**, 3, 1993, 365–409.
[208] J. C. O'Neill and E. C. Turner, *Test Elements and the Retract Theorem in Hyperbolic Groups*, **New York J. Math**, 6, 2000, 107–117.
[209] F. Paulin, *The Gromov toplogy on $\mathbb{R}$-trees*, **Topology Appl.**, 32, 1989, 197–221.
[210] N. Peczynski and W. Reiwer, *On cancellations in HNN groups*, **Math. Z.**, 158, 1978, 79–86.

[211] B. Plotkin, *Varieties of algebras and algebraic varieties. Categories of algebraic varieties*, preprint Hebrew University, Jerusalem, 1996.

[212] A. S. Rapinchuk, V. V. Benyash-Krivetz and V. I. Chernousov, *Representation varieties of the fundamental groups of compact orientable surfaces*, **Israel J. Math.**, 29, 1996.

[213] A. A. Razborov, *On systems of equations in free groups*, **Izv. Akad. Nauk SSSR**, 48, 1984, 779–832 English transl: Math, USSR Izv. 25, 115–162.

[214] V. N. Remeslennikov, *∃-free groups* **Siberian Mat. J.**, 30, 1989, 998–1001.

[215] V. N. Remeslennikov, *∃-free groups and groups with a length function*, **Cont. Math.**, 184, 1995, 369–376.

[216] V. N. Remeslennikov, R. Stöhr, **On algebraic sets over metabelian groups**, **J. Group Theory**, 8, 2005, 491–513.

[217] E. Rips and Z. Sela, *Cyclic Splittings of Finitely Presented Groups and the Canonical JSJ Decomposition*, **Ann. of Math. (2)**, 146, 1997, 53–109.

[218] N. S. Romanovskii, *On the elementary theory of an almost polycyclic group*, **Math.-USSR-Sb**, 39, 1981, 125–132.

[219] G. Rosenberger, *Applications of Nielsen's Reduction Method in the Solution of Combinatorial Problems in Group Theory*, **London Math. Soc. Lecture Notes**, 36, 1979, 339–358.

[220] G. Rosenberger, *Gleichungen in freien Produkten mit Amalgam*, **Math. Z.**, 173, 1980, 1–12. (Correction: Math. Z., 178, 1981, 579).

[221] G. Rosenberger, *Zum Rang und Isomorphieproblem für freie Produkte mit Amalgam*, Habilitationsschrift, Hamburg, 1974.

[222] G. Rosenberger, *Minimal generating systems for plane discontinuous groups and an equation in free groups*, **Proc. of Groups-Korea 1988**, Springer Verlag Lecture Notes in Mathematics, 1398, 1989, 170–186.

[223] G. Rosenberger, *Alternierende Produkte in freien Gruppen*, **Pacific J. of Math.**, 78, 1978, 243–250.

[224] G. Rosenberger, *Zum Isomorphieproblem für Gruppen mit einer definierenden Relation*, **Ill. J. Math.**, 20, 1976, 614–621.

[225] J. Rotman, **An Introduction to the Theory of Groups**, Third Edition, WC Brown, Dubuque, 1984.

[226] G. S. Sacerdote, *Elementary properties of free groups*, **Trans. Amer. Math. Soc.**, 178, 1972, 127–138.

[227] W. R. Scott, *Algebraically closed groups*, **Proc. American Math. Soc.**, 2, 1951, 118–126.

[228] G. P. Scott and C. T. C. Wall, *Topological methods in group theory*, in **Homological group theory** (eds. C. T. C. Wall and D. Johnson), London Math. Soc. lecture notes 36, 1979, 137–204.

[229] H. Schiek, *Ähnlichkeitsanalyse von Gruppenrelationen*, **Acta Math.**, 96, 1956, 157–252.

[230] M. Schutzenberger, **Comptes rendus des seances de l'academie des sciences**, 248, 1959, 248, 2435–2436.

[231] O. Schreier, *Die Untergruppen der freien Gruppen*, **Abh. Math. Sem. Hamburg Univ.**, 5, 1927, 161–183.

[232] Z. Sela, *The Isomorphism Problem for Hyperbolic Groups I*, **Ann. of Math**, 141, 2, 1995, 217–283.

[233] Z. Sela, *Diophantine Geometry over Groups I: Makanin–Razborov Diagrams*, **Publ. Math. de IHES**, 93, 2001, 31–105.

[234] Z. Sela, *Diophantine Geometry over Groups II: Completions, Closures and Fromal Solutions*, **Israel Jour. of Math.**, 104, 2003, 173–254.

[235] Z. Sela, *Diophantine Geometry over Groups III: Rigid and Solid Solutions*, **Israel Jour. of Math.**, 147, 2005, 1–73.

[236] Z. Sela, *Diophantine Geometry over Groups IV: An Iterative Procedure for Validation of a Sentence*, **Israel Jour. of Math.**, 143, 2004, 1–71.

[237] Z. Sela, *Diophantine Geometry over Groups V: Quantifier Elimination*, **Israel Jour. of Math.**, 150, 2005, 1–97.

[238] J. P. Serre, **Trees**, Springer-Verlag, 1980.

[239] P. Shalen, *Dendrology of groups: an Introduction*, in **Essays in group theory**, S. M. Gersten ed. Math. Sci. Res. Inst. Publ. 8, 1987, 265–319.

[240] P. Shalen, *Dendrology and its applications*, in **Group theory from a geometrical viewpoint**, E. Ghys, A. Haefliger, A. Verjovsky eds., World Scientific, 1991, 543–616.

[241] P. Shalen, *Linear representations of certain amalgamated products*, **J. Pure and Applied Algebra**, 15, 1979, 187–197.

[242] H. Short, *Notes on word hyperbolic groups*, in **Group theory from a geometrical viewpoint** (E. Ghys, A. Haefliger, A. Verjovsky eds.), World Scientific, 1991, 3–63.

[243] C. Simms, **Computations with Finitely Presented Groups**, Cambridge University Press, 1994.

[244] J. Stallings, *Topology of Finite Graphs*, **Invent. Math.**, 71, 3, 1983, 551–565.

[245] W. Szmielew, *Elementary properties of Abelian groups* **Fund. Math.**, 41, 1955, 203–271.

[246] V. A. Tartakovskii, *Solution of the word problem for groups with a k reduced basis for* $k > 6$, **Izv. Akad. Nauk SSSR Ser Math.**, 13, 1949, 483–494.

[247] J. Tits, *Sur le grupe de automorphismes* in **Essays in Topology and Related Subjects**, Springer-Verlag, 1970, 188–211.

[248] M. Tretkoff, *Covering Space Proofs in Combinatorial Group Theory*, **Comm. Alg.**, 3, 1975, 429–457.

[249] E. C. Turner, *Test words for automorphisms of free groups*, **Bull. London Math. Soc.**, 28, 1996, 255–263.

[250] M. Urbanski and L. Zamboni, *On free actions on Λ-trees*, **Proc. Cambridge Phil. Soc.**, 11, 1993, 535–542.

[251] E. R. Van Kampen, *On some lemmas in the theory of groups*, **Amer. J. Math.**, 55, 1933, 268–273.

[252] B. A. F. Wehrfritz, **Infinite linear groups**, Springer-Verlag New York Heidelberg Berlin, 1973.

[253] D. H. Wagner, *On free products of groups*, **Trans. Amer. Math. Soc.**, 84, 1957, 352–378.

[254] C. M. Weinbaum, *On relators and diagrams for groups with a single defining relator*, **Illinois J. Math.**, 16, 1972, 308–322.

[255] H. Wilton, *Solutions to Bestvina and Feighn's Exercises on Limit Groups*, preprint.

[256] Y.-F. Wu, *Groups in which commutativity is a transitive relation*, **J. Algebra**, 1998, 207, 165–181.

[257] H. Zieschang, *Über die Nielsensche Kürzungsmethode in freien Produkten mit Amalgam*, **Invent. Math.**, 10, 1970, 4–37.

[258] H. Zieschang, *Automorphismen ebener diskontinuierlicher Gruppen*, **Math. Ann.**, 166, 1966, 148–167.

[259] H. Zieschang, H. Vogt and E. Coldeway, **Surfaces and Planar Discontinuous Groups**, Springer Lecture Notes in Math., 876, 1981.

Index

De Gruyter Expositions in Mathematics

Volume 59
Friedrich Haslinger
The d-bar Neumann Problem and Schrödinger Operators, 2014
ISBN 978-3-11-031530-1, e-ISBN 978-3-11-031535-6, Set-ISBN 978-3-11-031536-3

Volume 58
Oleg K. Sheinman
Current Algebras on Riemann Surfaces, 2012
ISBN 978-3-11-026452-4, e-ISBN 978-3-11-026452-4, Set-ISBN 978-3-11-916387-3

Volume 57
Helmut Strade
Simple Lie Algebras, Completion of the Classification, 2012
ISBN 978-3-11-026298-8, e-ISBN 978-3-11-026301-5, Set-ISBN 978-3-11-916682-9

Volume 56
Yakov Berkovich, Zvonimir Janko
Groups of Prime Power Order 3, 2011
ISBN 978-3-11-020717-0, e-ISBN 978-3-11-025448-8, Set-ISBN 978-3-11-218909-2

Volume 55
Rainer Picard, Des McGhee
Partial Differential Equations, 2011
ISBN 978-3-11-025026-8, e-ISBN 978-3-11-025027-5, Set-ISBN 978-3-11-218895-8

Volume 54
Edgar E. Enochs, Overtoun M. G. Jenda
Relative Homological Algebra, 2011
ISBN 978-3-11-021522-9, e-ISBN 978-3-11-021523-6, Set-ISBN 978-3-11-173442-2

Volume 53
Adolfo Ballester-Bolinches, Ramon Esteban-Romero, Mohamed Asaad
Products of Finite Groups, 2010
ISBN 978-3-11-020417-9, e-ISBN 978-3-11-022061-2, Set-ISBN 978-3-11-173407-1

Bei Fragen zur Produktsicherheit wenden Sie sich bitte an:
If you have any questions regarding product safety,
please contact:

Walter de Gruyter GmbH
Genthiner Straße 13
10785 Berlin
productsafety@degruyterbrill.com